rms voltage for a half-wave rectif.

Power factor: $\text{pf} = \dfrac{P}{S} = \dfrac{P}{V_{rms} I_{rms}}$

Total harmonic distortion: $\text{THD} = \dfrac{\sqrt{\sum_{n=2}^{\infty} I_n^2}}{I_1}$

Distortion factor: $\text{DF} = \sqrt{\dfrac{1}{1 + (\text{THD})^2}}$

Form factor $= \dfrac{I_{rms}}{I_{avg}}$

Crest factor $= \dfrac{I_{peak}}{I_{rms}}$

Buck converter: $V_o = V_s D$

Boost converter: $V_o = \dfrac{V_s}{1 - D}$

Buck-boost and Ćuk converters: $V_o = -V_s \left(\dfrac{D}{1 - D}\right)$

SEPIC: $V_o = V_s \left(\dfrac{D}{1 - D}\right)$

Flyback converter: $V_o = V_s \left(\dfrac{D}{1 - D}\right)\left(\dfrac{N_2}{N_1}\right)$

Forward converter: $V_o = V_s D \left(\dfrac{N_2}{N_1}\right)$

Power Electronics

Daniel W. Hart
Valparaiso University
Valparaiso, Indiana

POWER ELECTRONICS

Published by McGraw-Hill, a business unit of The McGraw-Hill Companies, Inc., 1221 Avenue of the Americas, New York, NY 10020. Copyright © 2011 by The McGraw-Hill Companies, Inc. All rights reserved. No part of this publication may be reproduced or distributed in any form or by any means, or stored in a database or retrieval system, without the prior written consent of The McGraw-Hill Companies, Inc., including, but not limited to, in any network or other electronic storage or transmission, or broadcast for distance learning.

Some ancillaries, including electronic and print components, may not be available to customers outside the United States.

This book is printed on acid-free paper.

8 9 10 LHN 22 21 20

ISBN 978-0-07-338067-4
MHID 0-07-338067-9

Vice President & Editor-in-Chief: *Marty Lange*
Vice President, EDP: *Kimberly Meriwether-David*
Global Publisher: *Raghothaman Srinivasan*
Director of Development: *Kristine Tibbetts*
Developmental Editor: *Darlene M. Schueller*
Senior Marketing Manager: *Curt Reynolds*
Project Manager: *Erin Melloy*
Senior Production Supervisor: *Kara Kudronowicz*
Senior Media Project Manager: *Jodi K. Banowetz*
Design Coordinator: *Brenda A. Rolwes*
Cover Designer: *Studio Montage, St. Louis, Missouri*
(USE) Cover Image: *Figure 7.5a from interior*
Compositor: *Glyph International*
Typeface: *10.5/12 Times Roman*
Printer: *LSC Communications*

All credits appearing on page or at the end of the book are considered to be an extension of the copyright page.

This book was previously published by: Pearson Education, Inc.

Library of Congress Cataloging-in-Publication Data

Hart, Daniel W.
  Power electronics / Daniel W. Hart.
      p. cm.
  Includes bibliographical references and index.
  ISBN 978-0-07-338067-4 (alk. paper)
  1. Power electronics. I. Title.
  TK7881.15.H373 2010
  621.31'7—dc22
                                                                2009047266

www.mhhe.com

To my family, friends, and the many students
I have had the privilege and pleasure of guiding

BRIEF CONTENTS

Chapter 1
Introduction   1

Chapter 2
Power Computations   21

Chapter 3
Half-Wave Rectifiers   65

Chapter 4
Full-Wave Rectifiers   111

Chapter 5
AC Voltage Controllers   171

Chapter 6
DC-DC Converters   196

Chapter 7
DC Power Supplies   265

Chapter 8
Inverters   331

Chapter 9
Resonant Converters   387

Chapter 10
Drive Circuits, Snubber Circuits, and Heat Sinks   431

Appendix A Fourier Series for Some Common Waveforms   461

Appendix B State-Space Averaging   467

Index   473

CONTENTS

Chapter 1
Introduction 1

1.1 Power Electronics 1
1.2 Converter Classification 1
1.3 Power Electronics Concepts 3
1.4 Electronic Switches 5
The Diode 6
Thyristors 7
Transistors 8
1.5 Switch Selection 11
1.6 Spice, PSpice, and Capture 13
1.7 Switches in Pspice 14
The Voltage-Controlled Switch 14
Transistors 16
Diodes 17
Thyristors (SCRs) 18
Convergence Problems in PSpice 18
1.8 Bibliography 19
Problems 20

Chapter 2
Power Computations 21

2.1 Introduction 21
2.2 Power and Energy 21
Instantaneous Power 21
Energy 22
Average Power 22
2.3 Inductors and Capacitors 25
2.4 Energy Recovery 27
2.5 Effective Values: RMS 34
2.6 Apparent Power and Power Factor 42
Apparent Power S 42
Power Factor 43
2.7 Power Computations for Sinusoidal AC Circuits 43
2.8 Power Computations for Nonsinusoidal Periodic Waveforms 44
Fourier Series 45
Average Power 46
Nonsinusoidal Source and Linear Load 46
Sinusoidal Source and Nonlinear Load 48
2.9 Power Computations Using PSpice 51
2.10 Summary 58
2.11 Bibliography 59
Problems 59

Chapter 3
Half-Wave Rectifiers 65

3.1 Introduction 65
3.2 Resistive Load 65
Creating a DC Component Using an Electronic Switch 65
3.3 Resistive-Inductive Load 67
3.4 PSpice Simulation 72
Using Simulation Software for Numerical Computations 72

3.5   RL-Source Load   75
   Supplying Power to a DC Source from an AC Source   75
3.6   Inductor-Source Load   79
   Using Inductance to Limit Current   79
3.7   The Freewheeling Diode   81
   Creating a DC Current   81
   Reducing Load Current Harmonics   86
3.8   Half-Wave Rectifier With a Capacitor Filter   88
   Creating a DC Voltage from an AC Source   88
3.9   The Controlled Half-Wave Rectifier   94
   Resistive Load   94
   RL Load   96
   RL-Source Load   98
3.10   PSpice Solutions For Controlled Rectifiers   100
   Modeling the SCR in PSpice   100
3.11   Commutation   103
   The Effect of Source Inductance   103
3.12   Summary   105
3.13   Bibliography   106
   Problems   106

Chapter 4
Full-Wave Rectifiers   111

4.1   Introduction   111
4.2   Single-Phase Full-Wave Rectifiers   111
   The Bridge Rectifier   111
   The Center-Tapped Transformer Rectifier   114
   Resistive Load   115
   RL Load   115
   Source Harmonics   118
   PSpice Simulation   119
   RL-Source Load   120
   Capacitance Output Filter   122
   Voltage Doublers   125
   LC Filtered Output   126
4.3   Controlled Full-Wave Rectifiers   131
   Resistive Load   131
   RL Load, Discontinuous Current   133
   RL Load, Continuous Current   135
   PSpice Simulation of Controlled Full-Wave Rectifiers   139
   Controlled Rectifier with RL-Source Load   140
   Controlled Single-Phase Converter Operating as an Inverter   142
4.4   Three-Phase Rectifiers   144
4.5   Controlled Three-Phase Rectifiers   149
   Twelve-Pulse Rectifiers   151
   The Three-Phase Converter Operating as an Inverter   154
4.6   DC Power Transmission   156
4.7   Commutation: The Effect of Source Inductance   160
   Single-Phase Bridge Rectifier   160
   Three-Phase Rectifier   162
4.8   Summary   163
4.9   Bibliography   164
   Problems   164

Chapter 5
AC Voltage Controllers   171

5.1   Introduction   171
5.2   The Single-Phase AC Voltage Controller   171
   Basic Operation   171
   Single-Phase Controller with a Resistive Load   173
   Single-Phase Controller with an RL Load   177
   PSpice Simulation of Single-Phase AC Voltage Controllers   180

5.3 Three-Phase Voltage Controllers 183
- Y-Connected Resistive Load 183
- Y-Connected RL Load 187
- Delta-Connected Resistive Load 189

5.4 Induction Motor Speed Control 191
5.5 Static VAR Control 191
5.6 Summary 192
5.7 Bibliography 193
Problems 193

Chapter 6
DC-DC Converters 196

6.1 Linear Voltage Regulators 196
6.2 A Basic Switching Converter 197
6.3 The Buck (Step-Down) Converter 198
- Voltage and Current Relationships 198
- Output Voltage Ripple 204
- Capacitor Resistance—The Effect on Ripple Voltage 206
- Synchronous Rectification for the Buck Converter 207

6.4 Design Considerations 207
6.5 The Boost Converter 211
- Voltage and Current Relationships 211
- Output Voltage Ripple 215
- Inductor Resistance 218

6.6 The Buck-Boost Converter 221
- Voltage and Current Relationships 221
- Output Voltage Ripple 225

6.7 The Ćuk Converter 226
6.8 The Single-Ended Primary Inductance Converter (SEPIC) 231
6.9 Interleaved Converters 237
6.10 Nonideal Switches and Converter Performance 239
- Switch Voltage Drops 239
- Switching Losses 240

6.11 Discontinuous-Current Operation 241
- Buck Converter with Discontinuous Current 241
- Boost Converter with Discontinuous Current 244

6.12 Switched-Capacitor Converters 247
- The Step-Up Switched-Capacitor Converter 247
- The Inverting Switched-Capacitor Converter 249
- The Step-Down Switched-Capacitor Converter 250

6.13 PSpice Simulation of DC-DC Converters 251
- A Switched PSpice Model 252
- An Averaged Circuit Model 254

6.14 Summary 259
6.15 Bibliography 259
Problems 260

Chapter 7
DC Power Supplies 265

7.1 Introduction 265
7.2 Transformer Models 265
7.3 The Flyback Converter 267
- Continuous-Current Mode 267
- Discontinuous-Current Mode in the Flyback Converter 275
- Summary of Flyback Converter Operation 277

7.4 The Forward Converter 277
- Summary of Forward Converter Operation 283

7.5 The Double-Ended (Two-Switch) Forward Converter 285
7.6 The Push-Pull Converter 287
- Summary of Push-Pull Operation 290

7.7 Full-Bridge and Half-Bridge DC-DC Converters 291

- **7.8** Current-Fed Converters 294
- **7.9** Multiple Outputs 297
- **7.10** Converter Selection 298
- **7.11** Power Factor Correction 299
- **7.12** PSpice Simulation of DC Power Supplies 301
- **7.13** Power Supply Control 302
  - *Control Loop Stability 303*
  - *Small-Signal Analysis 304*
  - *Switch Transfer Function 305*
  - *Filter Transfer Function 306*
  - *Pulse-Width Modulation Transfer Function 307*
  - *Type 2 Error Amplifier with Compensation 308*
  - *Design of a Type 2 Compensated Error Amplifier 311*
  - *PSpice Simulation of Feedback Control 315*
  - *Type 3 Error Amplifier with Compensation 317*
  - *Design of a Type 3 Compensated Error Amplifier 318*
  - *Manual Placement of Poles and Zeros in the Type 3 Amplifier 323*
- **7.14** PWM Control Circuits 323
- **7.15** The AC Line Filter 323
- **7.16** The Complete DC Power Supply 325
- **7.17** Bibliography 326
  - Problems 327

Chapter 8
Inverters 331

- **8.1** Introduction 331
- **8.2** The Full-Bridge Converter 331
- **8.3** The Square-Wave Inverter 333
- **8.4** Fourier Series Analysis 337
- **8.5** Total Harmonic Distortion 339
- **8.6** PSpice Simulation of Square Wave Inverters 340
- **8.7** Amplitude and Harmonic Control 342
- **8.8** The Half-Bridge Inverter 346
- **8.9** Multilevel Inverters 348
  - *Multilevel Converters with Independent DC Sources 349*
  - *Equalizing Average Source Power with Pattern Swapping 353*
  - *Diode-Clamped Multilevel Inverters 354*
- **8.10** Pulse-Width-Modulated Output 357
  - *Bipolar Switching 357*
  - *Unipolar Switching 358*
- **8.11** PWM Definitions and Considerations 359
- **8.12** PWM Harmonics 361
  - *Bipolar Switching 361*
  - *Unipolar Switching 365*
- **8.13** Class D Audio Amplifiers 366
- **8.14** Simulation of Pulse-Width-Modulated Inverters 367
  - *Bipolar PWM 367*
  - *Unipolar PWM 370*
- **8.15** Three-Phase Inverters 373
  - *The Six-Step Inverter 373*
  - *PWM Three-Phase Inverters 376*
  - *Multilevel Three-Phase Inverters 378*
- **8.16** PSpice Simulation of Three-Phase Inverters 378
  - *Six-Step Three-Phase Inverters 378*
  - *PWM Three-Phase Inverters 378*
- **8.17** Induction Motor Speed Control 379
- **8.18** Summary 382
- **8.19** Bibliography 383
  - Problems 383

Chapter 9
Resonant Converters 387

- 9.1 Introduction 387
- 9.2 A Resonant Switch Converter: Zero-Current Switching 387
  - *Basic Operation* 387
  - *Output Voltage* 392
- 9.3 A Resonant Switch Converter: Zero-Voltage Switching 394
  - *Basic Operation* 394
  - *Output Voltage* 399
- 9.4 The Series Resonant Inverter 401
  - *Switching Losses* 403
  - *Amplitude Control* 404
- 9.5 The Series Resonant DC-DC Converter 407
  - *Basic Operation* 407
  - *Operation for $\omega_s > \omega_o$* 407
  - *Operation for $\omega_0/2 < \omega_s < \omega_0$* 413
  - *Operation for $\omega_s < \omega_0/2$* 413
  - *Variations on the Series Resonant DC-DC Converter* 414
- 9.6 The Parallel Resonant DC-DC Converter 415
- 9.7 The Series-Parallel DC-DC Converter 418
- 9.8 Resonant Converter Comparison 421
- 9.9 The Resonant DC Link Converter 422
- 9.10 Summary 426
- 9.11 Bibliography 426
  - Problems 427

Chapter 10
Drive Circuits, Snubber Circuits, and Heat Sinks 431

- 10.1 Introduction 431
- 10.2 MOSFET and IGBT Drive Circuits 431
  - *Low-Side Drivers* 431
  - *High-Side Drivers* 433
- 10.3 Bipolar Transistor Drive Circuits 437
- 10.4 Thyristor Drive Circuits 440
- 10.5 Transistor Snubber Circuits 441
- 10.6 Energy Recovery Snubber Circuits 450
- 10.7 Thyristor Snubber Circuits 450
- 10.8 Heat Sinks and Thermal Management 451
  - *Steady-State Temperatures* 451
  - *Time-Varying Temperatures* 454
- 10.9 Summary 457
- 10.10 Bibliography 457
  - Problems 458

Appendix A Fourier Series for Some Common Waveforms 461

Appendix B State-Space Averaging 467

Index 473

PREFACE

This book is intended to be an introductory text in power electronics, primarily for the undergraduate electrical engineering student. The text assumes that the student is familiar with general circuit analysis techniques usually taught at the sophomore level. The student should be acquainted with electronic devices such as diodes and transistors, but the emphasis of this text is on circuit topology and function rather than on devices. Understanding the voltage-current relationships for linear devices is the primary background required, and the concept of Fourier series is also important. Most topics presented in this text are appropriate for junior- or senior-level undergraduate electrical engineering students.

The text is designed to be used for a one-semester power electronics course, with appropriate topics selected or omitted by the instructor. The text is written for some flexibility in the order of the topics. It is recommended that Chap. 2 on power computations be covered at the beginning of the course in as much detail as the instructor deems necessary for the level of students. Chapters 6 and 7 on dc-dc converters and dc power supplies may be taken before Chaps. 3, 4, and 5 on rectifiers and voltage controllers. The author covers chapters in the order 1, 2 (introduction; power computations), 6, 7 (dc-dc converters; dc power supplies), 8 (inverters), 3, 4, 5 (rectifiers and voltage controllers), followed by coverage of selected topics in 9 (resonant converters) and 10 (drive and snubber circuits and heat sinks). Some advanced material, such as the control section in Chapter 7, may be omitted in an introductory course.

The student should use all the software tools available for the solution to the equations that describe power electronics circuits. These range from calculators with built-in functions such as integration and root finding to more powerful computer software packages such as MATLAB®, Mathcad®, Maple™, Mathematica®, and others. Numerical techniques are often suggested in this text. It is up to the student to select and adapt all the readily available computer tools to the power electronics situation.

Much of this text includes computer simulation using PSpice® as a supplement to analytical circuit solution techniques. Some prior experience with PSpice is helpful but not necessary. Alternatively, instructors may choose to use a different simulation program such as PSIM® or NI Multisim™ software instead of PSpice. Computer simulation is never intended to replace understanding of fundamental principles. It is the author's belief that using computer simulation for the instructional benefit of investigating the basic behavior of power electronics circuits adds a dimension to the student's learning that is not possible from strictly manipulating equations. Observing voltage and current waveforms from a computer simulation accomplishes some of the same objectives as those of a laboratory experience. In a computer simulation, all the circuit's voltages and currents can be investigated, usually much more efficiently than in a hardware lab. Variations in circuit performance for a change in components or operating parameters can be accomplished more easily with a computer simulation than in a laboratory. PSpice circuits presented in this text do not necessarily represent the most elegant way to simulate circuits. Students are encouraged to use their engineering skills to improve the simulation circuits wherever possible.

The website that accompanies this text can be found at www.mhhe.com/hart, and features Capture circuit files for PSpice simulation for students and instructors and a password-protected solutions manual and PowerPoint® lecture notes for instructors.

My sincere gratitude to reviewers and students who have made many valuable contributions to this project. Reviewers include

Ali Emadi
Illinois Institute of Technology

Shaahin Filizadeh
University of Manitoba

James Gover
Kettering University

Peter Idowu
Penn State, Harrisburg

Mehrdad Kazerani
University of Waterloo

Xiaomin Kou
University of Wisconsin-Platteville

Alexis Kwasinski
The University of Texas at Austin

Medhat M. Morcos
Kansas State University

Steve Pekarek
Purdue University

Wajiha Shireen
University of Houston

Hamid Toliyat
Texas A&M University

Zia Yamayee
University of Portland

Lin Zhao
Gannon University

A special thanks to my colleagues Kraig Olejniczak, Mark Budnik, and Michael Doria at Valparaiso University for their contributions. I also thank Nikke Ault for the preparation of much of the manuscript.

Complete Online Solutions Manual Organization System (COSMOS). Professors can benefit from McGraw-Hill's COSMOS electronic solutions manual. COSMOS enables instructors to generate a limitless supply of problem material for assignment, as well as transfer and integrate their own problems into the software. For additional information, contact your McGraw-Hill sales representative.

Electronic Textbook Option. This text is offered through CourseSmart for both instructors and students. CourseSmart is an online resource where students can purchase the complete text online at almost one-half the cost of a traditional text. Purchasing the eTextbook allows students to take advantage of CourseSmart's Web tools for learning, which include full text search, notes and highlighting, and e-mail tools for sharing notes among classmates. To learn more about CourseSmart options, contact your McGraw-Hill sales representative or visit www.CourseSmart.com.

Daniel W. Hart
Valparaiso University
Valparaiso, Indiana

CHAPTER 1

Introduction

1.1 POWER ELECTRONICS

Power electronics circuits convert electric power from one form to another using electronic devices. Power electronics circuits function by using semiconductor devices as switches, thereby controlling or modifying a voltage or current. Applications of power electronics range from high-power conversion equipment such as dc power transmission to everyday appliances, such as cordless screwdrivers, power supplies for computers, cell phone chargers, and hybrid automobiles. Power electronics includes applications in which circuits process milliwatts or megawatts. Typical applications of power electronics include conversion of ac to dc, conversion of dc to ac, conversion of an unregulated dc voltage to a regulated dc voltage, and conversion of an ac power source from one amplitude and frequency to another amplitude and frequency.

The design of power conversion equipment includes many disciplines from electrical engineering. Power electronics includes applications of circuit theory, control theory, electronics, electromagnetics, microprocessors (for control), and heat transfer. Advances in semiconductor switching capability combined with the desire to improve the efficiency and performance of electrical devices have made power electronics an important and fast-growing area in electrical engineering.

1.2 CONVERTER CLASSIFICATION

The objective of a power electronics circuit is to match the voltage and current requirements of the load to those of the source. Power electronics circuits convert one type or level of a voltage or current waveform to another and are hence called *converters*. Converters serve as an interface between the source and load (Fig. 1-1).

Figure 1-1 A source and load interfaced by a power electronics converter.

Converters are classified by the relationship between input and output:

ac input/dc output

The ac-dc converter produces a dc output from an ac input. Average power is transferred from an ac source to a dc load. The ac-dc converter is specifically classified as a *rectifier*. For example, an ac-dc converter enables integrated circuits to operate from a 60-Hz ac line voltage by converting the ac signal to a dc signal of the appropriate voltage.

dc input/ac output

The dc-ac converter is specifically classified as an *inverter*. In the inverter, average power flows from the dc side to the ac side. Examples of inverter applications include producing a 120-V rms 60-Hz voltage from a 12-V battery and interfacing an alternative energy source such as an array of solar cells to an electric utility.

dc input/dc output

The dc-dc converter is useful when a load requires a specified (often regulated) dc voltage or current but the source is at a different or unregulated dc value. For example, 5 V may be obtained from a 12-V source via a dc-dc converter.

ac input/ac output

The ac-ac converter may be used to change the level and/or frequency of an ac signal. Examples include a common light-dimmer circuit and speed control of an induction motor.

Some converter circuits can operate in different modes, depending on circuit and control parameters. For example, some rectifier circuits can be operated as inverters by modifying the control on the semiconductor devices. In such cases, it is the direction of average power flow that determines the converter classification. In Fig. 1-2, if the battery is charged from the ac power source, the converter is classified as a rectifier. If the operating parameters of the converter are changed and the battery acts as a source supplying power to the ac system, the converter is then classified as an inverter.

Power conversion can be a multistep process involving more than one type of converter. For example, an ac-dc-ac conversion can be used to modify an ac source by first converting it to direct current and then converting the dc signal to an ac signal that has an amplitude and frequency different from those of the original ac source, as illustrated in Fig. 1-3.

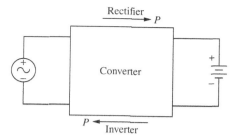

Figure 1-2 A converter can operate as a rectifier or an inverter, depending on the direction of average power P.

Figure 1-3 Two converters are used in a multistep process.

1.3 POWER ELECTRONICS CONCEPTS

To illustrate some concepts in power electronics, consider the design problem of creating a 3-V dc voltage level from a 9-V battery. The purpose is to supply 3 V to a load resistance. One simple solution is to use a voltage divider, as shown in Fig. 1-4. For a load resistor R_L, inserting a series resistance of $2R_L$ results in 3 V across R_L. A problem with this solution is that the power absorbed by the $2R_L$ resistor is twice as much as delivered to the load and is lost as heat, making the circuit only 33.3 percent efficient. Another problem is that if the value of the load resistance changes, the output voltage will change unless the $2R_L$ resistance changes proportionally. A solution to that problem could be to use a transistor in place of the $2R_L$ resistance. The transistor would be controlled such that the voltage across it is maintained at 6 V, thus regulating the output at 3 V. However, the same low-efficiency problem is encountered with this solution.

To arrive at a more desirable design solution, consider the circuit in Fig. 1-5a. In that circuit, a switch is opened and closed periodically. The switch is a short circuit when it is closed and an open circuit when it is open, making the voltage

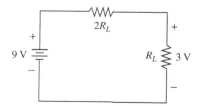

Figure 1-4 A simple voltage divider for creating 3 V from a 9-V source.

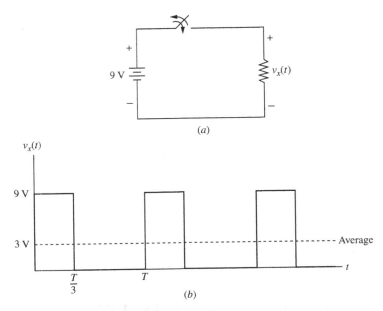

Figure 1-5 (a) A switched circuit; (b) a pulsed voltage waveform.

across R_L equal to 9 V when the switch is closed and 0 V when the switch is open. The resulting voltage across R_L will be like that of Fig. 1-5b. This voltage is obviously not a constant dc voltage, but if the switch is closed for one-third of the period, the average value of v_x (denoted as V_x) is one-third of the source voltage. Average value is computed from the equation

$$\text{avg}(v_x) = V_x = \frac{1}{T}\int_0^T v_x(t)\,dt = \frac{1}{T}\int_0^{T/3} 9\,dt + \frac{1}{T}\int_{T/3}^T 0\,dt = 3\text{ V} \quad (1\text{-}1)$$

Considering efficiency of the circuit, instantaneous power (see Chap. 2) absorbed by the switch is the product of voltage and current. When the switch is open, power absorbed by it is zero because the current in it is zero. When the switch is closed, power absorbed by it is zero because the voltage across it is zero. Since power absorbed by the switch is zero for both open and closed conditions, all power supplied by the 9-V source is delivered to R_L, making the circuit 100 percent efficient.

The circuit so far does not accomplish the design object of creating a dc voltage of 3 V. However, the voltage waveform v_x can be expressed as a Fourier series containing a dc term (the average value) plus sinusoidal terms at frequencies that are multiples of the pulse frequency. To create a 3-V dc voltage, v_x is applied to a low-pass filter. An ideal low-pass filter allows the dc component of voltage to pass through to the output while removing the ac terms, thus creating the desired dc output. If the filter is lossless, the converter will be 100 percent efficient.

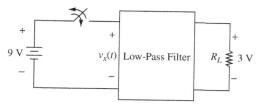

Figure 1-6 A low-pass filter allows just the average value of v_x to pass through to the load.

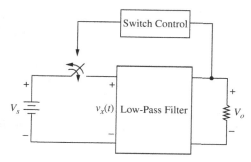

Figure 1-7 Feedback is used to control the switch and maintain the desired output voltage.

In practice, the filter will have some losses and will absorb some power. Additionally, the electronic device used for the switch will not be perfect and will have losses. However, the efficiency of the converter can still be quite high (more than 90 percent). The required values of the filter components can be made smaller with higher switching frequencies, making large switching frequencies desirable. Chaps. 6 and 7 describe the dc-dc conversion process in detail. The "switch" in this example will be some electronic device such as a metal-oxide field-effect transistors (MOSFET), or it may be comprised of more than one electronic device.

The power conversion process usually involves system control. Converter output quantities such as voltage and current are measured, and operating parameters are adjusted to maintain the desired output. For example, if the 9-V battery in the example in Fig. 1-6 decreased to 6 V, the switch would have to be closed 50 percent of the time to maintain an average value of 3 V for v_x. A feedback control system would detect if the output voltage were not 3 V and adjust the closing and opening of the switch accordingly, as illustrated in Fig. 1-7.

1.4 ELECTRONIC SWITCHES

An electronic switch is characterized by having the two states *on* and *off,* ideally being either a short circuit or an open circuit. Applications using switching devices are desirable because of the relatively small power loss in the device. If the switch is ideal, either the switch voltage or the switch current is zero, making the power absorbed by it zero. Real devices absorb some power when in the on state and when making transitions between the on and off states, but circuit efficiencies can still be quite high. Some electronic devices such as transistors can also operate in the active range where both voltage and current are nonzero, but it is desirable to use these devices as switches when processing power.

The emphasis of this textbook is on basic circuit operation rather than on device performance. The particular switching device used in a power electronics circuit depends on the existing state of device technology. The behaviors of power electronics circuits are often not affected significantly by the actual device used for switching, particularly if voltage drops across a conducting switch are small compared to other circuit voltages. Therefore, semiconductor devices are usually modeled as ideal switches so that circuit behavior can be emphasized. Switches are modeled as short circuits when on and open circuits when off. Transitions between states are usually assumed to be instantaneous, but the effects of nonideal switching are discussed where appropriate. A brief discussion of semiconductor switches is given in this section, and additional information relating to drive and snubber circuits is provided in Chap. 10. Electronic switch technology is continually changing, and thorough treatments of state-of-the-art devices can be found in the literature.

The Diode

A diode is the simplest electronic switch. It is uncontrollable in that the on and off conditions are determined by voltages and currents in the circuit. The diode is forward-biased (on) when the current i_d (Fig. 1-8a) is positive and reverse-biased (off) when v_d is negative. In the ideal case, the diode is a short circuit

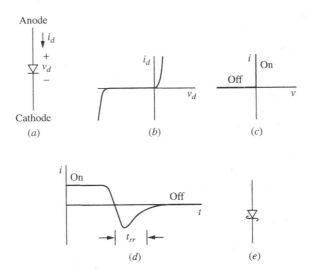

Figure 1-8 (a) Rectifier diode; (b) i-v characteristic; (c) idealized i-v characteristic; (d) reverse recovery time t_{rr}; (e) Schottky diode.

when it is forward-biased and is an open circuit when reverse-biased. The actual and idealized current-voltage characteristics are shown in Fig. 1-8b and c. The idealized characteristic is used in most analyses in this text.

An important dynamic characteristic of a nonideal diode is reverse recovery current. When a diode turns off, the current in it decreases and momentarily becomes negative before becoming zero, as shown in Fig. 1-8d. The time t_{rr} is the reverse recovery time, which is usually less than 1 μs. This phenomenon may become important in high-frequency applications. Fast-recovery diodes are designed to have a smaller t_{rr} than diodes designed for line-frequency applications. Silicon carbide (SiC) diodes have very little reverse recovery, resulting in more efficient circuits, especially in high-power applications.

Schottky diodes (Fig. 1-8e) have a metal-to-silicon barrier rather than a P-N junction. Schottky diodes have a forward voltage drop of typically 0.3 V. These are often used in low-voltage applications where diode drops are significant relative to other circuit voltages. The reverse voltage for a Schottky diode is limited to about 100 V. The metal-silicon barrier in a Schottky diode is not subject to recovery transients and turn-on and off faster than P-N junction diodes.

Thyristors

Thyristors are electronic switches used in some power electronic circuits where control of switch turn-on is required. The term *thyristor* often refers to a family of three-terminal devices that includes the silicon-controlled rectifier (SCR), the triac, the gate turnoff thyristor (GTO), the MOS-controlled thyristor (MCT), and others. *Thyristor* and *SCR* are terms that are sometimes used synonymously. The SCR is the device used in this textbook to illustrate controlled turn-on devices in the thyristor family. Thyristors are capable of large currents and large blocking voltages for use in high-power applications, but switching frequencies cannot be as high as when using other devices such as MOSFETs.

The three terminals of the SCR are the anode, cathode, and gate (Fig. 1-9a). For the SCR to begin to conduct, it must have a gate current applied while it has a positive anode-to-cathode voltage. After conduction is established, the gate signal is no longer required to maintain anode current. The SCR will continue to conduct as long as the anode current remains positive and above a minimum value called the holding level. Figs. 1-9a and b show the SCR circuit symbol and the idealized current-voltage characteristic.

The gate turnoff thyristor (GTO) of Fig. 1-9c, like the SCR, is turned on by a short-duration gate current if the anode-to-cathode voltage is positive. However, unlike the SCR, the GTO can be turned off with a negative gate current. The GTO is therefore suitable for some applications where control of both turn-on and turnoff of a switch is required. The negative gate turnoff current can be of brief duration (a few microseconds), but its magnitude must be very large compared to the turn-on current. Typically, gate turnoff current is one-third the on-state anode current. The idealized *i-v* characteristic is like that of Fig. 1-9b for the SCR.

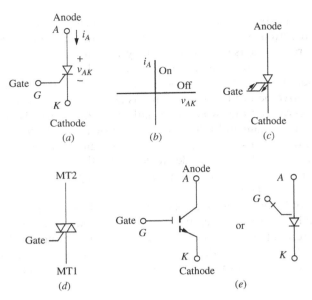

Figure 1-9 Thyristor devices: (*a*) silicon-controlled rectifier (SCR); (*b*) SCR idealized *i-v* characteristic; (*c*) gate turnoff (GTO) thyristor; (*d*) triac; (*e*) MOS-controlled thyristor (MCT).

The triac (Fig. 1-9*d*) is a thyristor that is capable of conducting current in either direction. The triac is functionally equivalent to two antiparallel SCRs (in parallel but in opposite directions). Common incandescent light-dimmer circuits use a triac to modify both the positive and negative half cycles of the input sine wave.

The MOS-controlled thyristor (MCT) in Fig. 1-9*e* is a device functionally equivalent to the GTO but without the high turnoff gate current requirement. The MCT has an SCR and two MOSFETs integrated into one device. One MOSFET turns the SCR on, and one MOSFET turns the SCR off. The MCT is turned on and off by establishing the proper voltage from gate to cathode, as opposed to establishing a gate current in the GTO.

Thyristors were historically the power electronics switch of choice because of high voltage and current ratings available. Thyristors are still used, especially in high-power applications, but ratings of power transistors have increased greatly, making the transistor more desirable in many applications.

Transistors

Transistors are operated as switches in power electronics circuits. Transistor drive circuits are designed to have the transistor either in the fully on or fully off state. This differs from other transistor applications such as in a linear amplifier circuit where the transistor operates in the region having simultaneously high voltage and current.

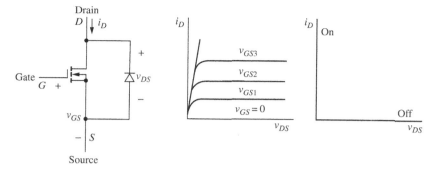

Figure 1-10 (*a*) MOSFET (N-channel) with body diode; (*b*) MOSFET characteristics; (*c*) idealized MOSFET characteristics.

Unlike the diode, turn-on and turnoff of a transistor are controllable. Types of transistors used in power electronics circuits include MOSFETs, bipolar junction transistors (BJTs), and hybrid devices such as insulated-gate bipolar junction transistors (IGBTs). Figs. 1-10 to 1-12 show the circuit symbols and the current-voltage characteristics.

The MOSFET (Fig. 1-10*a*) is a voltage-controlled device with characteristics as shown in Fig. 1-10*b*. MOSFET construction produces a parasitic (body) diode, as shown, which can sometimes be used to an advantage in power electronics circuits. Power MOSFETs are of the enhancement type rather than the depletion type. A sufficiently large gate-to-source voltage will turn the device on,

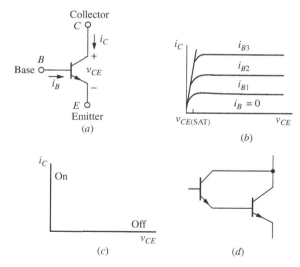

Figure 1-11 (*a*) BJT (NPN); (*b*) BJT characteristics; (*c*) idealized BJT characteristics; (*d*) Darlington configuration.

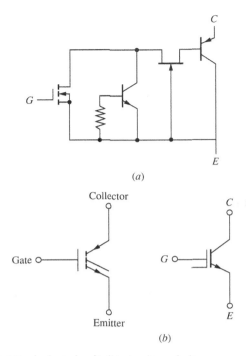

Figure 1-12 IGBT: (*a*) Equivalent circuit; (*b*) circuit symbols.

resulting in a small drain-to-source voltage. In the on state, the change in v_{DS} is linearly proportional to the change in i_D. Therefore, the on MOSFET can be modeled as an on-state resistance called $R_{DS(on)}$. MOSFETs have on-state resistances as low as a few milliohms. For a first approximation, the MOSFET can be modeled as an ideal switch with a characteristic shown in Fig. 1-10*c*. Ratings are to 1500 V and more than 600 A (although not simultaneously). MOSFET switching speeds are greater than those of BJTs and are used in converters operating into the megahertz range.

Typical BJT characteristics are shown in Fig. 1-11*b*. The on state for the transistor is achieved by providing sufficient base current to drive the BJT into saturation. The collector-emitter saturation voltage is typically 1 to 2 V for a power BJT. Zero base current results in an off transistor. The idealized *i-v* characteristic for the BJT is shown in Fig. 1-11*c*. The BJT is a current-controlled device, and power BJTs typically have low h_{FE} values, sometimes lower than 20. If a power BJT with $h_{FE} = 20$ is to carry a collector current of 60 A, for example, the base current would need to be more than 3 A to put the transistor into saturation. The drive circuit to provide a high base current is a significant power circuit in itself. Darlington configurations have two BJTs connected as shown in Fig. 1-11*d*. The effective current gain of the combination is approximately the product of individual gains and can thus reduce the current required from the drive circuit. The Darlington configuration can be constructed from two discrete transistors or can be obtained as a single integrated device. Power BJTs are rarely used in new applications, being surpassed by MOSFETs and IGBTs.

The IGBT of Fig. 1-12 is an integrated connection of a MOSFET and a BJT. The drive circuit for the IGBT is like that of the MOSFET, while the on-state characteristics are like those of the BJT. IGBTs have replaced BJTs in many applications.

1.5 SWITCH SELECTION

The selection of a power device for a particular application depends not only on the required voltage and current levels but also on its switching characteristics. Transistors and GTOs provide control of both turn-on and turnoff, SCRs of turn-on but not turnoff, and diodes of neither.

Switching speeds and the associated power losses are very important in power electronics circuits. The BJT is a minority carrier device, whereas the MOSFET is a majority carrier device that does not have minority carrier storage delays, giving the MOSFET an advantage in switching speeds. BJT switching times may be a magnitude larger than those for the MOSFET. Therefore, the MOSFET generally has lower switching losses and is preferred over the BJT.

When selecting a suitable switching device, the first consideration is the required operating point and turn-on and turnoff characteristics. Example 1-1 outlines the selection procedure.

EXAMPLE 1-1

Switch Selection

The circuit of Fig. 1-13a has two switches. Switch S_1 is on and connects the voltage source ($V_s = 24$ V) to the current source ($I_o = 2$ A). It is desired to open switch S_1 to disconnect V_s from the current source. This requires that a second switch S_2 close to provide a path for current I_o, as in Fig. 1-13b. At a later time, S_1 must reclose and S_2 must open to restore the circuit to its original condition. The cycle is to repeat at a frequency of 200 kHz. Determine the type of device required for each switch and the maximum voltage and current requirements of each.

■ **Solution**
The type of device is chosen from the turn-on and turnoff requirements, the voltage and current requirements of the switch for the on and off states, and the required switching speed.

The steady-state operating points for S_1 are at $(v_1, i_1) = (0, I_o)$ for S_1 closed and $(V_s, 0)$ for the switch open (Fig. 1-13c). The operating points are on the positive i and v axes, and S_1 must turn off when $i_1 = I_o > 0$ and must turn on when $v_1 = V_s > 0$. The device used for S_1 must therefore provide control of both turn-on and turnoff. The MOSFET characteristic

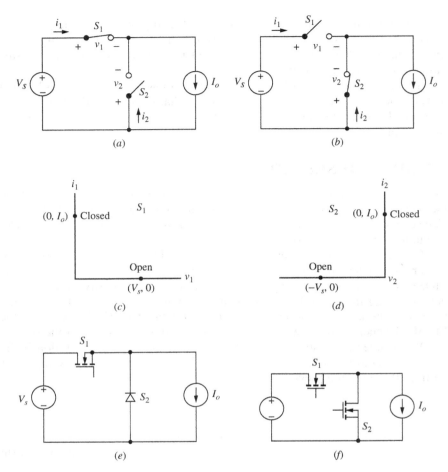

Figure 1-13 Circuit for Example 1-1. (a) S_1 closed, S_2 open; (b) S_1 open, S_2 closed; (c) operating points for S_1; (d) operating points for S_2; (e) switch implementation using a MOSFET and diode; (f) switch implementation using two MOSFETs (synchronous rectification).

of Fig. 1-10d or the BJT characteristic of Fig. 1-11c matches the requirement. A MOSFET would be a good choice because of the required switching frequency, simple gate-drive requirements, and relatively low voltage and current requirement (24 V and 2 A).

The steady-state operating points for S_2 are at $(v_2, i_2) = (-V_s, 0)$ in Fig. 1-13a and $(0, I_o)$ in Fig. 1-13b, as shown in Fig. 1-13d. The operating points are on the positive current axis and negative voltage axis. Therefore, a positive current in S_2 is the requirement to turn S_2 on, and a negative voltage exists when S_2 must turn off. Since the operating points match the diode (Fig. 1-8c) and no other control is needed for the device, a diode is an appropriate choice for S_2. Figure 1-13e shows the implementation of the switching circuit. Maximum current is 2 A, and maximum voltage in the blocking state is 24 V.

Although a diode is a sufficient and appropriate device for the switch S_2, a MOSFET would also work in this position, as shown in Fig. 1-13f. When S_2 is on and S_1 is off, current flows upward out of the drain of S_2. The advantage of using a MOSFET is that it has a much lower voltage drop across it when conducting compared to a diode, resulting in lower power loss and a higher circuit efficiency. The disadvantage is that a more complex control circuit is required to turn on S_2 when S_1 is turned off. However, several control circuits are available to do this. This control scheme is known as synchronous rectification or synchronous switching.

In a power electronics application, the current source in this circuit could represent an inductor that has a nearly constant current in it.

1.6 SPICE, PSPICE, AND CAPTURE

Computer simulation is a valuable analysis and design tool that is emphasized throughout this text. SPICE is a circuit simulation program developed in the Department of Electrical Engineering and Computer Science at the University of California at Berkeley. PSpice is a commercially available adaptation of SPICE that was developed for the personal computer. Capture is a graphical interface program that enables a simulation to be done from a graphical representation of a circuit diagram. Cadence provides a product called OrCAD Capture, and a demonstration version at no cost.[1] Nearly all simulations described in this textbook can be run using the demonstration version.

Simulation can take on various levels of device and component modeling, depending on the objective of the simulation. Most of the simulation examples and exercises use idealized or default component models, making the results first-order approximations, much the same as the analytical work done in the first discussion of a subject in any textbook. After understanding the fundamental operation of a power electronics circuit, the engineer can include detailed device models to predict more accurately the behavior of an actual circuit.

Probe, the graphics postprocessor program that accompanies PSpice, is especially useful. In Probe, the waveform of any current or voltage in a circuit can be shown graphically. This gives the student a look at circuit behavior that is not possible with pencil-and-paper analysis. Moreover, Probe is capable of mathematical computations involving currents and/or voltages, including numerical determination of rms and average values. Examples of PSpice analysis and design for power electronics circuits are an integral part of this textbook.

The PSpice circuit files listed in this text were developed using version 16.0. Continuous revision of software necessitates updates in simulation techniques.

---

[1] https://www.cadence.com/products/orcad/pages/downloads.aspx#demo

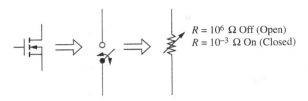

Figure 1-14 Implementing a switch with a resistance in PSpice.

1.7 SWITCHES IN PSPICE

The Voltage-Controlled Switch

The voltage-controlled switch Sbreak in PSpice can be used as an idealized model for most electronic devices. The voltage-controlled switch is a resistance that has a value established by a controlling voltage. Fig. 1-14 illustrates the concept of using a controlled resistance as a switch for PSpice simulation of power electronics circuits. A MOSFET or other switching device is ideally an open or closed switch. A large resistance approximates an open switch, and a small resistance approximates a closed switch. Switch model parameters are as follows:

Parameter	Description	Default Value
RON	"On" resistance	1 (reduce this to 0.001 or 0.01 Ω)
ROFF	"Off" resistance	10^6 Ω
VON	Control voltage for on state	1.0 V
VOFF	Control voltage for off state	0 V

The resistance is changed from large to small by the controlling voltage. The default off resistance is 1 MΩ, which is a good approximation for an open circuit in power electronics applications. The default on resistance of 1 Ω is usually too large. If the switch is to be ideal, the on resistance in the switch model should be changed to something much lower, such as 0.001 or 0.01 Ω.

EXAMPLE 1-2

A Voltage-Controlled Switch in PSpice

The Capture diagram of a switching circuit is shown in Fig. 1-15a. The switch is implemented with the voltage-controlled switch Sbreak, located in the Breakout library of devices. The control voltage is VPULSE and uses the characteristics shown. The rise and fall times, TR and TF, are made small compared to the pulse width and period, PW and PER. V1 and V2 must span the on and off voltage levels for the switch, 0 and 1 V by default. The switching period is 25 ms, corresponding to a frequency of 40 kHz.

The PSpice model for Sbreak is accessed by clicking *edit*, then *PSpice model*. The model editor window is shown in Fig 1-15b. The on resistance Ron is changed to 0.001 Ω

1.7 Switches in PSpice

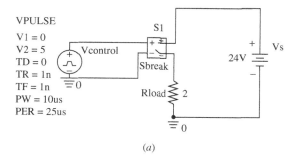

(a)

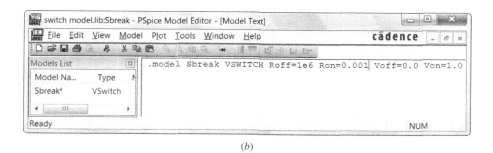

(b)

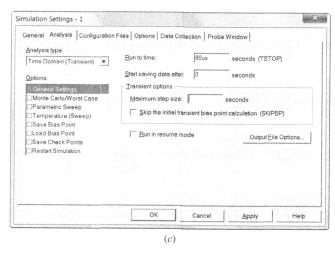

(c)

Figure 1-15 (a) Circuit for Example 1-2; (b) editing the PSpice Sbreak switch model to make Ron = 0.001Ω; (c) the transient analysis setup; (d) the Probe output.

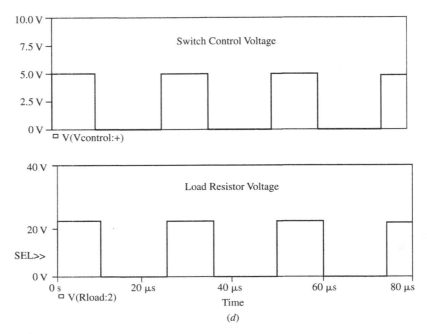

Figure 1-15 (*continued*)

to approximate an ideal switch. The Transient Analysis menu is accessed from Simulation Settings. This simulation has a run time of 80 μs, as shown in Fig. 1-15c.

Probe output showing the switch control voltage and the load resistor voltage waveforms is seen in Fig. 1-15d.

Transistors

Transistors used as switches in power electronics circuits can be idealized for simulation by using the voltage-controlled switch. As in Example 1-2, an ideal transistor can be modeled as very small on resistance. An on resistance matching the MOSFET characteristics can be used to simulate the conducting resistance $R_{DS(ON)}$ of a MOSFET to determine the behavior of a circuit with nonideal components. If an accurate representation of a transistor is required, a model may be available in the PSpice library of devices or from the manufacturer's website. The IRF150 and IRF9140 models for power MOSFETs are in the demonstration version library. The default MOSFET MbreakN or MbreakN3 model must have parameters for the threshold voltage VTO and the constant KP added to the PSpice device model for a meaningful simulation. Manufacturer's websites, such as International Rectifier at www.irf.com, have SPICE models available for their

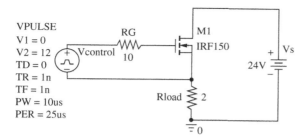

Figure 1-16 An idealized MOSFET drive circuit in PSpice.

products. The default BJT QbreakN can be used instead of a detailed transistor model for a rudimentary simulation.

Transistors in PSpice must have drive circuits, which can be idealized if the behavior of a specific drive circuit is not required. Simulations with MOSFETs can have drive circuits like that in Fig. 1-16. The voltage source VPULSE establishes the gate-to-source voltage of the MOSFET to turn it on and off. The gate resistor may not be necessary, but it sometimes eliminates numerical convergence problems.

Diodes

An ideal diode is assumed when one is developing the equations that describe a power electronics circuit, which is reasonable if the circuit voltages are much larger than the normal forward voltage drop across a conducting diode. The diode current is related to diode voltage by

$$i_d = I_S e^{v_d/nV_T} - 1 \qquad (1\text{-}2)$$

where n is the emission coefficient which has a default value of 1 in PSpice. An ideal diode can be approximated in PSpice by setting n to a small number such as 0.001 or 0.01. The nearly ideal diode is modeled with the part Dbreak with PSpice model

$$\text{model Dbreak D } n = 0.001$$

With the ideal diode model, simulation results will match the analytical results from the describing equations. A PSpice diode model that more accurately predicts diode behavior can be obtained from a device library. Simulations with a detailed diode model will produce more realistic results than the idealized case. However, if the circuit voltages are large, the difference between using an ideal diode and an accurate diode model will not affect the results in any significant way. The default diode model for Dbreak can be used as a compromise between the ideal and actual cases, often with little difference in the result.

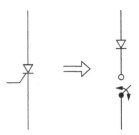

Figure 1-17 Simplified thyristor (SCR) model for PSpice.

Thyristors (SCRs)

An SCR model is available in the PSpice demonstration version part library and can be used in simulating SCR circuits. However, the model contains a relatively large number of components which imposes a size limit for the PSpice demonstration version. A simple SCR model that is used in several circuits in this text is a switch in series with a diode, as shown in Fig. 1-17. Closing the voltage-controlled switch is equivalent to applying a gate current to the SCR, and the diode prevents reverse current in the model. This simple SCR model has the significant disadvantage of requiring the voltage-controlled switch to remain closed during the entire on time of the SCR, thus requiring some prior knowledge of the behavior of a circuit that uses the device. Further explanation is included with the PSpice examples in later chapters.

Convergence Problems in PSpice

Some of the PSpice simulations in this book are subject to numerical convergence problems because of the switching that takes place in circuits with inductors and capacitors. All the PSpice files presented in this text have been designed to avoid convergence problems. However, sometimes changing a circuit parameter will cause a failure to converge in the transient analysis. In the event that there is a problem with PSpice convergence, the following remedies may be useful:

- Increase the iteration limit ITL4 from 10 to 100 or larger. This is an option accessed from the Simulation Profile Options, as shown in Fig. 1-18.
- Change the relative tolerance RELTOL to something other than the default value of 0.001.
- Change the device models to something that is less than ideal. For example, change the on resistance of a voltage-controlled switch to a larger value, or use a controlling voltage source that does not change as rapidly. An ideal diode could be made less ideal by increasing the value of n in the model. Generally, idealized device models will introduce more convergence problems than real device models.

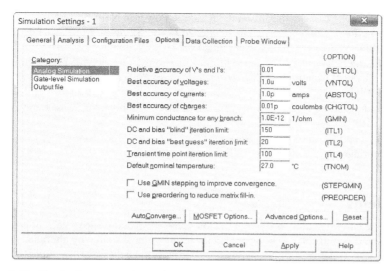

Figure 1-18 The Options menu for settings that can solve convergence problems. RELTOL and ITL4 have been changed here.

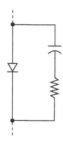

Figure 1-19 RC circuit to aid in PSpice convergence.

- Add an RC "snubber" circuit. A series resistance and capacitance with a small time constant can be placed across switches to prevent voltages from changing too rapidly. For example, placing a series combination of a 1-kΩ resistor and a 1-nF capacitor in parallel with a diode (Fig. 1-19) may improve convergence without affecting the simulation results.

1.8 BIBLIOGRAPHY

M. E. Balci and M. H. Hocaoglu, "Comparison of Power Definitions for Reactive Power Compensation in Nonsinusoidal Circuits," *International Conference on Harmonics and Quality of Power*, Lake Placid, N.Y. 2004.

L. S. Czarnecki, "Considerations on the Reactive Power in Nonsinusoidal Situations," *International Conference on Harmonics in Power Systems*, Worcester Polytechnic Institute, Worcester, Mass., 1984, pp 231–237.

A. E. Emanuel, "Powers in Nonsinusoidal Situations, A Review of Definitions and Physical Meaning," *IEEE Transactions on Power Delivery*, vol. 5, no. 3, July 1990.

G. T. Heydt, *Electric Power Quality*, Stars in a Circle Publications, West Lafayette, Ind., 1991.

W. Sheperd and P. Zand, *Energy Flow and Power Factor in Nonsinusoidal Circuits*, Cambridge University Press, 1979.

Problems

1-1. The current source in Example 1-1 is reversed so that positive current is upward. The current source is to be connected to the voltage source by alternately closing S_1 and S_2. Draw a circuit that has a MOSFET and a diode to accomplish this switching.

1-2. Simulate the circuit in Example 1-1 using PSpice. Use the voltage-controlled switch Sbreak for S_1 and the diode Dbreak for S_2. (*a*) Edit the PSpice models to idealize the circuit by using RON = 0.001 Ω for the switch and n = 0.001 for the diode. Display the voltage across the current source in Probe. (*b*) Use RON = 0.1 Ω in Sbreak and n = 1 (the default value) for the diode. How do the results of parts *a* and *b* differ?

1-3. The IRF150 power MOSFET model is in the EVAL library that accompanies the demonstration version of PSpice. Simulate the circuit in Example 1-1, using the IRF150 for the MOSFET and the default diode model Dbreak for S_2. Use an idealized gate drive circuit similar to that of Fig. 1-16. Display the voltage across the current source in Probe. How do the results differ from those using ideal switches?

1-4. Use PSpice to simulate the circuit of Example 1-1. Use the PSpice default BJT QbreakN for switch S_1. Use an idealized base drive circuit similar to that of the gate drive circuit for the MOSFET in Fig. 1-9. Choose an appropriate base resistance to ensure that the transistor turns on for a transistor h_{FE} of 100. Use the PSpice default diode Dbreak for switch S_2. Display the voltage across the current source. How do the results differ from those using ideal switches?

CHAPTER 2

Power Computations

2.1 INTRODUCTION

Power computations are essential in analyzing and designing power electronics circuits. Basic power concepts are reviewed in this chapter, with particular emphasis on power calculations for circuits with nonsinusoidal voltages and currents. Extra treatment is given to some special cases that are encountered frequently in power electronics. Power computations using the circuit simulation program PSpice are demonstrated.

2.2 POWER AND ENERGY

Instantaneous Power

The instantaneous power for any device is computed from the voltage across it and the current in it. *Instantaneous power* is

$$p(t) = v(t)i(t) \qquad (2\text{-}1)$$

This relationship is valid for any device or circuit. Instantaneous power is generally a time-varying quantity. If the passive sign convention illustrated in Fig. 2-1a is observed, the device is absorbing power if $p(t)$ is positive at a specified value of time t. The device is supplying power if $p(t)$ is negative. Sources frequently have an assumed current direction consistent with supplying power. With the convention of Fig. 2-1b, a positive $p(t)$ indicates the source is supplying power.

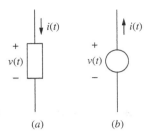

Figure 2-1 (*a*) Passive sign convention: $p(t) > 0$ indicates power is being absorbed; (*b*) $p(t) > 0$ indicates power is being supplied by the source.

Energy

Energy, or work, is the integral of instantaneous power. Observing the passive sign convention, energy absorbed by a component in the time interval from t_1 to t_2 is

$$W = \int_{t_1}^{t_2} p(t)\, dt \qquad (2\text{-}2)$$

If $v(t)$ is in volts and $i(t)$ is in amperes, power has units of watts and energy has units of joules.

Average Power

Periodic voltage and current functions produce a periodic instantaneous power function. Average power is the time average of $p(t)$ over one or more periods. *Average power P* is computed from

$$P = \frac{1}{T}\int_{t_0}^{t_0+T} p(t)\, dt = \frac{1}{T}\int_{t_0}^{t_0+T} v(t)i(t)\, dt \qquad (2\text{-}3)$$

where T is the period of the power waveform. Combining Eqs. (2-3) and (2-2), power is also computed from energy per period.

$$P = \frac{W}{T} \qquad (2\text{-}4)$$

Average power is sometimes called *real power* or *active power*, especially in ac circuits. The term *power* usually means average power. The total average power absorbed in a circuit equals the total average power supplied.

EXAMPLE 2-1

Power and Energy

Voltage and current, consistent with the passive sign convention, for a device are shown in Fig. 2-2a and b. (a) Determine the instantaneous power $p(t)$ absorbed by the device. (b) Determine the energy absorbed by the device in one period. (c) Determine the average power absorbed by the device.

■ **Solution**

(a) The instantaneous power is computed from Eq. (2-1). The voltage and current are expressed as

$$v(t) = \begin{cases} 20 \text{ V} & 0 < t < 10 \text{ ms} \\ 0 & 10 \text{ ms} < t < 20 \text{ ms} \end{cases}$$

$$i(t) = \begin{cases} 20 \text{ V} & 0 < t < 6 \text{ ms} \\ -15 \text{ A} & 6 \text{ ms} < t < 20 \text{ ms} \end{cases}$$

Instantaneous power, shown in Fig. 2-2c, is the product of voltage and current and is expressed as

$$p(t) = \begin{cases} 400 \text{ W} & 0 < t < 6 \text{ ms} \\ -300 \text{ W} & 6 \text{ ms} < t < 10 \text{ ms} \\ 0 & 10 \text{ ms} < t < 20 \text{ ms} \end{cases}$$

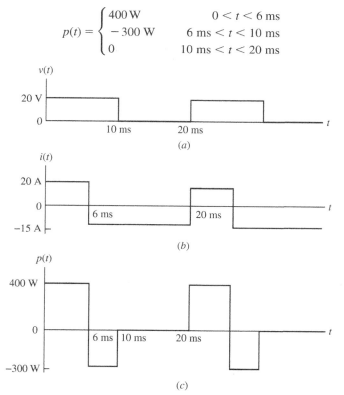

Figure 2-2 Voltage, current, and instantaneous power for Example 2-1.

(b) Energy absorbed by the device in one period is determined from Eq. (2-2).

$$W = \int_0^T p(t)\,dt = \int_0^{0.006} 400\,dt + \int_{0.006}^{0.010} -300\,dt + \int_{0.010}^{0.020} 0\,dt = 2.4 - 1.2 = 1.2 \text{ J}$$

(c) Average power is determined from Eq. (2-3).

$$P = \frac{1}{T}\int_0^T p(t)\,dt = \frac{1}{0.020}\left(\int_0^{0.006} 400\,dt + \int_{0.006}^{0.010} -300\,dt + \int_{0.010}^{0.020} 0\,dt\right)$$

$$= \frac{2.4 - 1.2 - 0}{0.020} = 60 \text{ W}$$

Average power could also be computed from Eq. (2-4) by using the energy per period from part (b).

$$P = \frac{W}{T} = \frac{1.2 \text{ J}}{0.020 \text{ s}} = 60 \text{ W}$$

A special case that is frequently encountered in power electronics is the power absorbed or supplied by a dc source. Applications include battery-charging circuits and dc power supplies. The average power absorbed by a dc voltage source $v(t) = V_{dc}$ that has a periodic current $i(t)$ is derived from the basic definition of average power in Eq. (2-3):

$$P_{dc} = \frac{1}{T}\int_{t_0}^{t_0+T} v(t)i(t)\,dt = \frac{1}{T}\int_{t_0}^{t_0+T} V_{dc}\,i(t)\,dt$$

Bringing the constant V_{dc} outside of the integral gives

$$P_{dc} = V_{dc}\left[\frac{1}{T}\int_{t_0}^{t_0+T} i(t)\,dt\right]$$

The term in brackets is the average of the current waveform. Therefore, *average power absorbed by a dc voltage source is the product of the voltage and the average current.*

$$\boxed{P_{dc} = V_{dc} I_{avg}} \qquad (2\text{-}5)$$

Similarly, average power absorbed by a dc source $i(t) = I_{dc}$ is

$$P_{dc} = I_{dc} V_{avg} \qquad (2\text{-}6)$$

2.3 INDUCTORS AND CAPACITORS

Inductors and capacitors have some particular characteristics that are important in power electronics applications. For periodic currents and voltages,

$$i(t + T) = i(t)$$
$$v(t + T) = v(t) \tag{2-7}$$

For an inductor, the stored energy is

$$w(t) = \frac{1}{2} Li^2(t) \tag{2-8}$$

If the inductor current is periodic, the stored energy at the end of one period is the same as at the beginning. No net energy transfer indicates that *the average power absorbed by an inductor is zero for steady-state periodic operation.*

$$\boxed{P_L = 0} \tag{2-9}$$

Instantaneous power is not necessarily zero because power may be absorbed during part of the period and returned to the circuit during another part of the period.

Furthermore, from the voltage-current relationship for the inductor

$$i(t_0 + T) = \frac{1}{L} \int_{t_0}^{t_0+T} v_L(t)\, dt + i(t_0) \tag{2-10}$$

Rearranging and recognizing that the starting and ending values are the same for periodic currents, we have

$$i(t_0 + T) - i(t_0) = \frac{1}{L} \int_{t_0}^{t_0+T} v_L(t)\, dt = 0 \tag{2-11}$$

Multiplying by L/T yields an expression equivalent to the average voltage across the inductor over one period.

$$\boxed{\mathrm{avg}[v_L(t)] = V_L = \frac{1}{T} \int_{t_0}^{t_0+T} v_L(t)\, dt = 0} \tag{2-12}$$

Therefore, *for periodic currents, the average voltage across an inductor is zero.* This is very important and will be used in the analysis of many circuits, including dc-dc converters and dc power supplies.

For a capacitor, stored energy is

$$w(t) = \frac{1}{2} Cv^2(t) \tag{2-13}$$

CHAPTER 2 Power Computations

If the capacitor voltage is periodic, the stored energy is the same at the end of a period as at the beginning. Therefore, *the average power absorbed by the capacitor is zero for steady-state periodic operation.*

$$\boxed{P_C = 0} \qquad (2\text{-}14)$$

From the voltage-current relationship for the capacitor,

$$v(t_0 + T) = \frac{1}{C}\int_{t_0}^{t_0+T} i_C(t)\, dt + v(t_0) \qquad (2\text{-}15)$$

Rearranging the preceding equation and recognizing that the starting and ending values are the same for periodic voltages, we get

$$v(t_0 + T) - v(t_0) = \frac{1}{C}\int_{t_0}^{t_0+T} i_C(t)\, dt = 0 \qquad (2\text{-}16)$$

Multiplying by C/T yields an expression for average current in the capacitor over one period.

$$\boxed{\text{avg}\,[i_C(t)] = I_C = \frac{1}{T}\int_{t_0}^{t_0+T} i_C(t)\, dt = 0} \qquad (2\text{-}17)$$

Therefore, *for periodic voltages, the average current in a capacitor is zero.*

EXAMPLE 2-2

Power and Voltage for an Inductor

The current in a 5-mH inductor of Fig. 2-3a is the periodic triangular wave shown in Fig. 2-3b. Determine the voltage, instantaneous power, and average power for the inductor.

■ **Solution**

The voltage across the inductor is computed from $v(t) = L(di/dt)$ and is shown in Fig. 2-3c. The average inductor voltage is zero, as can be determined from Fig. 2-3c by inspection. The instantaneous power in the inductor is determined from $p(t) = v(t)i(t)$ and is shown in Fig. 2-3d. When $p(t)$ is positive, the inductor is absorbing power, and when $p(t)$ is negative, the inductor is supplying power. The average inductor power is zero.

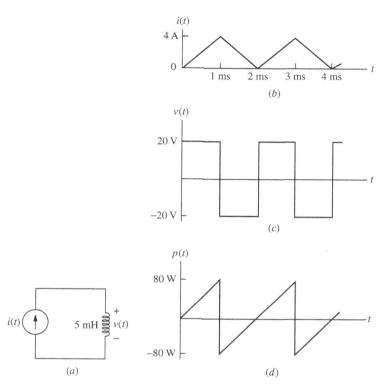

Figure 2.3 (a) Circuit for Example 2-2; (b) inductor current; (c) inductor voltage; (d) inductor instantaneous power.

2.4 ENERGY RECOVERY

Inductors and capacitors must be energized and deenergized in several applications of power electronics. For example, a fuel injector solenoid in an automobile is energized for a set time interval by a transistor switch. Energy is stored in the solenoid's inductance when current is established. The circuit must be designed to remove the stored energy in the inductor while preventing damage to the transistor when it is turned off. Circuit efficiency can be improved if stored energy can be transferred to the load or to the source rather than dissipated in circuit resistance. The concept of recovering stored energy is illustrated by the circuits described in this section.

Fig. 2-4a shows an inductor that is energized by turning on a transistor switch. The resistance associated with the inductance is assumed to be negligible, and the transistor switch and diode are assumed to be ideal. The diode-resistor path provides a means of opening the switch and removing the stored

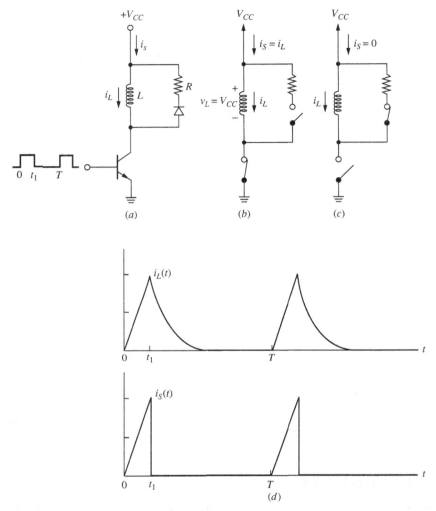

Figure 2-4 (*a*) A circuit to energize an inductance and then transfer the stored energy to a resistor; (*b*) Equivalent circuit when the transistor is on; (*c*) Equivalent circuit when the transistor is off and the diode is on; (*d*) Inductor and source currents.

energy in the inductor when the transistor turns off. Without the diode-resistor path, the transistor could be destroyed when it is turned off because a rapid decrease in inductor current would result in excessively high inductor and transistor voltages.

Assume that the transistor switch turns on at $t = 0$ and turns off at $t = t_1$. The circuit is analyzed first for the transistor switch on and then for the switch off.

Transistor on: $0 < t < t_1$

The voltage across the inductor is V_{CC}, and the diode is reverse-biased when the transistor is on (Fig. 2-4b).

$$v_L = V_{CC} \qquad (2\text{-}18)$$

An expression for inductor current is obtained from the voltage-current relationship:

$$i_L(t) = \frac{1}{L}\int_0^t v_L(\lambda)\,d\lambda + i_L(0) = \frac{1}{L}\int_0^t V_{CC}\,d\lambda + 0 = \frac{V_{CC}t}{L} \qquad (2\text{-}19)$$

Source current is the same as inductor current.

$$i_s(t) = i_L(t) \qquad (2\text{-}20)$$

Inductor and source currents thus increase linearly when the transistor is on.
The circuit is next analyzed for the transistor switch off.

Transistor off: $t_1 < t < T$

In the interval $t_1 < t < T$, the transistor switch is off and the diode is on (Fig. 2-4c). The current in the source is zero, and the current in the inductor and resistor is a decaying exponential with time constant L/R. The initial condition for inductor current is determined from Eq. (2-19):

$$i_L(t_1) = \frac{V_{CC}t_1}{L} \qquad (2\text{-}21)$$

Inductor current is then expressed as

$$i_L(t) = i_L(t_1)e^{-(t-t_1)/\tau} = \left(\frac{V_{CC}t_1}{L}\right)e^{-(t-t_1)/\tau} \quad t_1 < t < T \qquad (2\text{-}22)$$

where $\tau = L/R$. Source current is zero when the transistor is off.

$$i_S = 0 \qquad (2\text{-}23)$$

Average power supplied by the dc source during the switching period is determined from the product of voltage and average current [Eq. (2-5)].

$$P_S = V_S I_S = V_{CC}\left[\frac{1}{T}\int_0^T i_s(t)\,dt\right]$$

$$= V_{CC}\left[\frac{1}{T}\int_0^{t_1} \frac{V_{CC}t}{L}\,dt + \frac{1}{T}\int_{t_1}^T 0\,dt\right] = \frac{(V_{CC}t_1)^2}{2LT} \qquad (2\text{-}24)$$

Average power absorbed by the resistor could be determined by integrating an expression for instantaneous resistor power, but an examination of the circuit reveals an easier way. The average power absorbed by the inductor is zero, and power absorbed by the ideal transistor and diode is zero. Therefore, all power supplied by the source must be absorbed by the resistor:

$$P_R = P_S = \frac{(V_{CC} t_1)^2}{2LT} \qquad (2\text{-}25)$$

Another way to approach the problem is to determine the peak energy stored in the inductor,

$$W = \frac{1}{2} L i^2(t_1) = \frac{1}{2} L \left(\frac{V_{CC} t_1}{L} \right)^2 = \frac{(V_{CC} t_1)^2}{2L} \qquad (2\text{-}26)$$

The energy stored in the inductor is transferred to the resistor while the transistor switch is open. Power absorbed by the resistor can be determined from Eq. (2-4).

$$P_R = \frac{W}{T} = \frac{(V_{CC} t_1)^2}{2LT} \qquad (2\text{-}27)$$

which must also be the power supplied by the source. The function of the resistor in this circuit of Fig. 2-4a is to absorb the stored energy in the inductance and protect the transistor. This energy is converted to heat and represents a power loss in the circuit.

Another way to remove the stored energy in the inductor is shown in Fig. 2-5a. Two transistor switches are turned on and off simultaneously. The diodes provide a means of returning energy stored in the inductor back to the source. Assume that the transistors turn on at $t = 0$ and turn off at $t = t_1$. The analysis of the circuit of Fig. 2-5a begins with the transistors on.

Transistors on: $0 < t < t_1$

When the transistors are on, the diodes are reverse-biased, and the voltage across the inductor is V_{CC}. The inductor voltage is the same as the source when the transistors are on (Fig. 2-5b):

$$v_L = V_{CC} \qquad (2\text{-}28)$$

Inductor current is the function

$$i_L(t) = \frac{1}{L} \int_0^t v_L(\lambda)\, d\lambda + i_L(0) = \frac{1}{L} \int_0^t V_{CC}\, d\lambda + 0 = \frac{V_{CC} t}{L} \qquad (2\text{-}29)$$

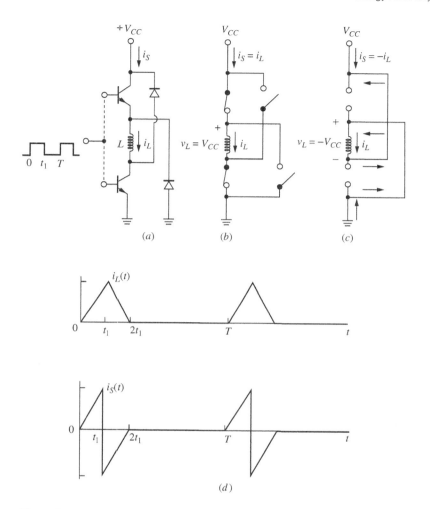

Figure 2-5 (*a*) A circuit to energize an inductance and recover the stored energy by transferring it back to the source; (*b*) Equivalent circuit when the transistors are on; (*c*) Equivalent circuit when the transistors are off and the diodes are on; (*d*) Inductor and source currents.

Source current is the same as inductor current.

$$i_S(t) = i_L(t) \qquad (2\text{-}30)$$

From the preceding equations, inductor and source currents increase linearly while the transistor switches are on, as was the case for the circuit of Fig. 2-4*a*.

The circuit is next analyzed for the transistors off.

Transistors off: $t_1 < t < T$

When the transistors are turned off, the diodes become forward-biased to provide a path for the inductor current (Fig. 2-5c). The voltage across the inductor then becomes the opposite of the source voltage:

$$v_L = -V_{CC} \tag{2-31}$$

An expression for inductor current is obtained from the voltage-current relationship.

$$i_L(t) = \frac{1}{L}\int_{t_1}^{t} v_L(\lambda)\,d\lambda + i_L(t_1) = \frac{1}{L}\int_{t_1}^{t}(-V_{CC})\,d\lambda + \frac{V_{CC}t_1}{L}$$

$$= \left(\frac{V_{CC}}{L}\right)[(t_1 - t) + t_1]$$

or,

$$i_L(t) = \left(\frac{V_{CC}}{L}\right)(2t_1 - t) \qquad t_1 < t < 2t_1 \tag{2-32}$$

Inductor current decreases and becomes zero at $t = 2t_1$, at which time the diodes turn off. Inductor current remains at zero until the transistors turn on again.

Source current is the opposite of inductor current when the transistors are off and the diodes are on:

$$i_S(t) = -i_L(t) \tag{2-33}$$

The source is absorbing power when the source current is negative. Average source current is zero, resulting in an average source power of zero.

The source supplies power while the transistors are on, and the source absorbs power while the transistors are off and the diodes are on. Therefore, the energy stored in the inductor is recovered by transferring it back to the source. Practical solenoids or other magnetic devices have equivalent resistances that represent losses or energy absorbed to do work, so not all energy will be returned to the source. The circuit of Fig. 2-5a has no energy losses inherent to the design and is therefore more efficient than that of Fig. 2-4a.

EXAMPLE 2-3

Energy Recovery

The circuit of Fig. 2-4a has $V_{CC} = 90$ V, $L = 200$ mH, $R = 20\ \Omega$, $t_1 = 10$ ms, and $T = 100$ ms. Determine (a) the peak current and peak energy storage in the inductor, (b) the average power absorbed by the resistor, and (c) the peak and average power supplied by the source. (d) Compare the results with what would happen if the inductor were energized using the circuit of Fig. 2-5a.

■ Solution

(a) From Eq. (2-19), when the transistor switch is on, inductor current is

$$i_L(t) = \left(\frac{V_{CC}}{L}\right)t = \left(\frac{90}{0.2}\right)t = 450t \text{ A} \quad 0 < t < 10 \text{ ms}$$

Peak inductor current and stored energy are

$$i_L(t_1) = 450(0.01) = 4.5 \text{ A}$$

$$W_L = \frac{1}{2}Li^2(t_1) = \frac{1}{2}(0.2)(4.5)^2 = 2.025 \text{ J}$$

(b) The time constant for the current when the switch is open is $L/R = 200 \text{ mH}/20\, \Omega = 10$ ms. The switch is open for 90 ms, which is 10 time constants, so essentially all stored energy in the inductor is transferred to the resistor:

$$W_R = W_L = 2.025 \text{ J}$$

Average power absorbed by the resistor is determined from Eq. (2-4):

$$P_R = \frac{W_R}{T} = \frac{2.025 \text{ J}}{0.1 \text{ s}} = 20.25 \text{ W}$$

(c) The source current is the same as the inductor current when the switch is closed and is zero when the switch is open. Instantaneous power supplied by the source is

$$p_S(t) = v_S(t)i_S(t) = \begin{cases} (90 \text{ V})(450t \text{ A}) = 40{,}500t \text{ W} & 0 < t < 10 \text{ ms} \\ 0 & 10 \text{ ms} < t < 100 \text{ ms} \end{cases}$$

which has a maximum value of 405 W at $t = 10$ ms. Average power supplied by the source can be determined from Eq. (2-3):

$$P_S = \frac{1}{T}\int_0^T p_S(t)\,dt = \frac{1}{0.1}\left(\int_0^{0.01} 40{,}500t\,dt + \int_{0.01}^{0.1} 0\,dt\right) = 20.25 \text{ W}$$

Average source power also can be determined from Eq. (2-5). Average of the triangular source current waveform over one period is

$$I_S = \frac{1}{2}\left[\frac{(0.01 \text{ s})(4.5 \text{ A})}{0.1 \text{ s}}\right] = 0.225 \text{ A}$$

and average source power is then

$$P_S = V_{CC}I_S = (90 \text{ V})(0.225 \text{ A}) = 20.25 \text{ W}$$

Still another computation of average source power comes from recognizing that the power absorbed by the resistor is the same as that supplied by the source.

$$P_S = P_R = 20.25 \text{ W}$$

(See Example 2-13 at the end of this chapter for the PSpice simulation of this circuit.)

(d) When the inductor is energized from the circuit of Fig. 2-5a, the inductor current is described by Eqs. (2-29) and (2-32).

$$i_L(t) = \begin{cases} 450t \text{ A} & 0 < t < 10 \text{ ms} \\ 9 - 450t \text{ A} & 10 \text{ ms} < t < 20 \text{ ms} \\ 0 & 20 \text{ ms} < t < 100 \text{ ms} \end{cases}$$

The peak current and peak energy storage are the same as for the circuit of Fig. 2-4a. The source current has the form shown in Fig. 2-5d and is expressed as

$$i_S(t) = \begin{cases} 450t \text{ A} & 0 < t < 10 \text{ ms} \\ 450t - 9 \text{ A} & 10 \text{ ms} < t < 20 \text{ ms} \\ 0 & 20 \text{ ms} < t < 100 \text{ ms} \end{cases}$$

Instantaneous power supplied by the source is

$$p_S(t) = 90 i_S(t) = \begin{cases} 40{,}500t \text{ W} & 0 < t < 10 \text{ ms} \\ 40{,}500t - 810 \text{ W} & 10 \text{ ms} < t < 20 \text{ ms} \\ 0 & 20 \text{ ms} < t < 100 \text{ ms} \end{cases}$$

Average source current is zero, and average source power is zero. Peak source power is peak current times voltage, which is 405 W as in part (c).

2.5 EFFECTIVE VALUES: RMS

The effective value of a voltage or current is also known as the root-mean-square (rms) value. The effective value of a periodic voltage waveform is based on the average power delivered to a resistor. For a dc voltage across a resistor,

$$P = \frac{V_{dc}^2}{R} \tag{2-34}$$

For a periodic voltage across a resistor, *effective voltage* is defined as the voltage that is as effective as the dc voltage in supplying average power. Effective voltage can be computed using the equation

$$P = \frac{V_{\text{eff}}^2}{R} \tag{2-35}$$

Computing average resistor power from Eq. (2-3) gives

$$P = \frac{1}{T}\int_0^T p(t)\,dt = \frac{1}{T}\int_0^T v(t)i(t)\,dt = \frac{1}{T}\int_0^T \frac{v^2(t)}{R}\,dt$$

$$= \frac{1}{R}\left[\frac{1}{T}\int_0^T v^2(t)\,dt\right] \tag{2-36}$$

2.5 Effective Values: RMS

Equating the expressions for average power in Eqs. (2-35) and (2-36) gives

$$P = \frac{V_{\text{eff}}^2}{R} = \frac{1}{R}\left[\frac{1}{T}\int_0^T v^2(t)\,dt\right]$$

or

$$V_{\text{eff}}^2 = \frac{1}{T}\int_0^T v^2(t)\,dt$$

resulting in the expression for effective or rms voltage

$$V_{\text{eff}} = V_{\text{rms}} = \sqrt{\frac{1}{T}\int_0^T v^2(t)\,dt} \qquad (2\text{-}37)$$

The effective value is the square *root* of the *mean* of the *square* of the voltage—hence the term *root mean square*.

Similarly, rms current is developed from $P = I_{\text{rms}}^2$ as

$$I_{\text{rms}} = \sqrt{\frac{1}{T}\int_0^T i^2(t)\,dt} \qquad (2\text{-}38)$$

The usefulness of the rms value of voltages and currents lies in the computing power absorbed by resistances. Additionally, ac power system voltages and currents are invariably given in rms values. Ratings of devices such as transformers are often specified in terms of rms voltage and current.

EXAMPLE 2-4

RMS Value of a Pulse Waveform

Determine the rms value of the periodic pulse waveform that has a duty ratio of D as shown in Fig. 2-6.

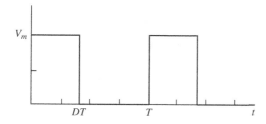

Figure 2-6 Pulse waveform for Example 2-4.

Solution

The voltage is expressed as

$$v(t) = \begin{cases} V_m & 0 < t < DT \\ 0 & DT < t < T \end{cases}$$

Using Eq. (2-37) to determine the rms value of the waveform gives

$$V_{rms} = \sqrt{\frac{1}{T}\int_0^T v^2(t)\,dt} = \sqrt{\frac{1}{T}\left(\int_0^{DT} V_m^2\,dt + \int_{DT}^T 0^2\,dt\right)} = \sqrt{\frac{1}{T}(V_m^2\,DT)}$$

yielding

$$V_{rms} = V_m\sqrt{D}$$

EXAMPLE 2-5

RMS Values of Sinusoids

Determine the rms values of (a) a sinusoidal voltage of $v(t) = V_m \sin(\omega t)$, (b) a full-wave rectified sine wave of $v(t) = |V_m \sin(\omega t)|$, and (c) a half-wave rectified sine wave of $v(t) = V_m \sin(\omega t)$ for $0 < t < T/2$ and zero otherwise.

Solution

(a) The rms value of the sinusoidal voltage is computed from Eq. (2-37):

$$V_{rms} = \sqrt{\frac{1}{T}\int_0^T V_m^2 \sin^2(\omega t)\,dt} \quad \text{where } T = \frac{2\pi}{\omega}$$

An equivalent expression uses ωt as the variable of integration. Without showing the details of the integration, the result is

$$V_{rms} = \sqrt{\frac{1}{2\pi}\int_0^{2\pi} V_m^2 \sin^2(\omega t)\,d(\omega t)} = \frac{V_m}{\sqrt{2}}$$

Note that the rms value is independent of the frequency.

(b) Equation (2-37) can be applied to the full-wave rectified sinusoid, but the results of part (a) can also be used to advantage. The rms formula uses the integral of the square of the function. The square of the sine wave is identical to the square of the full-wave rectified sine wave, so the rms values of the two waveforms are identical:

$$V_{rms} = \frac{V_m}{\sqrt{2}}$$

(c) Equation (2-37) can be applied to the half-wave rectified sinusoid.

$$V_{rms} = \sqrt{\frac{1}{2\pi}\left(\int_0^\pi V_m^2 \sin^2(\omega t)\, d(\omega t) + \int_\pi^{2\pi} 0^2\, d(\omega t)\right)} = \sqrt{\frac{1}{2\pi}\int_0^\pi V_m^2 \sin^2(\omega t)\, d(\omega t)}$$

The result of part (a) will again be used to evaluate this expression. The square of the function has one-half the area of that of the functions in (a) and (b). That is,

$$V_{rms} = \sqrt{\frac{1}{2\pi}\int_0^\pi V_m^2 \sin^2(\omega t)\, d(\omega t)} = \sqrt{\left(\frac{1}{2}\right)\frac{1}{2\pi}\int_0^{2\pi} V_m^2 \sin^2(\omega t)\, d(\omega t)}$$

Taking the 1/2 outside of the square root gives

$$V_{rms} = \left(\sqrt{\frac{1}{2}}\right)\sqrt{\frac{1}{2\pi}\int_0^{2\pi} V_m^2 \sin^2(\omega t)\, d(\omega t)}$$

The last term on the right is the rms value of a sine wave which is known to be $V_m/\sqrt{2}$, so the rms value of a half-wave rectified sine wave is

$$V_{rms} = \sqrt{\frac{1}{2}}\frac{V_m}{\sqrt{2}} = \frac{V_m}{2}$$

Figure 2-7 shows the waveforms.

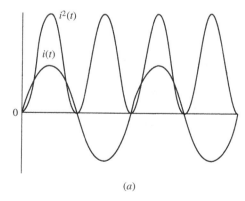

(a)

Figure 2-7 Waveforms and their squares for Example 2-5 (a) Sine wave; (b) full-wave rectified sine wave; (c) half-wave rectified sine wave.

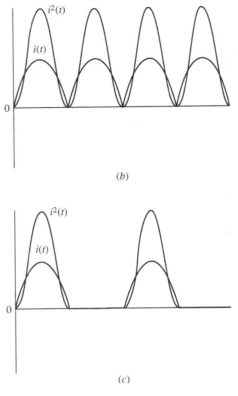

(b)

(c)

Figure 2-7 (*continued*)

EXAMPLE 2-6

Neutral Conductor Current in a Three-Phase System

An office complex is supplied from a three-phase four-wire voltage source (Fig. 2-8a). The load is highly nonlinear as a result of the rectifiers in the power supplies of the equipment, and the current in each of the three phases is shown in Fig. 2-8b. The neutral current is the sum of the phase currents. If the rms current in each phase conductor is known to be 20 A, determine the rms current in the neutral conductor.

■ **Solution**

Equation (2-38) may be applied to this case. Noting by inspection that the area of the square of the current function in the neutral i_n, is 3 times that of each of the phases i_a (Fig. 2-8c)

$$I_{n,\,\text{rms}} = \sqrt{\frac{1}{T}\int_0^T i_n^2(t)\,dt} = \sqrt{3\left(\frac{1}{T}\int_0^T i_a^2(t)\,dt\right)} = \sqrt{3}\,I_{a,\,\text{rms}}$$

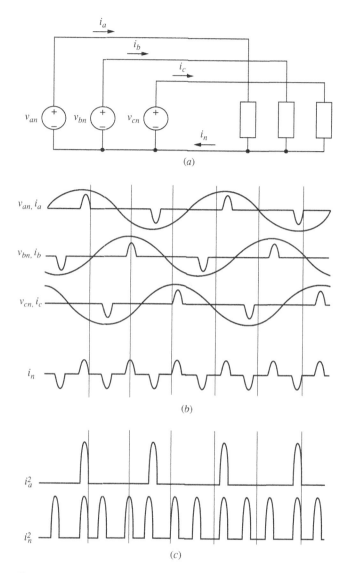

Figure 2-8 (a) Three-phase source supplying a balanced nonlinear three-phase load for Example 2-8; (b) phase and neutral currents; (c) squares of i_a and i_n.

The rms current in the neutral is therefore

$$I_{n,\,\mathrm{rms}} = \sqrt{3}\,(20) = 34.6 \text{ A}$$

Note that the rms neutral current is larger than the phase currents for this situation. This is much different from that for balanced linear loads where the line currents are sinusoids which are displaced by 120° and sum to zero. Three-phase distribution systems supplying highly nonlinear loads should have a neutral conductor capable of carrying $\sqrt{3}$ times as much current as the line conductor.

If a periodic voltage is the sum of two periodic voltage waveforms, $v(t) = v_1(t) + v_2(t)$, the rms value of $v(t)$ is determined from Eq. (2-37) as

$$V_{rms}^2 = \frac{1}{T}\int_0^T (v_1 + v_2)^2 \, dt = \frac{1}{T}\int_0^T \left(v_1^2 + 2v_1 v_2 + v_2^2\right) dt$$

or

$$V_{rms}^2 = \frac{1}{T}\int_0^T v_1^2 \, dt + \frac{1}{T}\int_0^T 2v_1 v_2 \, dt + \frac{1}{T}\int_0^T v_2^2 \, dt$$

The term containing the product $v_1 v_2$ in the above equation is zero if the functions v_1 and v_2 are orthogonal. A condition that satisfies that requirement occurs when v_1 and v_2 are sinusoids of different frequencies. For orthogonal functions,

$$V_{rms}^2 = \frac{1}{T}\int_0^T v_1^2(t) \, dt + \frac{1}{T}\int_0^T v_2^2(t) \, dt$$

Noting that

$$\frac{1}{T}\int_0^T v_1^2(t) \, dt = V_{1,rms}^2 \quad \text{and} \quad \frac{1}{T}\int_0^T v_2^2(t) \, dt = V_{2,rms}^2$$

then

$$V_{rms} = \sqrt{V_{1,rms}^2 + V_{2,rms}^2}$$

If a voltage is the sum of more than two periodic voltages, all orthogonal, the rms value is

$$V_{rms} = \sqrt{V_{1,rms}^2 + V_{2,rms}^2 + V_{3,rms}^2 + \cdots} = \sqrt{\sum_{n=1}^{N} V_{n,rms}^2} \quad (2\text{-}39)$$

Similarly,

$$I_{rms} = \sqrt{I_{1,rms}^2 + I_{2,rms}^2 + I_{3,rms}^2 + \cdots} = \sqrt{\sum_{n=1}^{N} I_{n,rms}^2} \quad (2\text{-}40)$$

Note that Eq. (2-40) can be applied to Example 2-6 to obtain the rms value of the neutral current.

2.5 Effective Values: RMS

EXAMPLE 2-7
RMS Value of the Sum of Waveforms

Determine the effective (rms) value of $v(t) = 4 + 8 \sin(\omega_1 t + 10°) + 5 \sin(\omega_2 t + 50°)$ for (a) $\omega_2 = 2\omega_1$ and (b) $\omega_2 = \omega_1$.

■ **Solution**

(a) The rms value of a single sinusoid is $V_m/\sqrt{2}$, and the rms value of a constant is the constant. When the sinusoids are of different frequencies, the terms are orthogonal and Eq. (2-39) applies.

$$V_{rms} = \sqrt{V_{1,rms}^2 + V_{2,rms}^2 + V_{3,rms}^2} = \sqrt{4^2 + \left(\frac{8}{\sqrt{2}}\right)^2 + \left(\frac{5}{\sqrt{2}}\right)^2} = 7.78 \text{ V}$$

(b) For sinusoids of the same frequency, Eq. (2-39) does not apply because the integral of the cross product over one period is not zero. First combine the sinusoids using phasor addition:

$$8\angle 10° + 5\angle 50° = 12.3\angle 25.2°$$

The voltage function is then expressed as

$$v(t) = 4 + 12.3 \sin(\omega_1 t + 25.2°) \text{ V}$$

The rms value of this voltage is determined from Eq. (2-39) as

$$V_{rms} = \sqrt{4^2 + \left(\frac{12.3}{\sqrt{2}}\right)^2} = 9.57 \text{ V}$$

EXAMPLE 2-8
RMS Value of Triangular Waveforms

(a) A triangular current waveform like that shown in Fig. 2-9a is commonly encountered in dc power supply circuits. Determine the rms value of this current.

(b) Determine the rms value of the offset triangular waveform in Fig. 2-9b.

■ **Solution**

(a) The current is expressed as

$$i(t) = \begin{cases} \dfrac{2I_m}{t_1} t - I_m & 0 < t < t_1 \\ \dfrac{-2I_m}{T - t_1} t + \dfrac{I_m(T + t_1)}{T - t_1} & t_1 < t < T \end{cases}$$

The rms value is determined from Eq. (2-38).

$$I_{rms}^2 = \frac{1}{T}\left[\int_0^{t_1} \left(\frac{2I_m}{t_1} t - I_m\right)^2 dt + \int_{t_1}^{T} \left(\frac{-2I_m}{T - t_1} t + \frac{I_m(T + t_1)}{T - t_1}\right)^2 dt\right]$$

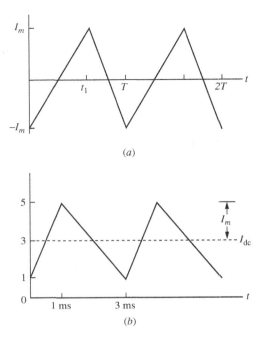

Figure 2-9 (a) Triangular waveform for Example 2-8; (b) offset triangular waveform.

The details of the integration are quite long, but the result is simple: The rms value of a triangular current waveform is

$$I_{\text{rms}} = \frac{I_m}{\sqrt{3}}$$

(b) The rms value of the offset triangular waveform can be determined by using the result of part (a). Since the triangular waveform of part (a) contains no dc component, the dc signal and the triangular waveform are orthogonal, and Eq. (2-40) applies.

$$I_{\text{rms}} = \sqrt{I_{1,\text{rms}}^2 + I_{2,\text{rms}}^2} = \sqrt{\left(\frac{I_m}{\sqrt{3}}\right)^2 + I_{\text{dc}}^2} = \sqrt{\left(\frac{2}{\sqrt{3}}\right)^2 + 3^2} = 3.22 \text{ A}$$

2.6 APPARENT POWER AND POWER FACTOR

Apparent Power S

Apparent power is the product of rms voltage and rms current magnitudes and is often used in specifying the rating of power equipment such as transformers. Apparent power is expressed as

$$S = V_{\text{rms}} I_{\text{rms}} \tag{2-41}$$

In ac circuits (linear circuits with sinusoidal sources), apparent power is the magnitude of complex power.

Power Factor

The *power factor* of a load is defined as the ratio of average power to apparent power:

$$\text{pf} = \frac{P}{S} = \frac{P}{V_{\text{rms}} I_{\text{rms}}} \qquad (2\text{-}42)$$

In sinusoidal ac circuits, the above calculation results in pf $= \cos\theta$ where θ is the phase angle between the voltage and current sinusoids. However, that is a special case and should be used only when both voltage and current are sinusoids. In general, power factor must be computed from Eq. (2-42).

2.7 POWER COMPUTATIONS FOR SINUSOIDAL AC CIRCUITS

In general, voltages and/or currents in power electronics circuits are not sinusoidal. However, a nonsinusoidal periodic waveform can be represented by a Fourier series of sinusoids. It is therefore important to understand thoroughly power computations for the sinusoidal case. The following discussion is a review of power computations for ac circuits.

For linear circuits that have sinusoidal sources, all steady-state voltages and currents are sinusoids. Instantaneous power and average power for ac circuits are computed using Eqs. (2-1) and (2-3) as follows:

For any element in an ac circuit, let

$$\begin{aligned} v(t) &= V_m \cos(\omega t + \theta) \\ i(t) &= I_m \cos(\omega t + \phi) \end{aligned} \qquad (2\text{-}43)$$

Then instantaneous power is

$$p(t) = v(t)i(t) = [V_m \cos(\omega t + \theta)][I_m \cos(\omega t + \phi)] \qquad (2\text{-}44)$$

Using the trigonometric identity gives

$$(\cos A)(\cos B) = \frac{1}{2}[\cos(A+B) + \cos(A-B)] \qquad (2\text{-}45)$$

$$p(t) = \left(\frac{V_m I_m}{2}\right)[\cos(2\omega t + \theta + \phi) + \cos(\theta - \phi)] \qquad (2\text{-}46)$$

Average power is

$$P = \frac{1}{T}\int_0^T p(t)\,dt = \left(\frac{V_m I_m}{2}\right)\int_0^T [\cos(2\omega t + \theta + \phi) + \cos(\theta - \phi)]\,dt \qquad (2\text{-}47)$$

The result of the above integration can be obtained by inspection. Since the first term in the integration is a cosine function, the integral over one period is zero because of equal areas above and below the time axis. The second term in the integration is the constant $\cos(\theta - \phi)$, which has an average value of $\cos(\theta - \phi)$. Therefore, the average power in any element in an ac circuit is

$$P = \left(\frac{V_m I_m}{2}\right) \cos(\theta - \phi) \tag{2-48}$$

This equation is frequently expressed as

$$P = V_{rms} I_{rms} \cos(\theta - \phi) \tag{2-49}$$

where $V_{rms} = V_m/\sqrt{2}$, $I_{rms} = I_m/\sqrt{2}$, and $\theta - \phi$ is the phase angle between voltage and current. The power factor is determined to be $\cos(\theta - \phi)$ by using Eq. (2-42).

In the steady state, no net power is absorbed by an inductor or a capacitor. The term *reactive power* is commonly used in conjunction with voltages and currents for inductors and capacitors. Reactive power is characterized by energy storage during one-half of the cycle and energy retrieval during the other half. Reactive power is computed with a relationship similar to Eq. (2-49):

$$Q = V_{rms} I_{rms} \sin(\theta - \phi) \tag{2-50}$$

By convention, inductors absorb positive reactive power and capacitors absorb negative reactive power.

Complex power combines real and reactive powers for ac circuits:

$$\mathbf{S} = P + jQ = (\mathbf{V}_{rms})(\mathbf{I}_{rms})^* \tag{2-51}$$

In the above equation, $\mathbf{V}_{rms}$ and $\mathbf{I}_{rms}$ are complex quantities often expressed as phasors (magnitude and angle), and $(\mathbf{I}_{rms})^*$ is the complex conjugate of phasor current, which gives results consistent with the convention that inductance, or lagging current, absorbs reactive power. Apparent power in ac circuits is the magnitude of complex power:

$$S = |\mathbf{S}| = \sqrt{P^2 + Q^2} \tag{2-52}$$

It is important to note that the complex power in Eq. (2-52) and power factor of $\cos(\theta - \phi)$ for sinusoidal ac circuits are special cases and are not applicable to nonsinusoidal voltages and currents.

2.8 POWER COMPUTATIONS FOR NONSINUSOIDAL PERIODIC WAVEFORMS

Power electronics circuits typically have voltages and/or currents that are periodic but not sinusoidal. For the general case, the basic definitions for the power terms described at the beginning of this chapter must be applied. A common error that is made when doing power computations is to attempt to apply some special relationships for sinusoids to waveforms that are not sinusoids.

2.8 Power Computations for Nonsinusoidal Periodic Waveforms

The Fourier series can be used to describe nonsinusoidal periodic waveforms in terms of a series of sinusoids. The power relationships for these circuits can be expressed in terms of the components of the Fourier series.

Fourier Series

A nonsinusoidal periodic waveform that meets certain conditions can be described by a Fourier series of sinusoids. The Fourier series for a periodic function $f(t)$ can be expressed in trigonometric form as

$$f(t) = a_0 + \sum_{n=1}^{\infty} [a_n \cos(n\omega_0 t) + b_n \sin(n\omega_0 t)] \qquad (2\text{-}53)$$

where

$$a_0 = \frac{1}{T} \int_{-T/2}^{T/2} f(t)\, dt$$

$$a_n = \frac{2}{T} \int_{-T/2}^{T/2} f(t) \cos(n\omega_0 t)\, dt \qquad (2\text{-}54)$$

$$b_n = \frac{2}{T} \int_{-T/2}^{T/2} f(t) \sin(n\omega_0 t)\, dt$$

Sines and cosines of the same frequency can be combined into one sinusoid, resulting in an alternative expression for a Fourier series:

$$f(t) = a_0 + \sum_{n=1}^{\infty} C_n \cos(n\omega_0 t + \theta_n)$$

where $\qquad (2\text{-}55)$

$$C_n = \sqrt{a_n^2 + b_n^2} \quad \text{and} \quad \theta_n = \tan^{-1}\left(\frac{-b_n}{a_n}\right)$$

or

$$f(t) = a_0 + \sum_{n=1}^{\infty} C_n \sin(n\omega_0 t + \theta_n)$$

where $\qquad (2\text{-}56)$

$$C_n = \sqrt{a_n^2 + b_n^2} \quad \text{and} \quad \theta_n = \tan^{-1}\left(\frac{a_n}{b_n}\right)$$

The term a_0 is a constant that is the average value of $f(t)$ and represents a dc voltage or current in electrical applications. The coefficient C_1 is the amplitude of the term at the fundamental frequency ω_0. Coefficients $C_2, C_3, \ldots$ are the amplitudes of the harmonics that have frequencies $2\omega_0, 3\omega_0, \ldots$.

The rms value of $f(t)$ can be computed from the Fourier series:

$$F_{\text{rms}} = \sqrt{\sum_{n=0}^{\infty} F_{n,\text{rms}}^2} = \sqrt{a_0^2 + \sum_{n=1}^{\infty} \left(\frac{C_n}{\sqrt{2}}\right)^2} \qquad (2\text{-}57)$$

Average Power

If periodic voltage and current waveforms represented by the Fourier series

$$v(t) = V_0 + \sum_{n=1}^{\infty} V_n \cos(n\omega_0 t + \theta_n)$$
$$i(t) = I_0 + \sum_{n=1}^{\infty} I_n \cos(n\omega_0 t + \phi_n) \qquad (2\text{-}58)$$

exist for a device or circuit, then average power is computed from Eq. (2-3).

$$P = \frac{1}{T} \int_0^T v(t)i(t)\,dt$$

The average of the products of the dc terms is $V_0 I_0$. The average of voltage and current products at the same frequency is described by Eq. (2-49), and the average of voltage and current products of different frequencies is zero. Consequently, average power for nonsinusoidal periodic voltage and current waveforms is

$$P = \sum_{n=0}^{\infty} P_n = V_0 I_0 + \sum_{n=1}^{\infty} V_{n,\text{rms}} I_{n,\text{rms}} \cos(\theta_n - \phi_n)$$

or $\qquad (2\text{-}59)$

$$P = V_0 I_0 + \sum_{n=1}^{\infty} \left(\frac{V_{n,\text{max}} I_{n,\text{max}}}{2}\right) \cos(\theta_n - \phi_n)$$

Note that total average power is the sum of the powers at the frequencies in the Fourier series.

Nonsinusoidal Source and Linear Load

If a nonsinusoidal periodic voltage is applied to a load that is a combination of linear elements, the power absorbed by the load can be determined by using superposition. A nonsinusoidal periodic voltage is equivalent to the series combination of the Fourier series voltages, as illustrated in Fig. 2-10. The current in the load can be determined using superposition, and Eq. (2-59) can be applied to compute average power. Recall that superposition for power is not valid when the sources are of the same frequency. The technique is demonstrated in Example 2-9.

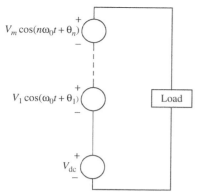

Figure 2.10 Equivalent circuit for Fourier analysis.

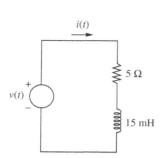

Figure 2.11 Circuit for Example 2-9.

EXAMPLE 2-9

Nonsinusoidal Source and Linear Load

A nonsinusoidal periodic voltage has a Fourier series of $v(t) = 10 + 20 \cos(2\pi 60 t - 25°) + 30 \cos(4\pi 60 t + 20°)$ V. This voltage is connected to a load that is a 5-Ω resistor and a 15-mH inductor connected in series as in Fig. 2-11. Determine the power absorbed by the load.

■ **Solution**

Current at each source frequency is computed separately. The dc current term is

$$I_0 = \frac{V_0}{R} = \frac{10}{5} = 2 \text{ A}$$

The amplitudes of the ac current terms are computed from phasor analysis:

$$I_1 = \frac{V_1}{R + j\omega_1 L} = \frac{20 \angle (-25°)}{5 + j(2\pi 60)(0.015)} = 2.65 \angle (-73.5°) \text{ A}$$

$$I_2 = \frac{V_2}{R + j\omega_2 L} = \frac{30 \angle 20°}{5 + j(4\pi 60)(0.015)} = 2.43 \angle (-46.2°) \text{ A}$$

Load current can then be expressed as

$$i(t) = 2 + 2.65 \cos(2\pi 60 t - 73.5°) + 2.43 \cos(4\pi 60 t - 46.2°) \text{ A}$$

Power at each frequency in the Fourier series is determined from Eq. (2-59):

dc term: $\quad P_0 = (10 \text{ V})(2 \text{ A}) = 20 \text{ W}$

$\omega = 2\pi 60: \quad P_1 = \dfrac{(20)(2.65)}{2} \cos(-25° + 73.5°) = 17.4 \text{ W}$

$\omega = 4\pi 60: \quad P_2 = \dfrac{(30)(2.43)}{2} \cos(20° + 46°) = 14.8 \text{ W}$

Total power is then

$$P = 20 + 17.4 + 14.8 = 52.2 \text{ W}$$

Power absorbed by the load can also be computed from $I_{\text{rms}}^2 R$ in this circuit because the average power in the inductor is zero.

$$P = I_{\text{rms}}^2 R = \left[2^2 + \left(\frac{2.65}{\sqrt{2}}\right)^2 + \left(\frac{2.43}{\sqrt{2}}\right)^2\right] 5 = 52.2 \text{ W}$$

Sinusoidal Source and Nonlinear Load

If a sinusoidal voltage source is applied to a nonlinear load, the current waveform will not be sinusoidal but can be represented as a Fourier series. If voltage is the sinusoid

$$v(t) = V_1 \sin(\omega_0 t + \theta_1) \tag{2-60}$$

and current is represented by the Fourier series

$$i(t) = I_0 + \sum_{n=1}^{\infty} I_n \sin(n\omega_0 t + \phi_n) \tag{2-61}$$

then average power absorbed by the load (or supplied by the source) is computed from Eq. (2-59) as

$$P = V_0 I_0 + \sum_{n=1}^{\infty} \left(\frac{V_{n,\max} I_{n,\max}}{2}\right) \cos(\theta_n - \phi_n)$$

$$= (0)(I_0) + \left(\frac{V_1 I_1}{2}\right) \cos(\theta_1 - \phi_1) + \sum_{n=2}^{\infty} \frac{(0)(I_{n,\max})}{2} \cos(\theta_n - \phi_n) \tag{2-62}$$

$$= \left(\frac{V_1 I_1}{2}\right) \cos(\theta_1 - \phi_1) = V_{1,\text{rms}} I_{1,\text{rms}} \cos(\theta_1 - \phi_1)$$

Note that the only nonzero power term is at the frequency of the applied voltage. The power factor of the load is computed from Eq. (2-42).

$$\text{pf} = \frac{P}{S} = \frac{P}{V_{\text{rms}} I_{\text{rms}}}$$

$$\text{pf} = \frac{V_{1,\text{rms}} I_{1,\text{rms}} \cos(\theta_1 - \phi_1)}{V_{1,\text{rms}} I_{\text{rms}}} = \left(\frac{I_{1,\text{rms}}}{I_{\text{rms}}}\right) \cos(\theta_1 - \phi_1) \tag{2-63}$$

where rms current is computed from

$$I_{\text{rms}} = \sqrt{\sum_{n=0}^{\infty} I_{n,\text{rms}}^2} = \sqrt{I_0^2 + \sum_{n=1}^{\infty} \left(\frac{I_n}{\sqrt{2}}\right)^2} \tag{2-64}$$

Note also that for a sinusoidal voltage and a sinusoidal current, pf $= \cos(\theta_1 - \phi_1)$, which is the power factor term commonly used in linear circuits and is called the *displacement power factor*. The ratio of the rms value of the fundamental frequency to the total rms value, $I_{1,\text{rms}}/I_{\text{rms}}$ in Eq. (2-63), is the *distortion factor* (DF).

$$\text{DF} = \frac{I_{1,\text{rms}}}{I_{\text{rms}}} \qquad (2\text{-}65)$$

The distortion factor represents the reduction in power factor due to the nonsinusoidal property of the current. Power factor is also expressed as

$$\text{pf} = [\cos(\theta_1 - \phi_1)]\,\text{DF} \qquad (2\text{-}66)$$

Total harmonic distortion (THD) is another term used to quantify the nonsinusoidal property of a waveform. THD is the ratio of the rms value of all the nonfundamental frequency terms to the rms value of the fundamental frequency term.

$$\text{THD} = \sqrt{\frac{\sum_{n \neq 1} I_{n,\text{rms}}^2}{I_{1,\text{rms}}^2}} = \frac{\sqrt{\sum_{n \neq 1} I_{n,\text{rms}}^2}}{I_{1,\text{rms}}} \qquad (2\text{-}67)$$

THD is equivalently expressed as

$$\text{THD} = \sqrt{\frac{I_{\text{rms}}^2 - I_{1,\text{rms}}^2}{I_{1,\text{rms}}^2}} \qquad (2\text{-}68)$$

Total harmonic distortion is often applied in situations where the dc term is zero, in which case THD may be expressed as

$$\text{THD} = \frac{\sqrt{\sum_{n=2}^{\infty} I_n^2}}{I_1} \qquad (2\text{-}69)$$

Another way to express the distortion factor is

$$\text{DF} = \sqrt{\frac{1}{1 + (\text{THD})^2}} \qquad (2\text{-}70)$$

Reactive power for a sinusoidal voltage and a nonsinusoidal current can be expressed as in Eq. (2-50). The only nonzero term for reactive power is at the voltage frequency:

$$Q = \frac{V_1 I_1}{2} \sin(\theta_1 - \phi_1) \qquad (2\text{-}71)$$

With P and Q defined for the nonsinusoidal case, apparent power S must include a term to account for the current at frequencies which are different from the voltage frequency. The term *distortion volt-amps D* is traditionally used in the computation of S,

$$S = \sqrt{P^2 + Q^2 + D^2} \tag{2-72}$$

where

$$D = V_{1,\text{rms}} \sqrt{\sum_{n \neq 1}^{\infty} I_{n,\text{rms}}^2} = \frac{V_1}{2}\sqrt{\sum_{n \neq 1}^{\infty} I_n} \tag{2-73}$$

Other terms that are sometimes used for nonsinusoidal current (or voltages) are *form factor* and *crest factor*.

$$\text{Form factor} = \frac{I_{\text{rms}}}{I_{\text{avg}}} \tag{2-74}$$

$$\text{Crest factor} = \frac{I_{\text{peak}}}{I_{\text{rms}}} \tag{2-75}$$

EXAMPLE 2-10

Sinusoidal Source and a Nonlinear Load

A sinusoidal voltage source of $v(t) = 100 \cos(377t)$ V is applied to a nonlinear load, resulting in a nonsinusoidal current which is expressed in Fourier series form as

$$i(t) = 8 + 15 \cos(377t + 30°) + 6 \cos[2(377)t + 45°] + 2 \cos[3(377)t + 60°]$$

Determine (*a*) the power absorbed by the load, (*b*) the power factor of the load, (*c*) the distortion factor of the load current, (*d*) the total harmonic distortion of the load current.

■ **Solution**

(*a*) The power absorbed by the load is determined by computing the power absorbed at each frequency in the Fourier series [Eq. (2-59)].

$$P = (0)(8) + \left(\frac{100}{\sqrt{2}}\right)\left(\frac{15}{\sqrt{2}}\right)\cos 30° + (0)\left(\frac{6}{\sqrt{2}}\right)\cos 45° + (0)\left(\frac{2}{\sqrt{2}}\right)\cos 60°$$

$$P = \left(\frac{100}{\sqrt{2}}\right)\left(\frac{15}{\sqrt{2}}\right)\cos 30° = 650 \text{ W}$$

(*b*) The rms voltage is

$$V_{\text{rms}} = \frac{100}{\sqrt{2}} = 70.7 \text{ V}$$

and the rms current is computed from Eq. (2-64):

$$I_{rms} = \sqrt{8^2 + \left(\frac{15}{\sqrt{2}}\right)^2 + \left(\frac{6}{\sqrt{2}}\right)^2 + \left(\frac{2}{\sqrt{2}}\right)^2} = 14.0 \text{ A}$$

The power factor is

$$pf = \frac{P}{S} = \frac{P}{V_{rms}I_{rms}} = \frac{650}{(70.7)(14.0)} = 0.66$$

Alternatively, power factor can be computed from Eq. (2-63):

$$pf = \frac{I_{1,rms} \cos(\theta_1 - \phi_1)}{I_{rms}} = \frac{\left(\frac{15}{\sqrt{2}}\right) \cos(0 - 30°)}{14.0} = 0.66$$

(c) The distortion factor is computed from Eq. (2-65) as

$$DF = \frac{I_{1,rms}}{I_{rms}} = \frac{\frac{15}{\sqrt{2}}}{14.0} = 0.76$$

(d) The total harmonic distortion of the load current is obtained from Eq. (2-68).

$$THD = \sqrt{\frac{I_{rms}^2 - I_{1,rms}^2}{I_{1,rms}^2}} = \sqrt{\frac{14^2 - \left(\frac{15}{\sqrt{2}}\right)^2}{\left(\frac{15}{\sqrt{2}}\right)^2}} = 0.86 = 86\%.$$

2.9 POWER COMPUTATIONS USING PSPICE

PSpice can be used to simulate power electronics circuits to determine voltages, currents, and power quantities. A convenient method is to use the numerical analysis capabilities of the accompanying graphics postprocessor program Probe to obtain power quantities directly. Probe is capable of

- Displaying voltage and current waveforms $(v)(t)$ and $i(t)$
- Displaying instantaneous power $p(t)$
- Computing energy absorbed by a device
- Computing average power P
- Computing average voltage and current
- Computing rms voltages and currents
- Determining the Fourier series of a periodic waveform

The examples that follow illustrate the use of PSpice to do power computations.

EXAMPLE 2-11

Instantaneous Power, Energy, and Average Power Using PSpice

PSpice can be used to display instantaneous power and to compute energy. A simple example is a sinusoidal voltage across a resistor. The voltage source has amplitude $V_m = 10$ V and frequency 60 Hz, and the resistor is 5 Ω. Use VSIN for the source, and select Time Domain (Transient) in the Simulation Setup. Enter a Run Time (Time to Stop) of 16.67 ms for one period of the source.

The circuit is shown in Fig. 2-12a. The top node is labeled as 1. When placing the resistor, rotate it 3 times so that the first node is upward. After running the simulation, the Netlist should look like this:

```
*source EXAMPLE 2-11
V_V1    1 0
+SIN 0 10 60 0 0 0
R_R1    1 0 5
```

When the simulation is completed, the Probe screen appears. The waveforms of voltage and current for the resistor are obtained by entering V(1) and I(R1). Instantaneous

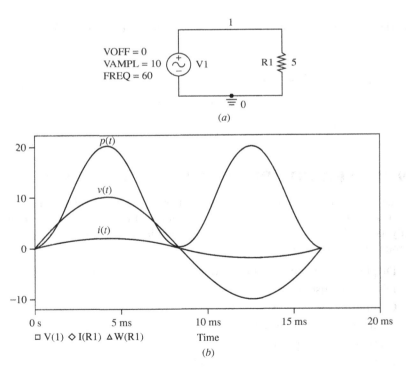

Figure 2.12 (a) PSpice circuit for Example 2-11; (b) voltage, current, and instantaneous power for the resistor; (c) energy absorbed by the resistor; (d) average power absorbed by the resistor.

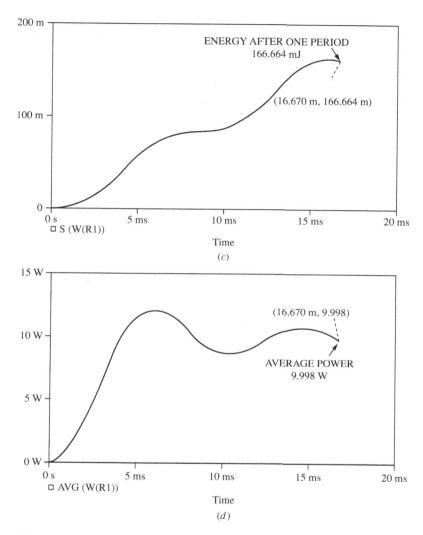

Figure 2.12 (*continued*)

power $p(t) = v(t)i(t)$ absorbed by the resistor is obtained from Probe by entering the expression V(1)*I(R1) or by selecting W(R1). The resulting display showing V(1), I(R1), and $p(t)$ is in Fig. 2-12b.

Energy can be computed using the definition of Eq. (2-2). When in Probe, enter the expression S(V(1)*I(R1)) or S(W(R1)), which computes the integral of instantaneous power. The result is a trace that shows that the energy absorbed increases with time. The energy absorbed by the resistor after one period of the source is determined by placing the cursor at the end of the trace, revealing $W_R = 166.66$ mJ (Fig. 2-12c).

The Probe feature of PSpice can also be used to determine the average value of power directly. For the circuit in the above example, average power is obtained by entering the expression AVG(V(1)*I(R1)) or AVG(W(R1)). The result is a "running" value of average power as computed in Eq. (2-3). Therefore, the average value of the power waveform must be obtained *at the end* of one or more periods of the waveform. Figure. 2-12*d* shows the output from Probe. The cursor option is used to obtain a precise value of average power. This output shows 9.998 W, very slightly different from the theoretical value of 10 W. Keep in mind that the integration is done numerically from discrete data points.

PSpice can also be used to determine power in an ac circuit containing an inductor or capacitor, but *the simulation must represent steady-state response* to be valid for steady-state operation of the circuit.

EXAMPLE 2-12

RMS and Fourier Analysis Using PSpice

Fig. 2-13*a* shows a periodic pulse voltage that is connected to a series R-L circuit with $R = 10\ \Omega$ and $L = 10$ mH. PSpice is used to determine the steady-state rms current and the Fourier components of the current.

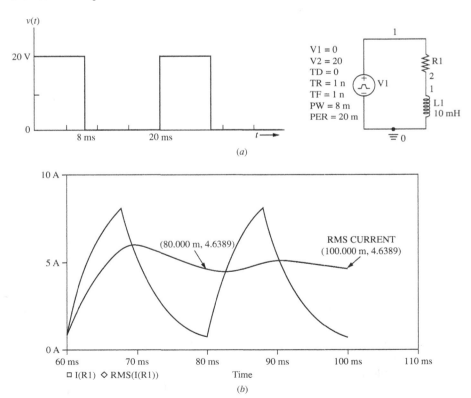

Figure 2-13 (*a*) A pulse waveform voltage source is applied to a series R-L circuit; (*b*) Probe output showing the steady-state current and the rms value.

In PSpice power calculations, *it is extremely important that the output being analyzed represent steady-state voltages and currents*. Steady-state current is reached after several periods of the pulse waveform. Therefore, the Simulation Settings have the Run Time (Time to Stop) at 100 ms and the Start Saving Data set at 60 ms. The 60-ms delay allows for the current to reach steady state. A maximum step size is set at 10 μs to produce a smooth waveform.

Current is displayed in Probe by entering I(R1), and steady state is verified by noting that the starting and ending values are the same for each period. The rms current is obtained by entering the expression RMS(I(R1)). The value of rms current, 4.6389 A, is obtained at the end of any period of the current waveform. Fig. 2-13*b* shows the Probe output.

The Fourier series of a waveform can be determined using PSpice. Fourier analysis is entered under Output File Options in the Transient Analysis menu. The Fast Fourier Transform (FFT) on the waveforms of the source voltage and

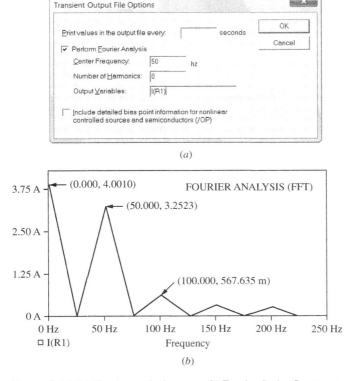

Figure 2-14 (*a*) Fourier analysis setup; (*b*) Fourier Series Spectrum from Probe using FFT.

the load current will appear in the output file. The fundamental frequency (Center Frequency) of the Fourier series is 50 Hz (1/20 mS). In this example, five periods of the waveform are simulated to ensure steady-state current for this L/R time constant.

A portion of the output file showing the Fourier components of source voltage and resistor current is as follows:

```
FOURIER COMPONENTS OF TRANSIENT RESPONSE I(R_R1)

DC COMPONENT = 4.000000E+00
```

HARMONIC NO	FREQUENCY (HZ)	FOURIER COMPONENT	NORMALIZED COMPONENT	PHASE (DEG)	NORMALIZED PHASE (DEG)
1	5.000E+01	3.252E+00	1.000E+00	−3.952E+01	0.000E+00
2	1.000E+02	5.675E−01	1.745E−01	−1.263E+02	−4.731E+01
3	1.500E+02	2.589E−01	7.963E−02	−2.402E+01	9.454E+01
4	2.000E+02	2.379E−01	7.317E−02	−9.896E+01	5.912E+01
5	2.500E+02	1.391E−07	4.277E−08	5.269E+00	2.029E+02
6	3.000E+02	1.065E−01	3.275E−02	−6.594E+01	1.712E+02
7	3.500E+02	4.842E−02	1.489E−02	−1.388E+02	1.378E+02
8	4.000E+02	3.711E−02	1.141E−02	−3.145E+01	2.847E+02
9	4.500E+02	4.747E−02	1.460E−02	−1.040E+02	2.517E+02

```
TOTAL HARMONIC DISTORTION = 2.092715E+01 PERCENT
```

When you use PSpice output for the Fourier series, remember that the values are listed as amplitudes (zero-to-peak), and conversion to rms by dividing by $\sqrt{2}$ is required for power computations. The phase angles are referenced to the sine rather than the cosine. The numerically computed Fourier components in PSpice may not be exactly the same as analytically computed values. Total harmonic distortion (THD) is listed at the end of the Fourier output. [*Note*: The THD computed in PSpice uses Eq. (2-69) and assumes that the dc component of the waveform is zero, which is not true in this case.]

The rms value of the load current can be computed from the Fourier series in the output file from Eq. (2-43).

$$I_{\text{rms}} = \sqrt{(4.0)^2 + \left(\frac{3.252}{\sqrt{2}}\right)^2 + \left(\frac{0.5675}{\sqrt{2}}\right)^2 + \cdots} \approx 4.63 \text{ A}$$

A graphical representation of the Fourier series can be produced in Probe. To display the Fourier series of a waveform, click the FFT button on the toolbar. Upon entering the variable to be displayed, the spectrum of the Fourier series will appear. It will be desirable to adjust the range of frequencies to obtain a useful graph. Fig. 2-14*b* shows the result for this example. Fourier component magnitudes are represented by the peaks of the graph, which can be determined precisely by using the cursor option.

EXAMPLE 2-13

PSpice Solution of Example 2-3

Use PSpice to simulate the inductor circuit of Fig. 2-4a with the parameters of Example 2-3.

■ **Solution**

Fig. 2-15 shows the circuit used in the PSpice simulation. The transistor is used as a switch, so a voltage-controlled switch (Sbreak) can be used in the PSpice circuit. The switch is idealized by setting the on resistance to $R_{on} = 0.001 \, \Omega$. The control for the switch is a pulse voltage source which has a pulse width of 10 ms and period of 100 ms. The diode Dbreak is used.

Some of the possible results that can be obtained from the Probe output are listed below. All traces except the maximum inductor current and the stored inductor energy are read at the end of the Probe trace, which is after one complete period. Note the agreement between the results of Example 2-3 and the PSpice results.

Desired Quantity	Probe Entry	Result
Inductor current	I(L1)	max ≈ 4.5 A
Energy stored in inductor	0.5*0.2*I(L1)*I(L1)	max ≈ 2.025 J
Average switch power	AVG(W(S1))	0.010 W
Average source power (absorbed)	AVG(W(VCC))	−20.3 W
Average diode power	AVG(W(D1))	0.464 W
Average inductor power	AVG(W(L1))	≈ 0
Average inductor voltage	AVG(V(1,2))	≈ 0
Average resistor power	AVG(W(R1))	19.9 W
Energy absorbed by resistor	S(W(R1))	1.99 J
Energy absorbed by diode	S(W(D1))	0.046 J
Energy absorbed by inductor	S(W(L1))	≈ 0
RMS resistor current	RMS(I(R1))	0.998 A

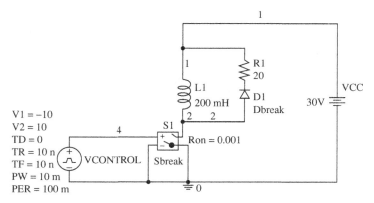

Figure 2-15 Circuit for Example 2-13, a PSpice simulation of the circuit in Example 2-4.

2.10 Summary

- Instantaneous power is the product of voltage and current at a particular time:

$$p(t) = v(t)i(t)$$

  Using the passive sign convention, the device is absorbing power if $(p)(t)$ is positive, and the device is supplying power if $(p)(t)$ is negative.

- *Power* usually refers to average power, which is the time average of periodic instantaneous power:

$$P = \frac{1}{T}\int_{t_0}^{t_0+T} v(t)i(t)\,dt = \frac{1}{T}\int_{t_0}^{t_0+T} p(t)\,dt$$

- The rms value is the root-mean-square or effective value of a voltage or current waveform.

$$V_{rms} = \sqrt{\frac{1}{T}\int_0^T v^2(t)\,dt}$$

$$I_{rms} = \sqrt{\frac{1}{T}\int_0^T i^2(t)\,dt}$$

- Apparent power is the product of rms voltage and current.

$$S = V_{rms}I_{rms}$$

- Power factor is the ratio of average power to apparent power.

$$\text{pf} = \frac{P}{S} = \frac{P}{V_{rms}I_{rms}}$$

- For inductors and capacitors that have periodic voltages and currents, the average power is zero. Instantaneous power is generally not zero because the device stores energy and then returns energy to the circuit.
- For periodic currents, the average voltage across an inductor is zero.
- For periodic voltages, the average current in a capacitor is zero.
- For nonsinusoidal periodic waveforms, average power may be computed from the basic definition, or the Fourier series method may be used. The Fourier series method treats each frequency in the series separately and uses superposition to compute total power.

$$P = \sum_{n=0}^{\infty} P_n = V_0 I_0 + \sum_{n=1}^{\infty} V_{n,\,rms}\,I_{n,\,rms}\cos(\theta_n - \phi_n)$$

- A simulation using the program PSpice may be used to obtain not only voltage and current waveforms but also instantaneous power, energy, rms values, and average power by using the numerical capabilities of the graphic postprocessor program

Probe. For numerical computations in Probe to be accurate, the simulation must represent steady-state voltages and currents.
- Fourier series terms are available in PSpice by using the Fourier Analysis in the Simulation Settings or by using the FFT option in Probe.

2.11 Bibliography

M. E. Balci and M. H. Hocaoglu, "Comparison of Power Definitions for Reactive Power Compensation in Nonsinusoidal Circuits," *International Conference on Harmonics and Quality of Power*, Lake Placid, New York, 2004.

L. S. Czarnecki, "Considerations on the Reactive Power in Nonsinusoidal Situations," *International Conference on Harmonics in Power Systems*, Worcester Polytechnic Institute, Worcester, Mass., 1984, pp. 231–237.

A. E. Emanuel, "Powers in Nonsinusoidal Situations, A Review of Definitions and Physical Meaning," *IEEE Transactions on Power Delivery*, vol. 5, no. 3, July 1990.

G. T. Heydt, *Electric Power Quality*, Stars in a Circle Publications, West Lafayette, Ind., 1991.

W. Sheperd and P. Zand, *Energy Flow and Power Factor in Nonsinusoidal Circuits*, Cambridge University Press, 1979.

Problems

Instantaneous and Average Power

2-1. Average power generally is *not* the product of average voltage and average current. Give an example of periodic waveforms for $v(t)$ and $i(t)$ that have zero average values and average power absorbed by the device is not zero. Sketch $v(t)$, $i(t)$, and $p(t)$.

2-2. The voltage across a 10-Ω resistor is $v(t) = 170 \sin(377t)$ V. Determine (*a*) an expression for instantaneous power absorbed by the resistor, (*b*) the peak power, and (*c*) the average power.

2-3. The voltage across an element is $v(t) = 5 \sin(2\pi t)$ V. Use graphing software to graph instantaneous power absorbed by the element, and determine the average power if the current, using the passive sign convention, is (*a*) $i(t) = 4 \sin(2\pi t)$ A and (*b*) $i(t) = 3 \sin(4\pi t)$ A.

2-4. The voltage and current for a device (using the passive sign convention) are periodic functions with $T = 100$ ms described by

$$v(t) = \begin{cases} 10 \text{ V} & 0 < t < 70 \text{ ms} \\ 0 & 70 \text{ ms} < t < 100 \text{ ms} \end{cases}$$

$$i(t) = \begin{cases} 0 & 0 < t < 50 \text{ ms} \\ 4 \text{ A} & 50 \text{ ms} < t < 100 \text{ ms} \end{cases}$$

Determine (*a*) the instantaneous power, (*b*) the average power, and (*c*) the energy absorbed by the device in each period.

2-5. The voltage and current for a device (using the passive sign convention) are periodic functions with $T = 20$ ms described by

$$v(t) = \begin{cases} 10 \text{ V} & 0 < t < 14 \text{ ms} \\ 0 & 14 \text{ ms} < t < 20 \text{ ms} \end{cases}$$

$$i(t) = \begin{cases} 7 \text{ A} & 0 < t < 6 \text{ ms} \\ -5 \text{ A} & 6 \text{ ms} < t < 10 \text{ ms} \\ 4 \text{ A} & 10 \text{ ms} < t < 20 \text{ ms} \end{cases}$$

Determine (a) the instantaneous power, (b) the average power, and (c) the energy absorbed by the device in each period.

2-6. Determine the average power absorbed by a 12-V dc source when the current into the positive terminal of the source is that given in (a) Prob. 2-4 and (b) Prob. 2-5.

2-7. A current of $5 \sin(2\pi 60 t)$ A enters an element. Sketch the instantaneous power and determine the average power absorbed by the load element when the element is (a) a 5-Ω resistor, (b) a 10-mH inductor, and (c) a 12-V source (current into the positive terminal).

2-8. A current source of $i(t) = 2 + 6 \sin(2\pi 60 t)$ A is connected to a load that is a series combination of a resistor, an inductor, and a dc voltage source (current into the positive terminal). If $R = 4\ \Omega$, $L = 15$ mH, and $V_{dc} = 6$ V, determine the average power absorbed by each element.

2-9. An electric resistance space heater rated at 1500 W for a voltage source of $v(t) = 120\sqrt{2} \sin(2\pi 60 t)$ V has a thermostatically controlled switch. The heater periodically switches on for 5 min and off for 7 min. Determine (a) the maximum instantaneous power, (b) the average power over the 12-min cycle, and (c) the electric energy converted to heat in each 12-min cycle.

Energy Recovery

2-10. An inductor is energized as in the circuit of Fig. 2-4a. The circuit has $L = 100$ mH, $R = 20\ \Omega$, $V_{CC} = 90$ V, $t_1 = 4$ ms, and $T = 40$ ms. Assuming the transistor and diode are ideal, determine (a) the peak energy stored in the inductor, (b) the energy absorbed by the resistor in each switching period, and (c) the average power supplied by the source. (d) If the resistor is changed to 40 Ω, what is the average power supplied by the source?

2-11. An inductor is energized as in the circuit of Fig. 2-4a. The circuit has $L = 10$ mH and $V_{CC} = 14$ V. (a) Determine the required on time of the switch such that the peak energy stored in the inductor is 1.2 J. (b) Select a value for R such that the switching cycle can be repeated every 20 ms. Assume the switch and the diode are ideal.

2-12. An inductor is energized as in the circuit of Fig. 2-5a. The circuit has $L = 50$ mH, $V_{CC} = 90$ V, $t_1 = 4$ ms, and $T = 50$ ms. (a) Determine the peak energy stored in the inductor. (b) Graph the inductor current, source current, inductor instantaneous power, and source instantaneous power versus time. Assume the transistors are ideal.

2-13. An alternative circuit for energizing an inductor and removing the stored energy without damaging a transistor is shown in Fig. P2-13. Here $V_{CC} = 12$ V, $L = 75$ mH, and the zener breakdown voltage is $V_Z = 20$ V. The transistor switch opens and closes periodically with $t_{on} = 20$ ms and $t_{off} = 50$ ms. (a) Explain how the zener diode allows the switch to open. (b) Determine and sketch the inductor current $i_L(t)$ and the zener diode current $i_Z(t)$ for one switching period. (c) Sketch $(p)(t)$ for the inductor and the zener diode. (d) Determine the average power absorbed by the inductor and by the zener diode.

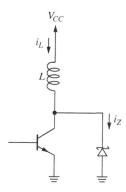

2-14. Repeat Prob. 2-13 with $V_{CC} = 20$ V, $L = 50$ mH, $V_Z = 30$ V, $t_{on} = 15$ ms, and $t_{off} = 60$ ms.

Effective Values: RMS

2-15. The rms value of a sinusoid is the peak value divided by $\sqrt{2}$. Give two examples to show that this is generally not the case for other periodic waveforms.

2-16. A three-phase distribution system is connected to a nonlinear load that has line and neutral currents like those of Fig. 2-8. The rms current in each phase is 12 A, and the resistance in each of the line and neutral conductors is 0.5 Ω. Determine the total power absorbed by the conductors. What should the resistance of the neutral conductor be such that it absorbs the same power as one of the phase conductors?

2-17. Determine the rms values of the voltage and current waveforms in Prob. 2-4.

2-18. Determine the rms values of the voltage and current waveforms in Prob. 2-5.

Nonsinusoidal Waveforms

2-19. The voltage and current for a circuit element are $v(t) = 2 + 5\cos(2\pi 60 t) - 3\cos(4\pi 60 t + 45°)$ V and $i(t) = 1.5 + 2\cos(2\pi 60 t + 20°) + 1.1\cos(4\pi 60 t - 20°)$ A. (a) Determine the rms values of voltage and current. (b) Determine the power absorbed by the element.

2-20. A current source $i(t) = 3 + 4\cos(2\pi 60 t) + 6\cos(4\pi 60 t)$ A is connected to a parallel RC load with $R = 100\ \Omega$ and $C = 50\ \mu\text{F}$. Determine the average power absorbed by the load.

2-21. In Fig. P2-21, $R = 4\ \Omega$, $L = 10$ mH, $V_{dc} = 12$ V, and $v_s(t) = 50 + 30\cos(4\pi 60 t) + 10\cos(8\pi 60 t)$ V. Determine the power absorbed by each component.

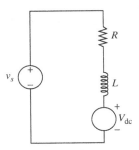

Figure P2-21

2-22. A nonsinusoidal periodic voltage has a Fourier series of $v(t) = 6 + 5\cos(2\pi 60 t) + 3\cos(6\pi 60 t)$. This voltage is connected to a load that is a 16-Ω resistor in series with a 25-mH inductor as in Fig. 2-11. Determine the power absorbed by the load.

2-23. Voltage and current for a device (using the passive sign convention) are

$$v(t) = 20 + \sum_{n=1}^{\infty}\left(\frac{20}{n}\right)\cos(n\pi t)\ \text{V}$$

$$i(t) = 5 + \sum_{n=1}^{\infty}\left(\frac{5}{n^2}\right)\cos(n\pi t)\ \text{A}$$

Determine the average power based on the terms through $n = 4$.

2-24. Voltage and current for a device (using the passive sign convention) are

$$v(t) = 50 + \sum_{n=1}^{\infty}\left(\frac{50}{n}\right)\cos(n\pi t)\ \text{V}$$

$$i(t) = 10 + \sum_{n=1}^{\infty}\left(\frac{10}{n^2}\right)\cos\left(n\pi t - \tan^{-1} n/2\right)$$

Determine the average power based on the terms through $n = 4$.

2-25. In Fig. P2-21, $R = 20\ \Omega$, $L = 25$ mH, and $V_{dc} = 36$ V. The source is a periodic voltage that has the Fourier series

$$v_s(t) = 50 + \sum_{n=1}^{\infty}\left(\frac{400}{n\pi}\right)\sin(200 n\pi t)$$

Using the Fourier series method, determine the average power absorbed by R, L, and V_{dc} when the circuit is operating in the steady state. Use as many terms in the Fourier series as necessary to obtain a reasonable estimate of power.

2-26. A sinusoidal current of 10 A rms at a 60-Hz fundamental frequency is contaminated with a ninth harmonic current. The current is expressed as

$$i(t) = 10\sqrt{2} \sin(2\pi 60 t) + I_9\sqrt{2} \sin(18\pi 60 t) \text{ A}$$

Determine the value of the ninth harmonic rms current I_9 if the THD is (a) 5 percent, (b) 10 percent, (c) 20 percent, and (d) 40 percent. Use graphing software or PSpice to show $i(t)$ for each case.

2-27. A sinusoidal voltage source of $v(t) = 170 \cos(2\pi 60 t)$ V is applied to a nonlinear load, resulting in a nonsinusoidal current that is expressed in Fourier series form as $i(t) = 10 \cos(2\pi 60 t + 30°) + 6 \cos(4\pi 60 t + 45°) + 3 \cos(8\pi 60 t + 20°)$ A. Determine (a) the power absorbed by the load, (b) the power factor of the load, (c) the distortion factor, and (d) the total harmonic distortion of the load current.

2-28. Repeat Prob. 2-27 with $i(t) = 12 \cos(2\pi 60 t - 40°) + 5 \sin(4\pi 60 t) + 4 \cos(8\pi 60 t)$ A.

2-29. A sinusoidal voltage source of $v(t) = 240\sqrt{2} \sin(2\pi 60 t)$ V is applied to a nonlinear load, resulting in a current $i(t) = 8 \sin(2\pi 60 t) + 4 \sin(4\pi 60 t)$ A. Determine (a) the power absorbed by the load, (b) the power factor of the load, (c) the THD of the load current, (d) the distortion factor of the load current, and (e) the crest factor of the load current.

2-30. Repeat Prob. 2-29 with $i(t) = 12 \sin(2\pi 60 t) + 9 \sin(4\pi 60 t)$ A.

2-31. A voltage source of $v(t) = 5 + 25 \cos(1000 t) + 10 \cos(2000 t)$ V is connected to a series combination of a 2-Ω resistor, a 1-mH inductor, and a 1000-μF capacitor. Determine the rms current in the circuit, and determine the power absorbed by each component.

PSpice

2-32. Use PSpice to simulate the circuit of Example 2-1. Define voltage and current with PULSE sources. Determine instantaneous power, energy absorbed in one period, and average power.

2-33. Use PSpice to determine the instantaneous and average power in the circuit elements of Prob. 2-7.

2-34. Use PSpice to determine the rms values of the voltage and current waveforms in (a) Prob. 2-5 and (b) Prob. 2-6.

2-35. Use PSpice to simulate the circuit of Prob. 2-10. (a) Idealize the circuit by using a voltage-controlled switch that has $R_{on} = 0.001$ Ω and a diode with $n = 0.001$. (b) Use $R_{on} = 0.5$ Ω and use the default diode.

2-36. Use PSpice to simulate the circuit of Fig. 2-5a. The circuit has $V_{CC} = 75$ V, $t_0 = 40$ ms, and $T = 100$ ms. The inductance is 100 mH and has an internal resistance of 20 Ω. Use a voltage-controlled switch with $R_{on} = 1$ Ω for the transistors, and use the PSpice default diode model. Determine the average power absorbed by each circuit element. Discuss the differences between the behavior of this circuit and that of the ideal circuit.

2-37. Use PSpice to simulate the circuit of Prob. 2-13. Use $R_{on} = 0.001\,\Omega$ for the switch model and use $n = 0.001$, BV = 20 V for the breakdown voltage and $I_{BV} = 10$ A for the current at breakdown for the zener diode model. (a) Display $i_L(t)$ and $i_Z(t)$. Determine the average power in the inductor and in the zener diode. (b) Repeat part (a) but include a 1.5-Ω series resistance with the inductor and use $R_{on} = 0.5\,\Omega$ for the switch.

2-38. Repeat Prob. 2-37, using the circuit of Prob. 2-14.

2-39. Use PSpice to determine the power absorbed by the load in Example 2-10. Model the system as a voltage source and four current sources in parallel.

2-40. Modify the switch model so $R_{on} = 1\,\Omega$ in the PSpice circuit file in Example 2-13. Determine the effect on each of the quantities obtained from Probe in the example.

2-41. Demonstrate with PSpice that a triangular waveform like that of Fig. 2-9a has an rms value of $V_m/\sqrt{3}$. Choose an arbitrary period T, and use at least three values of t_1. Use a VPULSE source with the rise and fall times representing the triangular wave.

CHAPTER 3

Half-Wave Rectifiers
The Basics of Analysis

3.1 INTRODUCTION

A rectifier converts ac to dc. The purpose of a rectifier may be to produce an output that is purely dc, or the purpose may be to produce a voltage or current waveform that has a specified dc component.

In practice, the half-wave rectifier is used most often in low-power applications because the average current in the supply will not be zero, and nonzero average current may cause problems in transformer performance. While practical applications of this circuit are limited, it is very worthwhile to analyze the half-wave rectifier in detail. A thorough understanding of the half-wave rectifier circuit will enable the student to advance to the analysis of more complicated circuits with a minimum of effort.

The objectives of this chapter are to introduce general analysis techniques for power electronics circuits, to apply the power computation concepts of the previous chapter, and to illustrate PSpice solutions.

3.2 RESISTIVE LOAD

Creating a DC Component Using an Electronic Switch

A basic half-wave rectifier with a resistive load is shown in Fig. 3-1a. The source is ac, and the objective is to create a load voltage that has a nonzero dc component. The diode is a basic electronic switch that allows current in one direction only. For the positive half-cycle of the source in this circuit, the diode is on (forward-biased).

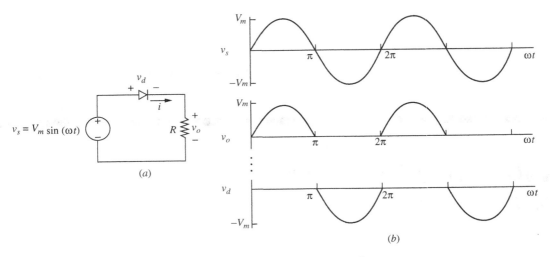

Figure 3-1 (a) Half-wave rectifier with resistive load; (b) Voltage waveforms.

Considering the diode to be ideal, the voltage across a forward-biased diode is zero and the current is positive.

For the negative half-cycle of the source, the diode is reverse-biased, making the current zero. The voltage across the reverse-biased diode is the source voltage, which has a negative value.

The voltage waveforms across the source, load, and diode are shown in Fig. 3-1b. Note that the units on the horizontal axis are in terms of angle (ωt). This representation is useful because the values are independent of frequency. The dc component V_o of the output voltage is the average value of a half-wave rectified sinusoid

$$V_o = V_{avg} = \frac{1}{2\pi} \int_0^{\pi} V_m \sin(\omega t) \, d(\omega t) = \frac{V_m}{\pi} \tag{3-1}$$

The dc component of the current for the purely resistive load is

$$I_o = \frac{V_o}{R} = \frac{V_m}{\pi R} \tag{3-2}$$

Average power absorbed by the resistor in Fig. 3-1a can be computed from $P = I_{rms}^2 R = V_{rms}^2/R$. When the voltage and current are half-wave rectified sine waves,

$$V_{rms} = \sqrt{\frac{1}{2\pi} \int_0^{\pi} [V_m \sin(\omega t)]^2 \, d(\omega t)} = \frac{V_m}{2} \tag{3-3}$$

$$I_{rms} = \frac{V_m}{2R}$$

In the preceding discussion, the diode was assumed to be ideal. For a real diode, the diode voltage drop will cause the load voltage and current to be reduced, but not appreciably if V_m is large. For circuits that have voltages much larger than the typical diode drop, the improved diode model may have only second-order effects on the load voltage and current computations.

EXAMPLE 3-1

Half-Wave Rectifier with Resistive Load

For the half-wave rectifier of Fig. 3-1a, the source is a sinusoid of 120 V rms at a frequency of 60 Hz. The load resistor is 5 Ω. Determine (a) the average load current, (b) the average power absorbed by the load and (c) the power factor of the circuit.

■ **Solution**

(a) The voltage across the resistor is a half-wave rectified sine wave with peak value $V_m = 120\sqrt{2} = 169.7$ V. From Eq. (3-2), the average voltage is V_m/π, and average current is

$$I_o = \frac{V_o}{R} = \frac{V_m}{\pi R} = \frac{\sqrt{2}(120)}{5\pi} = 10.8 \text{ A}$$

(b) From Eq. (3-3), the rms voltage across the resistor for a half-wave rectified sinusoid is

$$V_{rms} = \frac{V_m}{2} = \frac{\sqrt{2}(120)}{2} = 84.9 \text{ V}$$

The power absorbed by the resistor is

$$P = \frac{V_{rms}^2}{R} = \frac{84.9^2}{4} = 1440 \text{ W}$$

The rms current in the resistor is $V_m/(2R) = 17.0$ A, and the power could also be calculated from $I_{rms}^2 R = (17.0)^2(5) = 1440$ W.

(c) The power factor is

$$\text{pf} = \frac{P}{S} = \frac{P}{V_{s,rms} I_{s,rms}} = \frac{1440}{(120)(17)} = 0.707$$

3.3 RESISTIVE-INDUCTIVE LOAD

Industrial loads typically contain inductance as well as resistance. As the source voltage goes through zero, becoming positive in the circuit of Fig. 3-2a, the diode becomes forward-biased. The Kirchhoff voltage law equation that describes the current in the circuit for the forward-biased ideal diode is

$$V_m \sin(\omega t) = Ri(t) + L\frac{di(t)}{dt} \qquad (3\text{-}4)$$

The solution can be obtained by expressing the current as the sum of the forced response and the natural response:

$$i(t) = i_f(t) + i_n(t) \qquad (3\text{-}5)$$

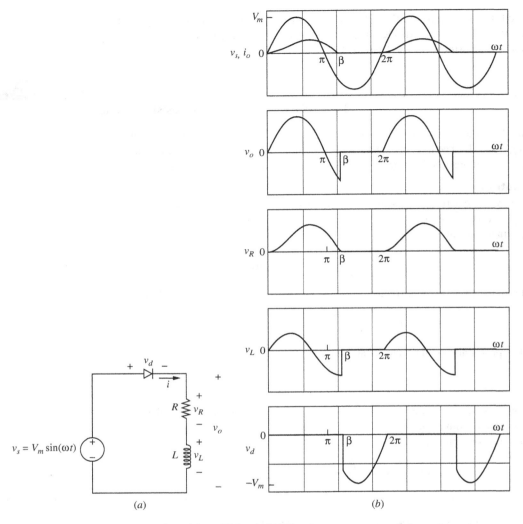

Figure 3-2 (a) Half-wave rectifier with an RL load; (b) Waveforms.

The forced response for this circuit is the current that exists after the natural response has decayed to zero. In this case, the forced response is the steady-state sinusoidal current that would exist in the circuit if the diode were not present. This steady-state current can be found from phasor analysis, resulting in

$$i_f(t) = \frac{V_m}{Z} \sin(\omega t - \theta) \tag{3-6}$$

where $\quad Z = \sqrt{R^2 + (\omega L)^2} \quad$ and $\quad \theta = \tan^{-1}\left(\frac{\omega L}{R}\right)$

The natural response is the transient that occurs when the load is energized. It is the solution to the homogeneous differential equation for the circuit without the source or diode:

$$Ri(t) + L\frac{di(t)}{dt} = 0 \qquad (3\text{-}7)$$

For this first-order circuit, the natural response has the form

$$i_n(t) = Ae^{-t/\tau} \qquad (3\text{-}8)$$

where τ is the time constant L/R and A is a constant that is determined from the initial condition. Adding the forced and natural responses gets the complete solution.

$$i(t) = i_f(t) + i_n(t) = \frac{V_m}{Z}\sin(\omega t - \theta) + Ae^{-t/\tau} \qquad (3\text{-}9)$$

The constant A is evaluated by using the initial condition for current. The initial condition of current in the inductor is zero because it was zero before the diode started conducting and it cannot change instantaneously.

Using the initial condition and Eq. (3-9) to evaluate A yields

$$i(0) = \frac{V_m}{Z}\sin(0 - \theta) + Ae^0 = 0$$

$$A = -\frac{V_m}{Z}\sin(-\theta) = \frac{V_m}{Z}\sin\theta \qquad (3\text{-}10)$$

Substituting for A in Eq. (3-9) gives

$$i(t) = \frac{V_m}{Z}\sin(\omega t - \theta) + \frac{V_m}{Z}\sin(\theta)e^{-t/\tau}$$

$$= \frac{V_m}{Z}\left[\sin(\omega t - \theta) + \sin(\theta)e^{-t/\tau}\right] \qquad (3\text{-}11)$$

It is often convenient to write the function in terms of the angle ωt rather than time. This merely requires ωt to be the variable instead of t. To write the above equation in terms of angle, t in the exponential must be written as ωt, which requires τ to be multiplied by ω also. The result is

$$i(\omega t) = \frac{V_m}{Z}\left[\sin(\omega t - \theta) + \sin(\theta)e^{-\omega t/\omega\tau}\right] \qquad (3\text{-}12)$$

A typical graph of circuit current is shown in Fig. 3-2b. Equation (3-12) is valid for positive currents only because of the diode in the circuit, so current is zero when the function in Eq. (3-12) is negative. When the source voltage again becomes positive, the diode turns on, and the positive part of the waveform in Fig. 3-2b is repeated. This occurs at every positive half-cycle of the source. The voltage waveforms for each element are shown in Fig. 3-2b.

Note that the diode remains forward-biased longer than π rad and that the source is negative for the last part of the conduction interval. This may seem unusual, but an examination of the voltages reveals that Kirchhoff's voltage law is satisfied and there is no contradiction. Also note that the inductor voltage is negative when the current is decreasing ($v_L = L\,di/dt$).

The point when the current reaches zero in Eq. (3-12) occurs when the diode turns off. The first positive value of ωt in Eq. (3-12) that results in zero current is called the extinction angle β.

Substituting $\omega t = \beta$ in Eq. (3-12), the equation that must be solved is

$$i(\beta) = \frac{V_m}{Z}\left[\sin(\beta - \theta) + \sin(\theta)e^{-\beta/\omega\tau}\right] = 0 \quad (3\text{-}13)$$

which reduces to

$$\boxed{\sin(\beta - \theta) + \sin(\theta)e^{-\beta/\omega\tau} = 0} \quad (3\text{-}14)$$

There is no closed-form solution for β, and some numerical method is required. To summarize, the current in the half-wave rectifier circuit with RL load (Fig. 3-2) is expressed as

$$i(\omega t) = \begin{cases} \dfrac{V_m}{Z}\left[\sin(\omega t - \theta) + \sin(\theta)e^{-\omega t/\omega\tau}\right] & \text{for } 0 \leq \omega t \leq \beta \\ 0 & \text{for } \beta \leq \omega t \leq 2\pi \end{cases} \quad (3\text{-}15)$$

where $\quad Z = \sqrt{R^2 + (\omega L)^2} \quad \theta = \tan^{-1}\left(\dfrac{\omega L}{R}\right) \quad$ and $\quad \tau = \dfrac{L}{R}$

The average power absorbed by the load is $I^2_{rms}R$, since the average power absorbed by the inductor is zero. The rms value of the current is determined from the current function of Eq. (3-15).

$$I_{rms} = \sqrt{\frac{1}{2\pi}\int_0^{2\pi} i^2(\omega t)\,d(\omega t)} = \sqrt{\frac{1}{2\pi}\int_0^{\beta} i^2(\omega t)\,d(\omega t)} \quad (3\text{-}16)$$

Average current is

$$I_o = \frac{1}{2\pi}\int_0^{\beta} i(\omega t)\,d(\omega t) \quad (3\text{-}17)$$

EXAMPLE 3-2

Half-Wave Rectifier with RL Load

For the half-wave rectifier of Fig. 3-2a, $R = 100\,\Omega$, $L = 0.1$ H, $\omega = 377$ rad/s, and $V_m = 100$ V. Determine (a) an expression for the current in this circuit, (b) the average current, (c) the rms current, (d) the power absorbed by the RL load, and (e) the power factor.

■ Solution

For the parameters given,

$$Z = [R^2 + (\omega L)^2]^{0.5} = 106.9\,\Omega$$
$$\theta = \tan^{-1}(\omega L/R) = 20.7° = 0.361 \text{ rad}$$
$$\omega\tau = \omega L/R = 0.377 \text{ rad}$$

(a) Equation (3-15) for current becomes

$$i(\omega t) = 0.936 \sin(\omega t - 0.361) + 0.331 e^{-\omega t/0.377} \quad \text{A} \quad \text{for } 0 \le \omega t \le \beta$$

Beta is found from Eq. (3-14).

$$\sin(\beta - 0.361) + \sin(0.361) e^{-\beta/0.377} = 0$$

Using a numerical root-finding program, β is found to be 3.50 rad, or 201°

(b) Average current is determined from Eq. (3-17).

$$I_o = \frac{1}{2\pi} \int_0^{3.50} \left[0.936\sin(\omega t - 0.361) + 0.331 e^{-\omega t/0.377} \right] d(\omega t) = 0.308 \text{ A}$$

(A numerical integration program is recommended.)

(c) The rms current is found from Eq. (3-16) to be

$$I_{rms} = \sqrt{\frac{1}{2\pi} \int_0^{3.50} \left[0.936\sin(\omega t - 0.361) + 0.331 e^{-\omega t/0.377} \right]^2 d(\omega t)} = 0.474 \text{ A}$$

(d) The power absorbed by the resistor is

$$P = I_{rms}^2 R = (0.474)^2 (100) = 22.4 \text{ W}$$

The average power absorbed by the inductor is zero. Also P can be computed from the definition of average power:

$$P = \frac{1}{2\pi} \int_0^{2\pi} p(\omega t) d(\omega t) = \frac{1}{2\pi} \int_0^{2\pi} v(\omega t) i(\omega t) d(\omega t)$$

$$= \frac{1}{2\pi} \int_0^{3.50} [100 \sin(\omega t)] \left[0.936 \sin(\omega t - 0.361) + 0.331 e^{-\omega t/0.377} \right] d(\omega t)$$

$$= 22.4 \text{ W}$$

(e) The power factor is computed from the definition pf $= P/S$, and P is power supplied by the source, which must be the same as that absorbed by the load.

$$\text{pf} = \frac{P}{S} = \frac{P}{V_{s,rms} I_{rms}} = \frac{22.4}{(100/\sqrt{2})0.474} = 0.67$$

Note that the power factor is *not* $\cos\theta$.

3.4 PSPICE SIMULATION

Using Simulation Software for Numerical Computations

A computer simulation of the half-wave rectifier can be performed using PSpice. PSpice offers the advantage of having the postprocessor program Probe which can display the voltage and current waveforms in the circuit and perform numerical computations. Quantities such as the rms and average currents, average power absorbed by the load, and power factor can be determined directly with PSpice. Harmonic content can be determined from the PSpice output.

A transient analysis produces the desired voltages and currents. One complete period is a sufficient time interval for the transient analysis.

EXAMPLE 3-3

PSpice Analysis

Use PSpice to analyze the circuit of Example 3-2.

■ **Solution**

The circuit of Fig. 3-2a is created using VSIN for the source and Dbreak for the diode. In the simulation settings, choose Time Domain (transient) for the analysis type, and set the Run Time to 16.67 ms for one period of the source. Set the Maximum Step Size to 10 μs to get adequate sampling of the waveforms. A transient analysis with a run time of 16.67 ms (one period for 60 Hz) and a maximum step size of 10 μs is used for the simulation settings.

If a diode model that approximates an ideal diode is desired for the purpose of comparing the simulation with analytical results, editing the PSpice model and using $n = 0.001$ will make the voltage drop across the forward-biased diode close to zero. Alternatively, a model for a power diode may be used to obtain a better representation of a real rectifier circuit. For many circuits, voltages and currents will not be affected significantly when different diode models are used. Therefore, it may be convenient to use the Dbreak diode model for a preliminary analysis.

When the transient analysis is performed and the Probe screen appears, display the current waveform by entering the expression I(R1). A method to display angle instead of time on the x axis is to use the x-variable option within the x-axis menu, entering TIME*60*360. The factor of 60 converts the axis to periods ($f = 60$ Hz), and the factor 360 converts the axis to degrees. Entering TIME*60*2*3.14 for the x variable converts the x axis to radians. Figure 3-3a shows the result. The extinction angle β is found to be 200° using the cursor option. Note that using the default diode model in PSpice resulted in a value of β very close to the 201° in Example 3-2.

Probe can be used to determine numerically the rms value of a waveform. While in Probe, enter the expression RMS(I(R1)) to obtain the rms value of the resistor current. Probe displays a "running" value of the integration in Eq. (3-16), so the appropriate value is *at the end of one or more complete periods* of the waveform. Figure 3-3b shows how to obtain the rms current. The rms current is read as approximately 468 mA. This compares

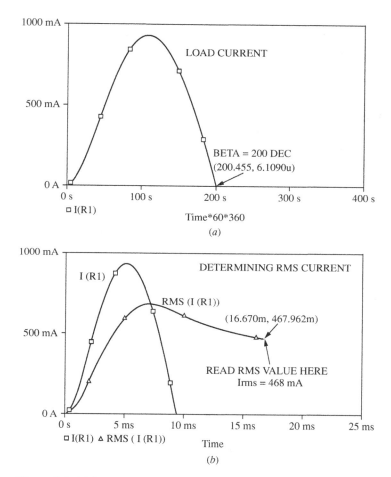

Figure 3-3 (a) Determining the extinction angle β in Probe. The time axis is changed to angle using the x-variable option and entering Time*60*360; (b) Determining the rms value of current in Probe.

very well with the 474 mA calculated in Example 3-2. Remember that the default diode model is used in PSpice and an ideal diode was used in Example 3-2. The average current is found by entering AVG(I(R1)), resulting in $I_o = 304$ mA.

PSpice is also useful in the design process. For example, the objective may be to design a half-wave rectifier circuit to produce a specified value of average current by selecting the proper value of L in an RL load. Since there is no closed-form solution, a trial-and-error iterative method must be used. A PSpice simulation that includes a parametric sweep is used to try several values of L. Example 3-4 illustrates this method.

EXAMPLE 3-4

Half-Wave Rectifier Design Using PSpice

Design a circuit to produce an average current of 2.0 A in a 10-Ω resistance. The source is 120 V rms at 60 Hz.

■ **Solution**

A half-wave rectifier is one circuit that can be used for this application. If a simple half-wave rectifier with the 10-Ω resistance were used, the average current would be $(120\sqrt{2}/\pi)/8 = 6.5$ A. Some means must be found to reduce the average current to the specified 2 A. A series resistance could be added to the load, but resistances absorb power. An added series inductance will reduce the current without adding losses, so an

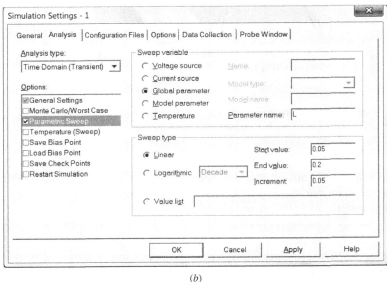

Figure 3-4 (a) PSpice circuit for Example 3-4; (b) A parametric sweep is established in the Simulation Settings box; (c) $L = 0.15$ H for an average current of approximately 2 A.

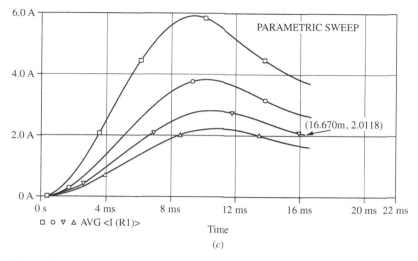

Figure 3-4 (*continued*)

inductor is chosen. Equations (3-15) and (3-17) describe the current function and its average for *RL* loads. There is no closed-form solution for *L*. A trial-and-error technique in PSpice uses the parameter (PARAM) part and a parametric sweep to try a series of values for *L*. The PSpice circuit and the Simulation Settings box are shown in Fig. 3-4.

Average current in the resistor is found by entering AVG(I(R1)) in Probe, yielding a family of curves for different inductance values (Fig. 3-4*c*). The third inductance in the sweep (0.15 H) results in an average current of 2.0118 A in the resistor, which is very close to the design objective. If further precision is necessary, subsequent simulations can be performed, narrowing the range of *L*.

3.5 *RL*-SOURCE LOAD

Supplying Power to a DC Source from an AC Source

Another variation of the half-wave rectifier is shown in Fig. 3-5*a*. The load consists of a resistance, an inductance, and a dc voltage. Starting the analysis at $\omega t = 0$ and assuming the initial current is zero, recognize that the diode will remain off as long as the voltage of the ac source is less than the dc voltage. Letting α be the value of ωt that causes the source voltage to be equal to V_{dc},

$$V_m \sin \alpha = V_{dc}$$

or

$$\alpha = \sin^{-1}\left(\frac{V_{dc}}{V_m}\right) \quad (3\text{-}18)$$

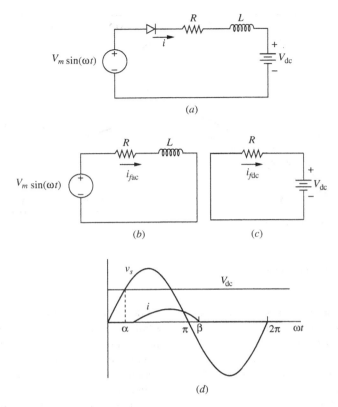

Figure 3-5 (a) Half-wave rectifier with RL source load; (b) Circuit for forced responce from ac source; (c) Circuit for forced responce from dc source; (d) Waveforms.

The diode starts to conduct at $\omega t = \alpha$. With the diode conducting, Kirchhoff's voltage law for the circuit yields the equation

$$V_m \sin(\omega t) = Ri(t) + L\frac{di(t)}{dt} + V_{dc} \quad (3\text{-}19)$$

Total current is determined by summing the forced and natural responses:

$$i(t) = i_f(t) + i_n(t)$$

The current $i_f(t)$ is determined using superposition for the two sources. The forced response from the ac source (Fig. 3-5b) is $(V_m/Z)\sin(\omega t - \theta)$. The forced response due to the dc source (Fig. 3-5c) is $-V_{dc}/R$. The entire forced response is

$$i_f(t) = \frac{V_m}{Z}\sin(\omega t - \theta) - \frac{V_{dc}}{R} \quad (3\text{-}20)$$

3.5 RL-Source Load

The natural response is

$$i_n(t) = Ae^{-t/\tau} \tag{3-21}$$

Adding the forced and natural responses gives the complete response.

$$i(\omega t) = \begin{cases} \dfrac{V_m}{Z}\sin(\omega t - \theta) - \dfrac{V_{dc}}{R} + Ae^{-\omega t/\omega\tau} & \text{for } \alpha \leq \omega t \leq \beta \\ 0 & \text{otherwise} \end{cases} \tag{3-22}$$

The extinction angle β is defined as the angle at which the current reaches zero, as was done earlier in Eq. (3-15). Using the initial condition of $i(\alpha) = 0$ and solving for A,

$$A = \left[-\dfrac{V_m}{Z}\sin(\alpha - \beta) + \dfrac{V_{dc}}{R}\right]e^{\alpha/\omega\tau} \tag{3-23}$$

Figure 3-5d shows voltage and current waveforms for a half-wave rectifier with RL-source load.

The average power absorbed by the resistor is $I_{rms}^2 R$, where

$$I_{rms} = \sqrt{\dfrac{1}{2\pi}\int_\alpha^\beta i^2(\omega t)\, d(\omega t)} \tag{3-24}$$

The average power absorbed by the dc source is

$$P_{dc} = I_o V_{dc} \tag{3-25}$$

where I_o is the average current, that is,

$$I_o = \dfrac{1}{2\pi}\int_\alpha^\beta i(\omega t)\, d(\omega t) \tag{3-26}$$

Assuming the diode and the inductor to be ideal, there is no average power absorbed by either. The power supplied by the ac source is equal to the sum of the power absorbed by the resistor and the dc source

$$P_{ac} = I_{rms}^2 R + I_o V_{dc} \tag{3-27}$$

or it can be computed from

$$P_{ac} = \dfrac{1}{2\pi}\int_0^{2\pi} v(\omega t) i(\omega t)\, d(\omega t) = \dfrac{1}{2\pi}\int_\alpha^\beta (V_m \sin \omega t) i(\omega t)\, d(\omega t) \tag{3-28}$$

EXAMPLE 3-5

Half-Wave Rectifier with *RL*-Source Load

For the circuit of Fig. 3-5a, $R = 2\,\Omega$, $L = 20$ mH, and $V_{dc} = 100$ V. The ac source is 120 V rms at 60 Hz. Determine (*a*) an expression for the current in the circuit, (*b*) the power absorbed by the resistor, (*c*) the power absorbed by the dc source, and (*d*) the power supplied by the ac source and the power factor of the circuit.

■ **Solution**

From the parameters given,

$$V_m = 120\sqrt{2} = 169.7 \text{ V}$$
$$Z = [R^2 + (\omega L)^2]^{0.5} = 7.80\,\Omega$$
$$\theta = \tan^{-1}(\omega L/R) = 1.31 \text{ rad}$$
$$\alpha = \sin^{-1}(100/169.7) = 36.1° = 0.630 \text{ rad}$$
$$\omega\tau = 377(0.02/2) = 3.77 \text{ rad}$$

(*a*) Using Eq. (3-22),

$$i(\omega t) = 21.8\sin(\omega t - 1.31) - 50 + 75.3e^{-\omega t/3.77} \quad \text{A}$$

The extinction angle β is found from the solution of

$$i(\beta) = 21.8\sin(\beta - 1.31) - 50 + 75.3e^{-\beta/3.77} = 0$$

which results in $\beta = 3.37$ rad (193°) using root-finding software.

(*b*) Using the preceding expression for $i(\omega t)$ in Eq. (3-24) and using a numerical integration program, the rms current is

$$I_{rms} = \sqrt{\frac{1}{2\pi}\int_{0.63}^{3.37} i^2(\omega t)\,d(\omega t)} = 3.98 \text{ A}$$

resulting in

$$P_R = I_{rms}^2 R = 3.98^2(2) = 31.7 \text{ W}$$

(*c*) The power absorbed by the dc source is $I_o V_{dc}$. Using Eq. (3-26),

$$I_o = \frac{1}{2\pi}\int_{0.63}^{3.37} i(\omega t)\,d(\omega t) = 2.25 \text{ A}$$

yielding

$$P_{dc} = I_o V_{dc} = (2.25)(100) = 225 \text{ W}$$

(*d*) The power supplied by the ac source is the sum of the powers absorbed by the load.

$$P_s = P_R + P_{dc} = 31.2 + 225 = 256 \text{ W}$$

The power factor is

$$\text{pf} = \frac{P}{S} = \frac{P}{V_{s,rms}I_{rms}} = \frac{256}{(120)(3.98)} = 0.54$$

■ PSpice Solution

The power quantities in this example can be determined from a PSpice simulation of this circuit. The circuit of Fig. 3-5a is created using VSIN, Dbreak, R, and L. In the simulation settings, choose Time Domain (transient) for the analysis type, and set the Run Time to 16.67 ms for one period of the source. Set the Maximum Step Size to 10 μs to get adequate sampling of the waveforms. A transient analysis with a run time of 16.67 ms (one period for 60 Hz) and a maximum step size of 10 μs is used for the simulation settings.

Average power absorbed by the 2-Ω resistor can be computed in Probe from the basic definition of the average of $p(t)$ by entering AVG(W(R1)), resulting in 29.7 W, or from $I_{rms}^2 R$ by entering RMS(I(R1))*RMS(I(R1))*2. The average power absorbed by the dc source is computed from the Probe expression AVG(W(Vdc)), yielding 217 W.

The PSpice values differ slightly from the values obtained analytically because of the diode model. However, the default diode is more realistic than the ideal diode in predicting actual circuit performance.

3.6 INDUCTOR-SOURCE LOAD

Using Inductance to Limit Current

Another variation of the half-wave rectifier circuit has a load that consists of an inductor and a dc source, as shown in Fig. 3-6. Although a practical implementation of this circuit would contain some resistance, the resistance may be negligible compared to other circuit parameters.

Starting at $\omega t = 0$ and assuming zero initial current in the inductor, the diode remains reverse-biased until the ac source voltage reaches the dc voltage. The value of ωt at which the diode starts to conduct is α, calculated using Eq. (3-18). With the diode conducting, Kirchhoff's voltage law for the circuit is

$$V_m \sin(\omega t) = L \frac{di(t)}{dt} + V_{dc} \quad (3\text{-}29)$$

or

$$V_m \sin(\omega t) = \frac{L}{\omega} \frac{di(\omega t)}{dt} + V_{dc} \quad (3\text{-}30)$$

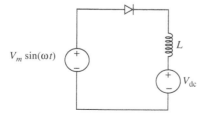

Figure 3-6 Half-wave rectifier with inductor source load.

Rearranging gives

$$\frac{di(\omega t)}{dt} = \frac{V_m \sin(\omega t) - V_{dc}}{\omega L} \tag{3-31}$$

Solving for $i(\omega t)$,

$$i(\omega t) = \frac{1}{\omega L}\int_\alpha^{\omega t} V_m \sin \lambda \, d(\lambda) - \frac{1}{\omega L}\int_\alpha^{\omega t} V_{dc} \, d(\lambda) \tag{3-32}$$

Performing the integration,

$$i(\omega t) = \begin{cases} \dfrac{V_m}{\omega L}(\cos \alpha - \cos \omega t) + \dfrac{V_{dc}}{\omega L}(\alpha - \omega t) & \text{for } \alpha \le \omega t \le \beta \\ 0 & \text{otherwise} \end{cases} \tag{3-33}$$

A distinct feature of this circuit is that the power supplied by the source is the same as that absorbed by the dc source, less any losses associated with a nonideal diode and inductor. If the objective is to transfer power from the ac source to the dc source, losses are kept to a minimum by using this circuit.

EXAMPLE 3-6

Half-Wave Rectifier with Inductor-Source Load

For the circuit of Fig. 3-6, the ac source is 120 V rms at 60 Hz, $L = 50$ mH, and $V_{dc} = 72$ V. Determine (*a*) an expression for the current, (*b*) the power absorbed by the dc source, and (*c*) the power factor.

■ **Solution**

For the parameters given,

$$\alpha = \sin^{-1}\left(\frac{72}{120\sqrt{2}}\right) = 25.1° = 0.438 \text{ rad}$$

(*a*) The equation for current is found from Eq. (3-33).

$$i(\omega t) = 9.83 - 9.00 \cos(\omega t) - 3.82 \, \omega t \quad \text{A} \quad \text{for } \alpha \le \omega t \le \beta$$

where β is found to be 4.04 rad from the numerical solution of $9.83 - 9.00 \cos \beta - 3.82 \beta = 0$.

(*b*) The power absorbed by the dc source is $I_o V_{dc}$, where

$$I_o = \frac{1}{2\pi}\int_\alpha^\beta i(\omega t) \, d(\omega t)$$

$$= \frac{1}{2\pi}\int_{0.438}^{4.04} [9.83 - 9.00 \cos(\omega t) - 3.82 \, \omega t] \, d(\omega t) = 2.46 \text{ A}$$

resulting in

$$P_{dc} = V_{dc} I_o = (2.46)(72) = 177 \text{ W}$$

(c) The rms current is found from

$$I_{rms} = \sqrt{\frac{1}{2\pi} \int_{\alpha}^{\beta} i^2(\omega t) \, d(\omega t)} = 3.81 \text{ A}$$

Therefore,

$$\text{pf} = \frac{P}{S} = \frac{P}{V_{rms} I_{rms}} = \frac{177}{(120)(3.81)} = 0.388$$

3.7 THE FREEWHEELING DIODE

Creating a DC Current

A freewheeling diode, D_2 in Fig. 3-7a, can be connected across an *RL* load as shown. The behavior of this circuit is somewhat different from that of the half-wave rectifier of Fig. 3-2. The key to the analysis of this circuit is to determine when each diode conducts. First, it is observed that both diodes cannot be forward-biased at the same time. Kirchhoff's voltage law around the path containing the source and the two diodes shows that one diode must be reverse-biased. Diode D_1 will be on when the source is positive, and diode D_2 will be on when the source is negative.

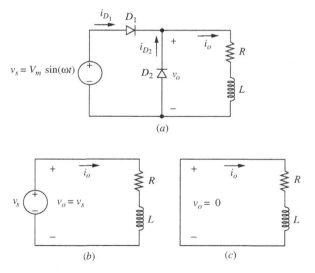

Figure 3-7 (a) Half-wave rectifier with freewheeling diode; (b) Equivalent circuit for $v_s > 0$; (c) Equivalent circuit for $v_s < 0$.

CHAPTER 3 Half-Wave Rectifiers

For a positive source voltage,

- D_1 is on.
- D_2 is off.
- The equivalent circuit is the same as that of Fig. 3-2, shown again in Fig. 3-7b.
- The voltage across the RL load is the same as the source.

For a negative source voltage,

- D_1 is off.
- D_2 is on.
- The equivalent circuit is the same at that of Fig. 3-7c.
- The voltage across the RL load is zero.

Since the voltage across the RL load is the same as the source voltage when the source is positive and is zero when the source is negative, the load voltage is a half-wave rectified sine wave.

When the circuit is first energized, the load current is zero and cannot change instantaneously. The current reaches periodic steady state after a few periods (depending on the L/R time constant), which means that the current at the end of a period is the same as the current at the beginning of the period, as shown in Fig. 3-8. The steady-state current is usually of greater interest than the transient that occurs when the circuit is first energized. Steady-state load, source, and diode currents are shown in Fig. 3-9.

The Fourier series for the half-wave rectified sine wave for the voltage across the load is

$$v(t) = \frac{V_m}{\pi} + \frac{V_m}{2}\sin(\omega_0 t) - \sum_{n=2,4,6...}^{\infty} \frac{2V_m}{(n^2-1)\pi}\cos(n\omega_0 t) \quad (3\text{-}34)$$

The current in the load can be expressed as a Fourier series by using superposition, taking each frequency separately. The Fourier series method is illustrated in Example 3-7.

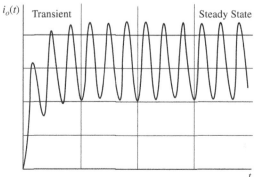

Figure 3-8 Load current reaching steady state after the circuit is energized.

3.7 The Freewheeling Diode

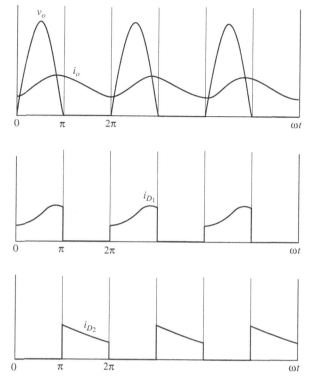

Figure 3-9 Steady-state load voltage and current waveforms with freewheeling diode.

EXAMPLE 3-7

Half-Wave Rectifier with Freewheeling Diode

Determine the average load voltage and current, and determine the power absorbed by the resistor in the circuit of Fig. 3-7a, where $R = 2\,\Omega$ and $L = 25$ mH, V_m is 100 V, and the frequency is 60 Hz.

■ **Solution**
The Fourier series for this half-wave rectified voltage that appears across the load is obtained from Eq. (3-34). The average load voltage is the dc term in the Fourier series:

$$V_o = \frac{V_m}{\pi} = \frac{100}{\pi} = 31.8 \text{ V}$$

Average load current is

$$I_o = \frac{V_o}{R} = \frac{31.8}{2} = 15.9 \text{ A}$$

Load power can be determined from $I_{rms}^2 R$, and rms current is determined from the Fourier components of current. The amplitudes of the ac current components are determined from phasor analysis:

$$I_n = \frac{V_n}{Z_n}$$

where
$$Z_n = |R + jn\omega_0 L| = |2 + jn377(0.025)|$$

The ac voltage amplitudes are determined from Eq. (3-34), resulting in

$$V_1 = \frac{V_m}{2} = \frac{100}{2} = 50 \text{ V}$$

$$V_2 = \frac{2V_m}{(2^2-1)\pi} = 21.2 \text{ V}$$

$$V_4 = \frac{2V_m}{(4^2-1)\pi} = 4.24 \text{ V}$$

$$V_6 = \frac{2V_m}{(6^2-1)\pi} = 1.82 \text{ V}$$

The resulting Fourier terms are as follows:

n	V_n (V)	Z_n (Ω)	I_n (A)
0	31.8	2.00	15.9
1	50.0	9.63	5.19
2	21.2	18.96	1.12
4	4.24	37.75	0.11
6	1.82	56.58	0.03

The rms current is obtained using Eq. (2-64).

$$I_{rms} = \sqrt{\sum_{k=0}^{\infty} I_{k,rms}^2} \approx \sqrt{15.9^2 + \left(\frac{5.19}{\sqrt{2}}\right)^2 + \left(\frac{1.12}{\sqrt{2}}\right)^2 + \left(\frac{0.11}{\sqrt{2}}\right)^2} = 16.34 \text{ A}$$

Notice that the contribution to rms current from the harmonics decreases as n increases, and higher-order terms are not significant. Power in the resistor is $I_{rms}^2 R = (16.34)^2 2 = 534$ W.

■ PSpice Solution

The circuit of Fig. 3-7a is created using VSIN, Dbreak, R, and L. The PSpice model for Dbreak is changed to make $n = 0.001$ to approximate an ideal diode. A transient analysis is run with a run time of 150 ms with data saved after 100 ms to eliminate the start-up transient from the data. A maximum step size of 10 μs gives a smooth waveform.

A portion of the output file is as follows:

```
****     FOURIER ANALYSIS          TEMPERATURE =   27.000 DEG C

FOURIER COMPONENTS OF TRANSIENT RESPONSE V(OUT)

DC COMPONENT =      3.183002E+01
```

HARMONIC NO	FREQUENCY (HZ)	FOURIER COMPONENT	NORMALIZED COMPONENT	PHASE (DEG)	NORMALIZED PHASE (DEG)
1	6.000E+01	5.000E+01	1.000E+00	-5.804E-05	0.000E+00
2	1.200E+02	2.122E+01	4.244E-01	-9.000E+01	-9.000E+01
3	1.800E+02	5.651E-05	1.130E-06	-8.831E+01	-8.831E+01
4	2.400E+02	4.244E+00	8.488E-02	-9.000E+01	-9.000E+01
5	3.000E+02	5.699E-05	1.140E-06	-9.064E+01	-9.064E+01
6	3.600E+02	1.819E+00	3.638E-02	-9.000E+01	-9.000E+01
7	4.200E+02	5.691E-05	1.138E-06	-9.111E+01	-9.110E+01
8	4.800E+02	1.011E+00	2.021E-02	-9.000E+01	-9.000E+01
9	5.400E+02	5.687E-05	1.137E-06	-9.080E+01	-9.079E+01

FOURIER COMPONENTS OF TRANSIENT RESPONSE I(R_R1)

DC COMPONENT =   1.591512E+01

HARMONIC NO	FREQUENCY (HZ)	FOURIER COMPONENT	NORMALIZED COMPONENT	PHASE (DEG)	NORMALIZED PHASE (DEG)
1	6.000E+01	5.189E+00	1.000E+00	-7.802E+01	0.000E+00
2	1.200E+02	1.120E+00	2.158E-01	-1.739E+02	-1.788E+01
3	1.800E+02	1.963E-04	3.782E-05	-3.719E+01	1.969E+02
4	2.400E+02	1.123E-01	2.164E-02	-1.770E+02	1.351E+02
5	3.000E+02	7.524E-05	1.450E-05	6.226E+01	4.524E+02
6	3.600E+02	3.217E-02	6.200E-03	-1.781E+02	2.900E+02
7	4.200E+02	8.331E-05	1.605E-05	1.693E+02	7.154E+02
8	4.800E+02	1.345E-02	2.592E-03	-1.783E+02	4.458E+02
9	5.400E+02	5.435E-05	1.047E-05	-1.074E+02	5.948E+02

Note the close agreement between the analytically obtained Fourier terms and the PSpice output. Average current can be obtained in Probe by entering AVG(I(R1)), yielding 15.9 A. Average power in the resistor can be obtained by entering AVG(W(R1)), yielding $P = 535$ W. It is important that the simulation represent steady-state periodic current for the results to be valid.

Reducing Load Current Harmonics

The average current in the *RL* load is a function of the applied voltage and the resistance but not the inductance. The inductance affects only the ac terms in the Fourier series. If the inductance is infinitely large, the impedance of the load to ac terms in the Fourier series is infinite, and the load current is purely dc. The load current is then

$$i_o(t) \approx I_o = \frac{V_o}{R} = \frac{V_m}{\pi R} \qquad \frac{L}{R} \to \infty \qquad (3\text{-}35)$$

A large inductor ($L/R \gg T$) with a freewheeling diode provides a means of establishing a nearly constant load current. Zero-to-peak fluctuation in load current can be estimated as being equal to the amplitude of the first ac term in the Fourier series. The peak-to-peak ripple is then

$$\Delta I_o \approx 2I_1 \qquad (3\text{-}36)$$

EXAMPLE 3-8

Half-Wave Rectifier with Freewheeling Diode: $L/R \to \infty$

For the half-wave rectifier with a freewheeling diode and *RL* load as shown in Fig. 3-7a, the source is 240 V rms at 60 Hz and $R = 8\ \Omega$. (*a*) Assume *L* is infinitely large. Determine the power absorbed by the load and the power factor as seen by the source. Sketch v_o, i_{D_1}, and i_{D_2}. (*b*) Determine the average current in each diode. (*c*) For a finite inductance, determine *L* such that the peak-to-peak current is no more than 10 percent of the average current.

■ **Solution**

(*a*) The voltage across the *RL* load is a half-wave rectified sine wave, which has an average value of V_m/π. The load current is

$$i(\omega t) = I_o = \frac{V_o}{R} = \frac{V_m/\pi}{R} = \frac{(240\sqrt{2})/\pi}{8} = 13.5\ \text{A} \approx I_{\text{rms}}$$

Power in the resistor is

$$P = (I_{\text{rms}})^2 R = (13.5)^2 8 = 1459\ \text{W}$$

Source rms current is computed from

$$I_{s,\text{rms}} = \sqrt{\frac{1}{2\pi} \int_0^\pi (13.5)^2 d(\omega t)} = 9.55\ \text{A}$$

The power factor is

$$\text{pf} = \frac{P}{V_{s,\text{rms}} I_{s,\text{rms}}} = \frac{1459}{(240)(9.55)} = 0.637$$

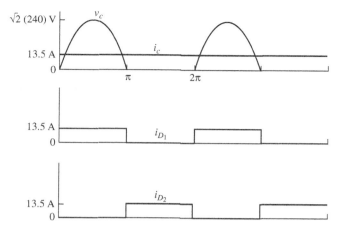

Figure 3-10 Waveforms for the half-wave rectifier with freewheeling diode of Example 3-8 with $L/R \to \infty$.

Voltage and current waveforms are shown in Fig. 3-10.

(b) Each diode conducts for one-half of the time. Average current for each diode is $I_o/2 = 13.5/2 = 6.75$ A.

(c) The value of inductance required to limit the variation in load current to 10 percent can be approximated from the fundamental frequency of the Fourier series. The voltage input to the load for $n = 1$ in Eq. (3-34) has amplitude $V_m/2 = \sqrt{2}(240)/2 = 170$ V the peak-to-peak current must be limited to

$$\Delta I_o = (0.10)(I_o) = (0.10)(13.5) = 1.35 \text{ A}$$

which corresponds to an amplitude of $1.35/2 = 0.675$ A. The load impedance at the fundamental frequency must then be

$$Z_1 = \frac{V_1}{I_1} = \frac{170}{0.675} = 251 \text{ }\Omega$$

The load impedance is

$$Z_1 = 251 = |R + j\omega L| = |8 + j377L|$$

Since the 8-Ω resistance is negligible compared to the total impedance, the inductance can be approximated as

$$L \approx \frac{Z_1}{\omega} = \frac{251}{377} = 0.67 \text{ H}$$

The inductance will have to be slightly larger than 0.67 H because Fourier terms higher than $n = 1$ were neglected in this estimate.

3.8 HALF-WAVE RECTIFIER WITH A CAPACITOR FILTER

Creating a DC Voltage from an AC Source

A common application of rectifier circuits is to convert an ac voltage input to a dc voltage output. The half-wave rectifier of Fig. 3-11a has a parallel *RC* load. The purpose of the capacitor is to reduce the variation in the output voltage, making it more like dc. The resistance may represent an external load, and the capacitor may be a filter which is part of the rectifier circuit.

Assuming the capacitor is initially uncharged and the circuit is energized at $\omega t = 0$, the diode becomes forward-biased as the source becomes positive. With the diode on, the output voltage is the same as the source voltage, and the capacitor charges. The capacitor is charged to V_m when the input voltage reaches its positive peak at $\omega t = \pi/2$.

As the source decreases after $\omega t = \pi/2$, the capacitor discharges into the load resistor. At some point, the voltage of the source becomes less than the output voltage, reverse-biasing the diode and isolating the load from the source. The output voltage is a decaying exponential with time constant RC while the diode is off.

The point when the diode turns off is determined by comparing the rates of change of the source and the capacitor voltages. The diode turns off when the

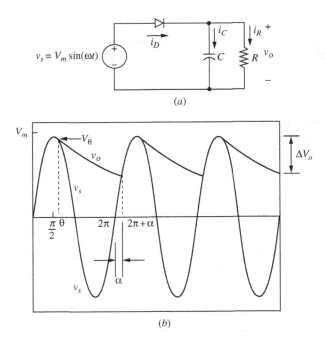

Figure 3-11 (*a*) Half-wave rectifier with *RC* load; (*b*) Input and output voltages.

downward rate of change of the source exceeds that permitted by the time constant of the RC load. The angle $\omega t = \theta$ is the point when the diode turns off in Fig. 3-11b. The output voltage is described by

$$v_o(\omega t) = \begin{cases} V_m \sin \omega t & \text{diode on} \\ V_\theta e^{-(\omega t - \theta)/\omega RC} & \text{diode off} \end{cases} \quad (3\text{-}37)$$

where
$$V_\theta = V_m \sin \theta \quad (3\text{-}38)$$

The slopes of these functions are

$$\frac{d}{d(\omega t)}[V_m \sin(\omega t)] = V_m \cos(\omega t) \quad (3\text{-}39)$$

and

$$\frac{d}{d(\omega t)}\left(V_m \sin \theta \, e^{-(\omega t - \theta)/\omega RC}\right) = V_m \sin \theta \left(-\frac{1}{\omega RC}\right) e^{-(\omega t - \theta)/\omega RC} \quad (3\text{-}40)$$

At $\omega t = \theta$, the slopes of the voltage functions are equal:

$$V_m \cos \theta = \left(\frac{V_m \sin \theta}{-\omega RC}\right) e^{-(\theta - \theta)/\omega RC} = \frac{V_m \sin \theta}{-\omega RC}$$

$$\frac{V_m \cos \theta}{V_m \sin \theta} = \frac{1}{-\omega RC}$$

$$\frac{1}{\tan \theta} = \frac{1}{-\omega RC}$$

Solving for θ and expressing θ so it is in the proper quadrant, we have

$$\boxed{\theta = \tan^{-1}(-\omega RC) = -\tan^{-1}(\omega RC) + \pi} \quad (3\text{-}41)$$

In practical circuits where the time constant is large,

$$\boxed{\theta \approx \frac{\pi}{2} \quad \text{and} \quad V_m \sin \theta \approx V_m} \quad (3\text{-}42)$$

When the source voltage comes back up to the value of the output voltage in the next period, the diode becomes forward-biased, and the output again is the same as the source voltage. The angle at which the diode turns on in the second period, $\omega t = 2\pi + \alpha$, is the point when the sinusoidal source reaches the same value as the decaying exponential output:

$$V_m \sin(2\pi + \alpha) = (V_m \sin \theta) e^{-(2\pi + \alpha - \theta)/\omega RC}$$

or

$$\sin\alpha - (\sin\theta)e^{-(2\pi+\alpha-\theta)/\omega RC} = 0 \qquad (3\text{-}43)$$

Equation (3-43) must be solved numerically for α.

The current in the resistor is calculated from $i_R = v_o/R$. The current in the capacitor is calculated from

$$i_C(t) = C\frac{dv_o(t)}{dt}$$

which can also be expressed, using ωt as the variable, as

$$i_C(\omega t) = \omega C\frac{dv_o(\omega t)}{d(\omega t)}$$

Using v_o from Eq. (3-37),

$$i_C(\omega t) = \begin{cases} -\left(\dfrac{V_m \sin\theta}{R}\right)e^{-(\omega t-\theta)/\omega RC} & \text{for } \theta \le \omega t \le 2\pi + \alpha \quad \text{(diode off)} \\ \omega C V_m \cos(\omega t) & \text{for } 2\pi + \alpha \le \omega t \le 2\pi + \theta \quad \text{(diode on)} \end{cases} \qquad (3\text{-}44)$$

The source current, which is the same as the diode current, is

$$i_S = i_D = i_R + i_C \qquad (3\text{-}45)$$

The average capacitor current is zero, so the average diode current is the same as the average load current. Since the diode is on for a short time in each cycle, the peak diode current is generally much larger than the average diode current. Peak capacitor current occurs when the diode turns on at $\omega t = 2\pi + \alpha$. From Eq. (3-44),

$$I_{C,\text{peak}} = \omega C V_m \cos(2\pi + \alpha) = \omega C V_m \cos\alpha \qquad (3\text{-}46)$$

Resistor current at $\omega t = 2\pi + \alpha$ is obtained from Eq. (3-37).

$$i_R(2\omega t + \alpha) = \frac{V_m \sin(2\omega t + \alpha)}{R} = \frac{V_m \sin\alpha}{R} \qquad (3\text{-}47)$$

Peak diode current is

$$I_{D,\text{peak}} = \omega C V_m \cos\alpha + \frac{V_m \sin\alpha}{R} = V_m\left(\omega C \cos\alpha + \frac{\sin\alpha}{R}\right) \qquad (3\text{-}48)$$

The effectiveness of the capacitor filter is determined by the variation in output voltage. This may be expressed as the difference between the maximum and minimum output voltage, which is the peak-to-peak ripple voltage. For the half-wave rectifier of Fig. 3-11a, the maximum output voltage is V_m. The minimum output voltage occurs at $\omega t = 2\pi + \alpha$, which can be computed from $V_m \sin \alpha$. The peak-to-peak ripple for the circuit of Fig. 3-11a is expressed as

$$\Delta V_o = V_m - V_m \sin \alpha = V_m(1 - \sin \alpha) \qquad (3\text{-}49)$$

In circuits where the capacitor is selected to provide for a nearly constant dc output voltage, the RC time constant is large compared to the period of the sine wave, and Eq. (3-42) applies. Moreover, the diode turns on close to the peak of the sine wave when $\alpha \approx \pi/2$. The change in output voltage when the diode is off is described in Eq. (3-37). In Eq. (3-37), if $V_\theta \approx V_m$ and $\theta \approx \pi/2$, then Eq. (3-37) evaluated at $\alpha = \pi/2$ is

$$v_o(2\pi + \alpha) = V_m e^{-(2\pi + \pi/2 - \pi/2)\omega RC} = V_m e^{-2\pi/\omega RC}$$

The ripple voltage can then be approximated as

$$\Delta V_o \approx V_m - V_m e^{-2\pi/\omega RC} = V_m\left(1 - e^{-2\pi/\omega RC}\right) \qquad (3\text{-}50)$$

Furthermore, the exponential in the above equation can be approximated by the series expansion:

$$e^{-2\pi/\omega RC} \approx 1 - \frac{2\pi}{\omega RC}$$

Substituting for the exponential in Eq. (3-50), the peak-to-peak ripple is approximately

$$\Delta V_o \approx V_m\left(\frac{2\pi}{\omega RC}\right) = \frac{V_m}{fRC} \qquad (3\text{-}51)$$

The output voltage ripple is reduced by increasing the filter capacitor C. As C increases, the conduction interval for the diode decreases. Therefore, increasing the capacitance to reduce the output voltage ripple results in a larger peak diode current.

EXAMPLE 3-9

Half-Wave Rectifier with RC Load

The half-wave rectifier of Fig. 3-11a has a 120-V rms source at 60 Hz, $R = 500\ \Omega$, and $C = 100\ \mu\text{F}$. Determine (a) an expression for output voltage, (b) the peak-to-peak voltage variation on the output, (c) an expression for capacitor current, (d) the peak diode current, and (e) the value of C such that ΔV_o is 1 percent of V_m.

■ **Solution**
From the parameters given,

$$V_m = 120\sqrt{2} = 169.7\ \text{V}$$
$$\omega RC = (2\pi 60)(500)(10)^{-6} = 18.85\ \text{rad}$$

The angle θ is determined from Eq. (3-41).
$$\theta = -\tan^{-1}(18.85) + \pi = 1.62 \text{ rad} = 93°$$
$$V_m \sin\theta = 169.5 \text{ V}$$

The angle α is determined from the numerical solution of Eq. (3-43).
$$\sin\alpha - \sin(1.62)e^{-(2\pi + \alpha - 1.62/18.85)} = 0$$

yielding
$$\alpha = 0.843 \text{ rad} = 48°$$

(a) Output voltage is expressed from Eq. (3-37).
$$v_o(\omega t) = \begin{cases} 169.7 \sin(\omega t) & 2\pi + \alpha \leq \omega t \leq 2\pi + \theta \\ 169.5 e^{-(\omega t - 1.62)/18.85} & \theta \leq \omega t \leq 2\pi + \alpha \end{cases}$$

(b) Peak-to-peak output voltage is described by Eq. (3-49).
$$\Delta V_o = V_m(1 - \sin\alpha) = 169.7(1 - \sin 0.843) = 43 \text{ V}$$

(c) The capacitor current is determined from Eq. (3-44).
$$i_C(\omega t) = \begin{cases} -0.339 e^{-(\omega t - 1.62)/18.85} \text{ A} & \theta \leq \omega t \leq 2\pi + \alpha \\ 6.4 \cos(\omega t) \text{ A} & 2\pi + \alpha \leq \omega t \leq 2\pi + \theta \end{cases}$$

(d) Peak diode current is determined from Eq. (3-48).
$$I_{D,\text{peak}} = \sqrt{2}(120)\left[377(10)^{-4} \cos 0.843 + \frac{\sin 8.43}{500} \right]$$
$$= 4.26 + 0.34 = 4.50 \text{ A}$$

(e) For $\Delta V_o = 0.01 V_m$, Eq. (3-51) can be used.
$$C \approx \frac{V_m}{fR(\Delta V_o)} = \frac{V_m}{(60)(500)(0.01 V_m)} = \frac{1}{300} \text{F} = 3333 \text{ μF}$$

Note that peak diode current can be determined from Eq. (3-48) using an estimate of α from Eq. (3-49).
$$\alpha \approx \sin^{-1}\left(1 - \frac{\Delta V_o}{V_m}\right) = \sin^{-1}\left(1 - \frac{1}{fRC}\right) = 81.9°$$

From Eq. (3-48), peak diode current is 30.4 A.

■ **PSpice Solution**

A PSpice circuit is created for Fig. 3-11a using VSIN, Dbreak, R, and C. The diode Dbreak used in this analysis causes the results to differ slightly from the analytic solution based on the ideal diode. The diode drop causes the maximum output voltage to be slightly less than that of the source.

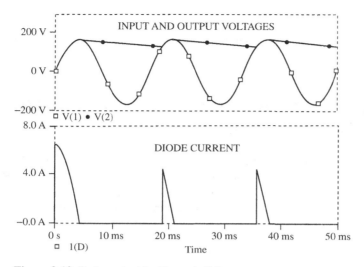

Figure 3-12 Probe output for Example 3-9.

The Probe output is shown in Fig. 3-12. Angles θ and α are determined directly by first modifying the x-variable to indicate degrees (x-variable = time*60*360) and then using the cursor option. The restrict data option is used to compute quantities based on steady-state values (16.67 to 50 ms). Steady state is characterized by waveforms beginning and ending a period at the same values. Note that the peak diode current is largest in the first period because the capacitor is initially uncharged.

■ **Results from the Probe Cursor**

Quantity	Result
α + 360°	408° (α = 48°)
θ	98.5°
V_o max	168.9 V
V_o min	126 V
ΔV_o	42.9 V
$I_{D,peak}$	4.42 A steady state; 6.36 A first period
$I_{C,peak}$	4.12 A steady state; 6.39 A first period

■ **Results after Restricting the Data to Steady State**

Quantity	Probe Expression	Result
$I_{D,avg}$	AVG(I(D1))	0.295 A
$I_{C,rms}$	RMS(I(C1))	0.905 A
$I_{R,avg}$	AVG(W(R1))	43.8 W
P_s	AVG(W(Vs))	−44.1 W
P_D	AVG(W(D1))	345 mW

In this example, the ripple, or variation in output voltage, is very large, and the capacitor is not an effective filter. In many applications, it is desirable to produce an output that is closer to dc. This requires the time constant RC to be large compared to the period of the input voltage, resulting in little decay of the output voltage. For an effective filter capacitor, the output voltage is essentially the same as the peak voltage of the input.

3.9 THE CONTROLLED HALF-WAVE RECTIFIER

The half-wave rectifiers analyzed previously in this chapter are classified as uncontrolled rectifiers. Once the source and load parameters are established, the dc level of the output and the power transferred to the load are fixed quantities.

A way to control the output of a half-wave rectifier is to use an SCR[1] instead of a diode. Figure 3-13a shows a basic controlled half-wave rectifier with a resistive load. Two conditions must be met before the SCR can conduct:

1. The SCR must be forward-biased ($v_{SCR} > 0$).
2. A current must be applied to the gate of the SCR.

Unlike the diode, the SCR will not begin to conduct as soon as the source becomes positive. Conduction is delayed until a gate current is applied, which is the basis for using the SCR as a means of control. Once the SCR is conducting, the gate current can be removed and the SCR remains on until the current goes to zero.

Resistive Load

Figure 3-13b shows the voltage waveforms for a controlled half-wave rectifier with a resistive load. A gate signal is applied to the SCR at $\omega t = \alpha$, where α is the delay angle. The average (dc) voltage across the load resistor in Fig. 3-13a is

$$V_o = \frac{1}{2\pi} \int_\alpha^\pi V_m \sin(\omega t) \, d(\omega t) = \frac{V_m}{2\pi}(1 + \cos \alpha) \quad (3\text{-}52)$$

The power absorbed by the resistor is V_{rms}^2/R, where the rms voltage across the resistor is computed from

$$V_{rms} = \sqrt{\frac{1}{2\pi} \int_0^{2\pi} v_o^2(\omega t) \, d(\omega t)}$$

$$= \sqrt{\frac{1}{2\pi} \int_\alpha^\pi [V_m \sin(\omega t)]^2 \, d(\omega t)}$$

$$= \frac{V_m}{2} \sqrt{1 - \frac{\alpha}{\pi} + \frac{\sin(2\alpha)}{2\pi}} \quad (3\text{-}53)$$

---

[1] Switching with other controlled turn-on devices such as transistors or IGBTs can be used to control the output of a converter.

3.9 The Controlled Half-Wave Rectifier

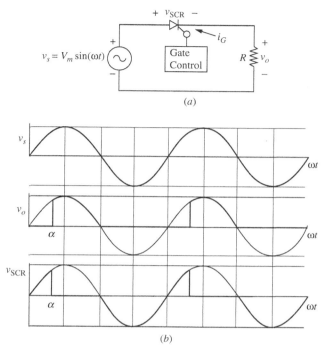

Figure 3-13 (a) A basic controlled rectifier; (b) Voltage waveforms.

EXAMPLE 3-10

Controlled Half-Wave Rectifier with Resistive Load

Design a circuit to produce an average voltage of 40 V across a 100-Ω load resistor from a 120-V rms 60-Hz ac source. Determine the power absorbed by the resistance and the power factor.

■ **Solution**

If an uncontrolled half-wave rectifier is used, the average voltage will be $V_m/\pi = 120\sqrt{2}/\pi = 54$ V. Some means of reducing the average resistor voltage to the design specification of 40 V must be found. A series resistance or inductance could be added to an uncontrolled rectifier, or a controlled rectifier could be used. The controlled rectifier of Fig. 3-13a has the advantage of not altering the load or introducing losses, so it is selected for this application.

Equation (3-52) is rearranged to determine the required delay angle:

$$\alpha = \cos^{-1}\left[V_o\left(\frac{2\pi}{V_m}\right) - 1\right]$$

$$= \cos^{-1}\left\{40\left[\frac{2\pi}{\sqrt{2}(120)}\right] - 1\right\} = 61.2° = 1.07 \text{ rad}$$

Equation (3-53) gives

$$V_{rms} = \frac{\sqrt{2}(120)}{2}\sqrt{1 - \frac{1.07}{\pi} + \frac{\sin[2(1.07)]}{2\pi}} = 75.6 \text{ V}$$

Load power is

$$P_R = \frac{V_{rms}^2}{R} = \frac{(75.6)^2}{100} = 57.1 \text{ W}$$

The power factor of the circuit is

$$\text{pf} = \frac{P}{S} = \frac{P}{V_{S,rms}I_{rms}} = \frac{57.1}{(120)(75.6/100)} = 0.63$$

RL Load

A controlled half-wave rectifier with an *RL* load is shown in Fig. 3-14a. The analysis of this circuit is similar to that of the uncontrolled rectifier. The current is the sum of the forced and natural responses, and Eq. (3-9) applies:

$$i(\omega t) = i_f(\omega t) + i_n(\omega t) = \frac{V_m}{Z}\sin(\omega t - \theta) + Ae^{-\omega t/\omega\tau}$$

The constant A is determined from the initial condition $i(\alpha) = 0$:

$$i(\alpha) = 0 = \frac{V_m}{Z}\sin(\alpha - \theta) + Ae^{-\alpha/\omega\tau}$$

$$A = \left[-\frac{V_m}{Z}\sin(\alpha - \theta)\right] = e^{\alpha/\omega\tau} \tag{3-54}$$

Substituting for A and simplifying,

$$i(\omega t) = \begin{cases} \frac{V_m}{Z}\left[\sin(\omega t - \theta) - \sin(\alpha - \theta)e^{(\alpha - \omega t)/\omega\tau}\right] & \text{for } \alpha \le \omega t \le \beta \\ 0 & \text{otherwise} \end{cases} \tag{3-55}$$

The *extinction angle* β is defined as the angle at which the current returns to zero, as in the case of the uncontrolled rectifier. When $\omega t = \beta$,

$$i(\beta) = 0 = \frac{V_m}{Z}\left[\sin(\beta - \theta) - \sin(\alpha - \theta)e^{(\alpha - \beta)/\omega\tau}\right] \tag{3-56}$$

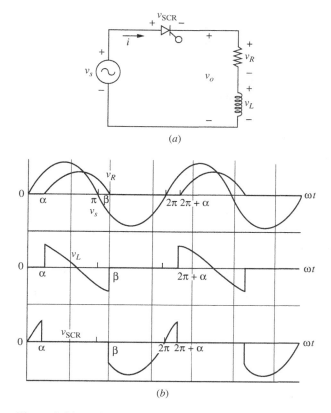

Figure 3-14 (a) Controlled half-wave rectifier with *RL* load; (b) Voltage waveforms.

which must be solved numerically for β. The angle $\beta - \alpha$ is called the *conduction angle* γ. Figure 3-14b shows the voltage waveforms.

The average (dc) output voltage is

$$V_o = \frac{1}{2\pi} \int_\alpha^\beta V_m \sin(\omega t)\, d(\omega t) = \frac{V_m}{2\pi}(\cos\alpha - \cos\beta) \tag{3-57}$$

The average current is computed from

$$I_o = \frac{1}{2\pi} \int_\alpha^\beta i(\omega t)\, d(\omega t) \tag{3-58}$$

where $i(\omega t)$ is defined in Eq. (3-55). Power absorbed by the load is $I_{rms}^2 R$, where the rms current is computed from

$$I_{rms} = \sqrt{\frac{1}{2\pi} \int_\alpha^\beta i^2(\omega t)\, d(\omega t)} \tag{3-59}$$

EXAMPLE 3-11

Controlled Half-Wave Rectifier with *RL* Load

For the circuit of Fig. 3-14a, the source is 120 V rms at 60 Hz, $R = 20\,\Omega$, $L = 0.04$ H, and the delay angle is 45°. Determine (a) an expression for $i(\omega t)$, (b) the average current, (c) the power absorbed by the load, and (d) the power factor.

■ **Solution**

(a) From the parameters given,

$$V_m = 120\sqrt{2} = 169.7\text{ V}$$
$$Z = [R^2 + (\omega L)^2]^{0.5} = [20^2 + (377*0.04)^2]^{0.5} = 25.0\,\Omega$$
$$\theta = \tan^{-1}(\omega L/R) = \tan^{-1}(377*0.04/20) = 0.646\text{ rad}$$
$$\omega\tau = \omega L/R = 377*0.04/20 = 0.754$$
$$\alpha = 45° = 0.785\text{ rad}$$

Substituting the preceding quantities into Eq. (3-55), current is expressed as

$$i(\omega t) = 6.78\sin(\omega t - 0.646) - 2.67 e^{-\omega t/0.754}\quad \text{A} \quad \text{for } \alpha \leq \omega t \leq \beta$$

The preceding equation is valid from α to β, where β is found numerically by setting the equation to zero and solving for ωt, with the result $\beta = 3.79$ rad (217°). The conduction angle is $\gamma = \beta - \alpha = 3.79 - 0.785 = 3.01$ rad $= 172°$.

(b) Average current is determined from Eq. (3-58).

$$I_o = \frac{1}{2\pi}\int_{0.785}^{3.79}\left[6.78\sin(\omega t - 0.646) - 2.67 e^{-\omega t/0.754}\right]d(\omega t) = 2.19\text{ A}$$

(c) The power absorbed by the load is computed from $I_{rms}^2 R$, where

$$I_{rms} = \sqrt{\frac{1}{2\pi}\int_{0.785}^{3.79}\left[6.78\sin(\omega t - 0.646) - 2.67 e^{-\omega t/0.754}\right]^2 d(\omega t)} = 3.26\text{ A}$$

yielding

$$P = I_{rms}^2 R = (3.26)^2(20) = 213\text{ W}$$

(d) The power factor is

$$\text{pf} = \frac{P}{S} = \frac{213}{(120)(3.26)} = 0.54$$

RL-Source Load

A controlled rectifier with a series resistance, inductance, and dc source is shown in Fig. 3-15. The analysis of this circuit is very similar to that of the uncontrolled half-wave rectifier discussed earlier in this chapter. The major difference is that for the uncontrolled rectifier, conduction begins as soon as the source voltage

3.9 The Controlled Half-Wave Rectifier

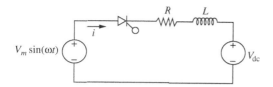

Figure 3-15 Controlled rectifier with *RL*-source load.

reaches the level of the dc voltage. For the controlled rectifier, conduction begins when a gate signal is applied to the SCR, provided the SCR is forward-biased. Thus, the gate signal may be applied at any time that the ac source is larger than the dc source:

$$\alpha_{min} = \sin^{-1}\left(\frac{V_{dc}}{V_m}\right) \qquad (3\text{-}60)$$

Current is expressed as in Eq. (3-22), with α specified within the allowable range:

$$i(\omega t) \begin{cases} \dfrac{V_m}{Z}\sin(\omega t - \theta) - \dfrac{V_{dc}}{R} + Ae^{-\omega t/\omega \tau} & \text{for } \alpha \leq \omega t \leq \beta \\ 0 & \text{otherwise} \end{cases} \qquad (3\text{-}61)$$

where A is determined from Eq. (3-61):

$$A = \left[-\frac{V_m}{Z}\sin(\alpha - \theta) + \frac{V_{dc}}{R}\right]e^{\alpha/\omega \tau}$$

EXAMPLE 3-12

Controlled Rectifier with *RL*-Source Load

The controlled half-wave rectifier of Fig. 3-15 has an ac input of 120 V rms at 60 Hz, $R = 2\,\Omega$, $L = 20$ mH, and $V_{dc} = 100$ V. The delay angle α is 45°. Determine (*a*) an expression for the current, (*b*) the power absorbed by the resistor, and (*c*) the power absorbed by the dc source in the load.

■ **Solution:**
From the parameters given,

$V_m = 120\sqrt{2} = 169.7$ V
$Z = [R^2 + (\omega L)^2]^{0.5} = [2^2 + (377*0.02)^2]^{0.5} = 7.80\,\Omega$
$\theta = \tan^{-1}(\omega L/R) = \tan^{-1}(377*0.02)/2) = 1.312$ rad
$\omega\tau = \omega L/R = 377*0.02/2 = 3.77$
$\alpha = 45° = 0.785$ rad

(a) First, use Eq. (3-60) to determine if $\alpha = 45°$ is allowable. The minimum delay angle is

$$\alpha_{min} = \sin^{-1}\left(\frac{100}{120\sqrt{2}}\right) = 36°$$

which indicates that 45° is allowable. Equation (3-61) becomes

$i(\omega t) = 21.8 \sin(\omega t - 1.312) - 50 + 75.0 e^{-\omega t/3.77}$ A   for $0.785 \leq \omega t \leq 3.37$ rad

where the extinction angle β is found numerically to be 3.37 rad from the equation $i(\beta) = 0$.

(b) Power absorbed by the resistor is $I_{rms}^2 R$, where I_{rms} is computed from Eq. (3-59) using the preceding expression for $i(\omega t)$.

$$I_{rms} = \sqrt{\frac{1}{2\pi}\int_\alpha^\beta i^2(\omega t)\, d(\omega t)} = 3.90 \text{ A}$$

$$P = (3.90)^2 (2) = 30.4 \text{ W}$$

(c) Power absorbed by the dc source is $I_o V_{dc}$, where I_o is computed from Eq. (3-58).

$$I_o = \frac{1}{2\pi}\int_\alpha^\beta i(\omega t)\, d(\omega t) = 2.19 \text{ A}$$

$$P_{dc} = I_o V_{dc} = (2.19)(100) = 219 \text{ W}$$

3.10 PSPICE SOLUTIONS FOR CONTROLLED RECTIFIERS

Modeling the SCR in PSpice

To simulate the controlled half-wave rectifier in PSpice, a model for the SCR must be selected. An SCR model available in a device library can be utilized in the simulation of a controlled half-wave rectifier. A circuit for Example 3-10 using the 2N1595 SCR in the PSpice demo version library of devices is shown in Fig. 3-16a. An alternative model for the SCR is a voltage-controlled switch and a diode as described in Chap. 1. The switch controls when the SCR begins to conduct, and the diode allows current in only one direction. The switch must be closed for at least the conduction angle of the current. An advantage of using this SCR model is that the device can be made ideal. The major disadvantage of the model is that the switch control must keep the switch closed for the entire conduction period and open the switch before the source becomes positive again. A circuit for the circuit in Example 3-11 is shown in Fig. 3-16b.

EXAMPLE 3-13

Controlled Half-Wave Rectifier Design Using PSpice

A load consists of a series-connected resistance, inductance, and dc voltage source with $R = 2 \,\Omega$, $L = 20$ mH, and $V_{dc} = 100$ V. Design a circuit that will deliver 150 W to the dc voltage source from a 120-V rms 60-Hz ac source.

Controlled Half-wave Rectifier with SCR 2N1595

Change the SCR model for a higher voltage rating

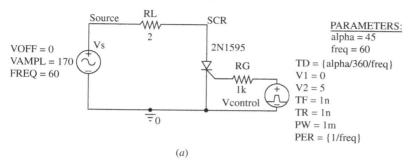

(a)

CONTROLLED HALFWAVE RECTIFIER
switch and diode for SCR

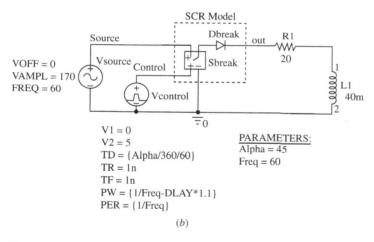

(b)

Figure 3-16 (a) A controlled half-wave rectifier using an SCR from the library of devices; (b) An SCR model using a voltage-controlled switch and a diode.

■ Solution

Power in the dc source of 150 W requires an average load current of 150 W/100 V = 1.5 A. An uncontrolled rectifier with this source and load will have an average current of 2.25 A and an average power in the dc source of 225 W, as was computed in Example 3-5 previously. A means of limiting the average current to 1.5 A must be found. Options include the addition of series resistance or inductance. Another option that is chosen for this application is the controlled half-wave rectifier of Fig. 3-15. The power delivered to the load components is determined by the delay angle α. Since there is no closed-form solution for α, a trial-and-error iterative method must be used. A PSpice simulation that includes a parametric sweep is used to try several values of alpha. The parametric sweep

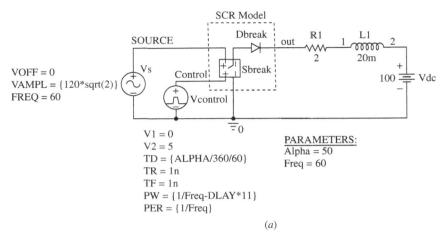

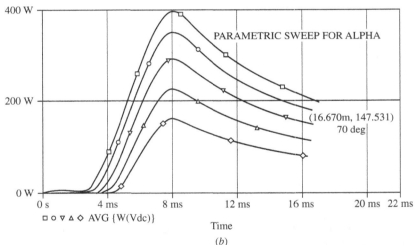

Figure 3-17 (*a*) PSpice circuit for Example 3-13; (*b*) Probe output for showing a family of curves for different delay angles.

is established in the Simulation Setting menu (see Example 3-4). A PSpice circuit is shown in Fig. 3-17*a*.

When the expression AVG(W(Vdc)) is entered, Probe produces a family of curves representing the results for a number of values of α, as shown in Fig. 3-17*b*. An α of 70°, which results in about 148 W delivered to the load, is the approximate solution.

The following results are obtained from Probe for $\alpha = 70°$:

Quantity	Expression	Result
DC source power	AVG(W(Vdc))	148 W (design objective of 150 W)
RMS current	RMS(I(R1))	2.87 A
Resistor power	AVG(W(R1))	16.5 W
Source apparent power	RMS(V(SOURCE))*RMS(I(Vs))	344 VA
Source average power	AVG(W(Vs))	166 W
Power factor (P/S)	166/344	0.48

3.11 COMMUTATION

The Effect of Source Inductance

The preceding discussion on half-wave rectifiers assumed an ideal source. In practical circuits, the source has an equivalent impedance which is predominantly inductive reactance. For the single-diode half-wave rectifiers of Figs. 3-1 and 3-2, the nonideal circuit is analyzed by including the source inductance with the load elements. However, the source inductance causes a fundamental change in circuit behavior for circuits like the half-wave rectifier with a freewheeling diode.

A half-wave rectifier with a freewheeling diode and source inductance L_s is shown in Fig. 3-18a. Assume that the load inductance is very large, making the load current constant. At $t = 0^-$, the load current is I_L, D_1 is off, and D_2 is on. As the source voltage becomes positive, D_1 turns on, but the source current does not instantly equal the load current because of L_s. Consequently, D_2 must remain on while the current in L_s and D_1 increases to that of the load. The interval when both D_1 and D_2 are on is called the commutation time or commutation angle. *Commutation is the process of turning off an electronic switch, which usually involves transferring the load current from one switch to another.*[2]

When both D_1 and D_2 are on, the voltage across L_s is

$$v_{Ls} = V_m \sin(\omega t) \tag{3-62}$$

and current in L_s and the source is

$$i_s = \frac{1}{\omega L_s} \int_0^{\omega t} v_{Ls} d(\omega t) + i_s(0) = \frac{1}{\omega L_s} \int_0^{\omega t} V_m \sin(\omega t) d(\omega t) + 0 \tag{3-63}$$

$$i_s = \frac{V_m}{\omega L_s}(1 - \cos \omega t)$$

---

[2] Commutation in this case is an example of *natural commutation* or *line commutation*, where the change in instantaneous line voltage results in a device turning off. Other applications may use *forced commutation*, where current in a device such as a thyristor is forced to zero by additional circuitry. *Load commutation* makes use of inherent oscillating currents produced by the load to turn a device off.

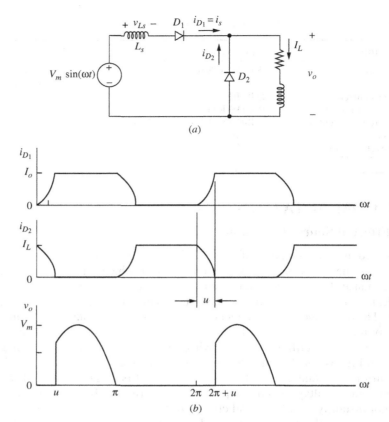

Figure 3-18 (*a*) Half-wave rectifier with freewheeling diode and source inductance; (*b*) Diode currents and load voltage showing the effects of Commutation.

Current in D_2 is

$$i_{D_2} = I_L - i_s = I_L - \frac{V_m}{\omega L_s}(1 - \cos \omega t)$$

The current in D_2 starts at I_L and decreases to zero. Letting the angle at which the current reaches zero be $\omega t = u$,

$$i_{D_2}(u) = I_L - \frac{V_m}{\omega L_s}(1 - \cos u) = 0$$

Solving for u,

$$\boxed{u = \cos^{-1}\left(1 - \frac{I_L \omega L_s}{V_m}\right) = \cos^{-1}\left(1 - \frac{I_L X_s}{V_m}\right)} \qquad (3\text{-}64)$$

where $X_s = \omega L_s$ is the reactance of the source. Figure 3-18b shows the effect of the source reactance on the diode currents. The commutation from D_1 to D_2 is analyzed similarly, yielding an identical result for the commutation angle u.

The commutation angle affects the voltage across the load. Since the voltage across the load is zero when D_2 is conducting, the load voltage remains at zero through the commutation angle, as shown in Fig. 3-17b. Recall that the load voltage is a half-wave rectified sinusoid when the source is ideal.

Average load voltage is

$$V_o = \frac{1}{2\pi} \int_u^\pi V_m \sin(\omega t) d(\omega t)$$

$$= \frac{V_m}{2\pi}[-\cos(\omega t)]\Big|_u^\pi = \frac{V_m}{2\pi}(1 + \cos u)$$

Using u from Eq. (3-64),

$$V_o = \frac{V_m}{\pi}\left(1 - \frac{I_L X_s}{2V_m}\right) \tag{3-65}$$

Recall that the average of a half-wave rectified sine wave is V_m/π. Source reactance thus reduces average load voltage.

3.12 Summary

- A rectifier converts ac to dc. Power transfer is from the ac source to the dc load.
- The half-wave rectifier with a resistive load has an average load voltage of V_m/π and an average load current of $V_m/\pi R$.
- The current in a half-wave rectifier with an RL load contains a natural and a forced response, resulting in

$$i(\omega t) = \begin{cases} \frac{V_m}{Z}\left[\sin(\omega t - \theta) + \sin(\theta)e^{-\omega t/\omega \tau}\right] & \text{for } 0 \leq \omega t \leq \beta \\ 0 & \text{for } \beta \leq \omega t \leq 2\pi \end{cases}$$

where $Z = \sqrt{R^2 + (\omega L)^2}$, $\theta = \tan^{-1}\left(\frac{\omega L}{R}\right)$ and $\tau = \frac{L}{R}$.

The diode remains on as long as the current is positive. Power in the RL load is $I_{rms}^2 R$.

- A half-wave rectifier with an RL-source load does not begin to conduct until the ac source reaches the dc voltage in the load. Power in the resistance is $I_{rms}^2 R$, and power absorbed by the dc source is $I_o V_{dc}$, where I_o is the average load current. The load current is expressed as

$$i(\omega t) = \begin{cases} \frac{V_m}{Z}\sin(\omega t - \theta) - \frac{V_{dc}}{R} + Ae^{-\omega t/\omega \tau} & \text{for } \alpha \leq \omega t \leq \beta \\ 0 & \text{otherwise} \end{cases}$$

where

$$A = \left[-\frac{V_m}{Z}\sin(\alpha - \beta) + \frac{V_{dc}}{R}\right]e^{\alpha/\omega\tau}$$

- A freewheeling diode forces the voltage across an *RL* load to be a half-wave rectified sine wave. The load current can be analyzed using Fourier analysis. A large load inductance results in a nearly constant load current.
- A large filter capacitor across a resistive load makes the load voltage nearly constant. Average diode current must be the same as average load current, making the peak diode current large.
- An SCR in place of the diode in a half-wave rectifier provides a means of controlling output current and voltage.
- PSpice simulation is an effective way of analyzing circuit performance. The parametric sweep in PSpice allows several values of a circuit parameter to be tried and is an aid in circuit design.

3.13 Bibliography

S. B. Dewan and A. Straughen, *Power Semiconductor Circuits*, Wiley, New York, 1975.

Y.-S. Lee and M. H. L. Chow, *Power Electronics Handbook*, edited by M. H. Rashid, Academic Press, New York, 2001, Chapter 10.

N. Mohan, T. M. Undeland, and W. P. Robbins, *Power Electronics: Converters, Applications, and Design,* 3d ed., Wiley, New York, 2003.

M. H. Rashid, *Power Electronics: Circuits, Devices, and Systems,* 3d ed., Prentice-Hall, Upper Saddle River, NJ., 2004.

R. Shaffer, *Fundamentals of Power Electronics with MATLAB,* Charles River Media, Boston, Mass., 2007.

B. Wu, *High-Power Converters and AC Drives,* Wiley, New York, 2006.

Problems

Half-Wave Rectifier with Resistive Load

3-1. The half-wave rectifier circuit of Fig. 3-1*a* has $v_s(t) = 170 \sin(377t)$ V and a load resistance $R = 15 \, \Omega$. Determine (*a*) the average load current, (*b*) the rms load current, (*c*) the power absorbed by the load, (*d*) the apparent power supplied by the source, and (*e*) the power factor of the circuit.

3-2. The half-wave rectifier circuit of Fig. 3-1*a* has a transformer inserted between the source and the remainder of the circuit. The source is 240 V rms at 60 Hz, and the load resistor is 20 Ω. (*a*) Determine the required turns ratio of the transformer such that the average load current is 12 A. (*b*) Determine the average current in the primary winding of the transformer.

3-3. For a half-wave rectifier with a resistive load, (*a*) show that the power factor is $1/\sqrt{2}$ and (*b*) determine the displacement power factor and the distortion factor as defined in Chap. 2. The Fourier series for the half-wave rectified voltage is given in Eq. (3-34).

Half-Wave Rectifier with *RL* Load

3-4. A half-wave rectifier has a source of 120 V rms at 60 Hz and an *RL* load with $R = 12\ \Omega$ and $L = 12$ mH. Determine (*a*) an expression for load current, (*b*) the average current, (*c*) the power absorbed by the resistor, and (*d*) the power factor.

3-5. A half-wave rectifier has a source of 120 V rms at 60 Hz and an *RL* load with $R = 10\ \Omega$ and $L = 15$ mH. Determine (*a*) an expression for load current, (*b*) the average current, (*c*) the power absorbed by the resistor, and (*d*) the power factor.

3-6. A half-wave rectifier has a source of 240 V rms at 60 Hz and an *RL* load with $R = 15\ \Omega$ and $L = 80$ mH. Determine (*a*) an expression for load current, (*b*) the average current, (*c*) the power absorbed by the resistor, and (*d*) the power factor. (*e*) Use PSpice to simulate the circuit. Use the default diode model and compare your PSpice results with analytical results.

3-7. The inductor in Fig. 3-2*a* represents an electromagnet modeled as a 0.1-H inductance. The source is 240 V at 60 Hz. Use PSpice to determine the value of a series resistance such that the average current is 2.0 A.

Half-Wave Rectifier with *RL*-Source Load

3-8. A half-wave rectifier of Fig. 3-5*a* has a 240 V rms, 60 Hz ac source. The load is a series inductance, resistance, and dc source, with $L = 75$ mH, $R = 10\ \Omega$, and $V_{dc} = 100$ V. Determine (*a*) the power absorbed by the dc voltage source, (*b*) the power absorbed by the resistance, and (*c*) the power factor.

3-9. A half-wave rectifier of Fig. 3-5*a* has a 120 V rms, 60 Hz ac source. The load is a series inductance, resistance, and dc source, with $L = 120$ mH, $R = 12\ \Omega$, and $V_{dc} = 48$ V. Determine (*a*) the power absorbed by the dc voltage source, (*b*) the power absorbed by the resistance, and (*c*) the power factor.

3-10. A half-wave rectifier of Fig. 3-6 has a 120 V rms, 60 Hz ac source. The load is a series inductance and dc voltage with $L = 100$ mH and $V_{dc} = 48$ V. Determine the power absorbed by the dc voltage source.

3-11. A half-wave rectifier with a series inductor-source load has an ac source of 240 V rms, 60 Hz. The dc source is 96 V. Use PSpice to determine the value of inductance which results in 150 W absorbed by the dc source. Use the default diode model.

3-12. A half-wave rectifier with a series inductor and dc source has an ac source of 120 V rms, 60 Hz. The dc source is 24 V. Use PSpice to determine the value of inductance which results in 50 W absorbed by the dc source. Use the default diode.

Freewheeling Diode

3-13. The half-wave rectifier with a freewheeling diode (Fig. 3-7*a*) has $R = 12\ \Omega$ and $L = 60$ mH. The source is 120 V rms at 60 Hz. (*a*) From the Fourier series of the half-wave rectified sine wave that appears across the load, determine the dc component of the current. (*b*) Determine the amplitudes of the first four nonzero ac terms in the Fourier series. Comment on the results.

3-14. In Example 3-8, the inductance required to limit the peak-to-peak ripple in load current was estimated by using the first ac term in the Fourier series. Use PSpice to determine the peak-to-peak ripple with this inductance, and compare it to the estimate. Use the ideal diode model ($n = 0.001$).

3-15. The half-wave rectifier with a freewheeling diode (Fig. 3-7a) has $R = 4\,\Omega$ and a source with $V_m = 50$ V at 60 Hz. (a) Determine a value of L such that the amplitude of the first ac current term in the Fourier series is less than 5 percent of the dc current. (b) Verify your results with PSpice, and determine the peak-to-peak current.

3-16. The circuit of Fig. P3-16 is similar to the circuit of Fig. 3-7a except that a dc source has been added to the load. The circuit has $v_s(t) = 170 \sin(377t)$ V, $R = 10\,\Omega$, and $V_{dc} = 24$ V. From the Fourier series, (a) determine the value of L such that the peak-to-peak variation in load current is no more than 1 A. (b) Determine the power absorbed by the dc source. (c) Determine the power absorbed by the resistor.

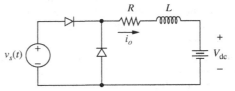

Figure P3-16

Half-Wave Rectifier with a Filter Capacitor

3-17. A half-wave rectifier with a capacitor filter has $V_m = 200$ V, $R = 1\,\text{k}\Omega$, $C = 1000\,\mu\text{F}$, and $\omega = 377$. (a) Determine the ratio of the RC time constant to the period of the input sine wave. What is the significance of this ratio? (b) Determine the peak-to-peak ripple voltage using the exact equations. (c) Determine the ripple using the approximate formula in Eq. (3-51).

3-18. Repeat Prob. 3-17 with (a) $R = 100\,\Omega$ and (b) $R = 10\,\Omega$. Comment on the results.

3-19. A half-wave rectifier with a 1-kΩ load has a parallel capacitor. The source is 120 V rms, 60 Hz. Determine the peak-to-peak ripple of the output voltage when the capacitor is (a) 4000 μF and (b) 20 μF. Is the approximation of Eq. (3-51) reasonable in each case?

3-20. Repeat Prob. 3-19 with $R = 500\,\Omega$.

3-21. A half-wave rectifier has a 120 V rms, 60 Hz ac source. The load is 750 Ω. Determine the value of a filter capacitor to keep the peak-to-peak ripple across the load to less than 2 V. Determine the average and peak values of diode current.

3-22. A half-wave rectifier has a 120 V rms 60 Hz ac source. The load is 50 W. (a) Determine the value of a filter capacitor to keep the peak-to-peak ripple across the load to less than 1.5 V. (b) Determine the average and peak values of diode current.

Controlled Half-Wave Rectifier

3-23. Show that the controlled half-wave rectifier with a resistive load in Fig. 3-13a has a power factor of

$$\text{pf} = \sqrt{\frac{1}{2} - \frac{\alpha}{2\pi} + \frac{\sin(2\alpha)}{4\pi}}$$

3-24. For the controlled half-wave rectifier with resistive load, the source is 120 V rms at 60 Hz. The resistance is 100 Ω, and the delay angle α is 45°. (a) Determine the average voltage across the resistor. (b) Determine the power absorbed by the resistor. (c) Determine the power factor as seen by the source.

3-25. A controlled half-wave rectifier has an ac source of 240 V rms at 60 Hz. The load is a 30-Ω resistor. (a) Determine the delay angle such that the average load current is 2.5 A. (b) Determine the power absorbed by the load. (c) Determine the power factor.

3-26. A controlled half-wave rectifier has a 120 V rms 60 Hz ac source. The series RL load has $R = 25$ ω and $L = 50$ mH. The delay angle is 30°. Determine (a) an expression for load current, (b) the average load current, and (c) the power absorbed by the load.

3-27. A controlled half-wave rectifier has a 120 V rms 60 Hz ac source. The series RL load has $R = 40$ Ω and $L = 75$ mH. The delay angle is 60°. Determine (a) an expression for load current, (b) the average load current, and (c) the power absorbed by the load.

3-28. A controlled half-wave rectifier has an RL load with $R = 20$ Ω and $L = 40$ mH. The source is 120 V rms at 60 Hz. Use PSpice to determine the delay angle required to produce an average current of 2.0 A in the load. Use the default diode in the simulation.

3-29. A controlled half-wave rectifier has an RL load with $R = 16$ Ω and $L = 60$ mH. The source is 120 V rms at 60 Hz. Use PSpice to determine the delay angle required to produce an average current of 1.8 A in the load. Use the default diode in the simulation.

3-30. A controlled half-wave rectifier has a 120 V, 60 Hz ac source. The load is a series inductance, resistance, and dc source, with $L = 100$ mH, $R = 12$ Ω, and $V_{dc} = 48$ V. The delay angle is 50°. Determine (a) the power absorbed by the dc voltage source, (b) the power absorbed by the resistance, and (c) the power factor.

3-31. A controlled half-wave rectifier has a 240 V rms 60 Hz ac source. The load is a series resistance, inductance, and dc source with $R = 100$ Ω, $L = 150$ mH, and $V_{dc} = 96$ V. The delay angle is 60°. Determine (a) the power absorbed by the dc voltage source, (b) the power absorbed by the resistance, and (c) the power factor.

3-32. Use PSpice to determine the delay angle required such that the dc source in Prob. 3-31 absorbs 35 W.

3-33. A controlled half-wave rectifier has a series resistance, inductance, and dc voltage source with $R = 2$ Ω, $L = 75$ mH, and $V_{dc} = 48$ V. The source is 120 V rms at 60 Hz. The delay angle is 50°. Determine (a) an expression for load current, (b) the power absorbed by the dc voltage source, and (c) the power absorbed by the resistor.

3-34. Use PSpice to determine the delay angle required such that the dc source in Prob. 3-33 absorbs 50 W.

3-35. Develop an expression for current in a controlled half-wave rectifier circuit that has a load consisting of a series inductance L and dc voltage V_{dc}. The source is $v_s = V_m \sin \omega t$, and the delay angle is α. (a) Determine the average current if $V_m = 100$ V, $L = 35$ mH, $V_{dc} = 24$ V, ω = 2π60 rad/s, and α = 75°. (b) Verify your result with PSpice.

3-36. A controlled half-wave rectifier has an RL load. A freewheeling diode is placed in parallel with the load. The inductance is large enough to consider the load current to be constant. Determine the load current as a function of the delay angle alpha. Sketch the current in the SCR and the freewheeling diode. Sketch the voltage across the load.

Commutation

3-37. The half-wave rectifier with freewheeling diode of Fig. 3-18a has a 120 V rms ac source that has an inductance of 1.5 mH. The load current is a constant 5 A. Determine the commutation angle and the average output voltage. Use PSpice to verify your results. Use ideal diodes in the simulation. Verify that the commutation angle for D_1 to D_2 is the same as for D_2 to D_1.

3-38. The half-wave rectifier with freewheeling diode of Fig. 3-18a has a 120 V rms ac source which has an inductance of 10 mH. The load is a series resistance-inductance with $R = 20\ \Omega$ and $L = 500$ mH. Use PSpice to determine (a) the steady-state average load current, (b) the average load voltage, and (c) the commutation angle. Use the default diode in the simulation. Comment on the results.

3-39. The half-wave rectifier with freewheeling diode of Fig. 3-18a has a 120 V rms ac source which has an inductance of 5 mH. The load is a series resistance-inductance with $R = 15\ \Omega$ and $L = 500$ mH. Use PSpice to determine (a) the steady-state average load current, (b) the average load voltage, and (c) the commutation angle. Use the default diode in the simulation.

3-40. The commutation angle given in Eq. (3-64) for the half-wave rectifier with a freewheeling diode was developed for commutation of load current from D_2 to D_1. Show that the commutation angle is the same for commutation from D_1 to D_2.

3-41. Diode D_1 in Fig. 3-18a is replaced with an SCR to make a controlled half-wave rectifier. Show that the angle for commutation from the diode to the SCR is

$$u = \cos^{-1}\left(\cos \alpha - \frac{I_L X_s}{V_m}\right) - \alpha$$

where α is the delay angle of the SCR.

Design Problems

3-42. A certain situation requires that either 160 or 75 W be supplied to a 48 V battery from a 120 V rms 60 Hz ac source. There is a two-position switch on a control panel set at either 160 or 75. Design a single circuit to deliver both values of power, and specify what the control switch will do. Specify the values of all the components in your circuit. The internal resistance of the battery is 0.1 Ω.

3-43. Design a circuit to produce an average current of 2 A in an inductance of 100 mH. The ac source available is 120 V rms at 60 Hz. Verify your design with PSpice. Give alternative circuits that could be used to satisfy the design specifications, and give reasons for your selection.

3-44. Design a circuit that will deliver 100 W to a 48 V dc source from a 120 V rms 60 Hz ac source. Verify your design with PSpice. Give alternative circuits that could be used to satisfy the design specifications, and give reasons for your selection.

3-45. Design a circuit which will deliver 150 W to a 100 V dc source from a 120 V rms 60 Hz ac source. Verify your design with PSpice. Give alternative circuits that could be used to satisfy the design specifications, and give reasons for your selection.

CHAPTER 4

Full-Wave Rectifiers
Converting ac to dc

4.1 INTRODUCTION

The objective of a full-wave rectifier is to produce a voltage or current that is purely dc or has some specified dc component. While the purpose of the full-wave rectifier is basically the same as that of the half-wave rectifier, full-wave rectifiers have some fundamental advantages. The average current in the ac source is zero in the full-wave rectifier, thus avoiding problems associated with nonzero average source currents, particularly in transformers. The output of the full-wave rectifier has inherently less ripple than the half-wave rectifier.

In this chapter, uncontrolled and controlled single-phase and three-phase full-wave converters used as rectifiers are analyzed for various types of loads. Also included are examples of controlled converters operating as inverters, where power flow is from the dc side to the ac side.

4.2 SINGLE-PHASE FULL-WAVE RECTIFIERS

The bridge rectifier and the center-tapped transformer rectifier of Figs. 4-1 and 4-2 are two basic single-phase full-wave rectifiers.

The Bridge Rectifier

For the bridge rectifier of Fig. 4-1, these are some basic observations:

1. Diodes D_1 and D_2 conduct together, and D_3 and D_4 conduct together. Kirchhoff's voltage law around the loop containing the source, D_1, and D_3

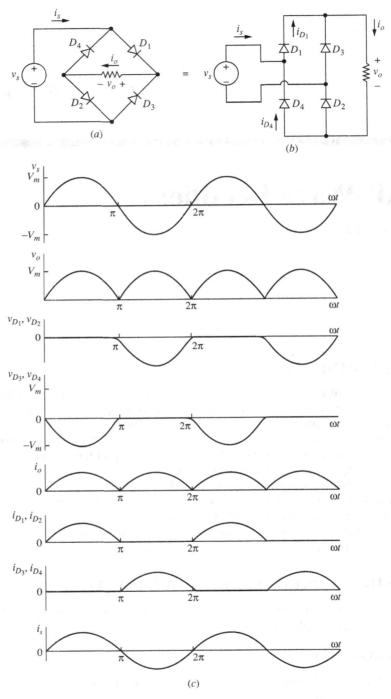

Figure 4-1 Full-wave bridge rectifier. (*a*) Circuit diagram. (*b*) Alternative representation. (*c*) Voltages and currents.

4.2 Single-Phase Full-Wave Rectifiers

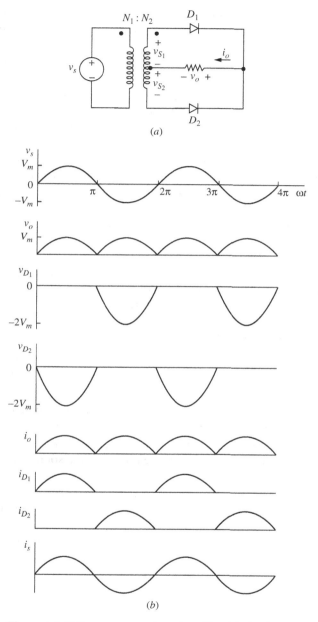

Figure 4-2 Full-wave center-tapped rectifier (*a*) circuit; (*b*) voltages and currents.

shows that D_1 and D_3 cannot be on at the same time. Similarly, D_2 and D_4 cannot conduct simultaneously. The load current can be positive or zero but can never be negative.

2. The voltage across the load is $+v_s$ when D_1 and D_2 are on. The voltage across the load is $-v_s$ when D_3 and D_4 are on.
3. The maximum voltage across a reverse-biased diode is the peak value of the source. This can be shown by Kirchhoff's voltage law around the loop containing the source, D_1, and D_3. With D_1 on, the voltage across D_3 is $-v_s$.
4. The current entering the bridge from the source is $i_{D_1} - i_{D_4}$, which is symmetric about zero. Therefore, the average source current is zero.
5. The rms source current is the same as the rms load current. The source current is the same as the load current for one-half of the source period and is the negative of the load current for the other half. The squares of the load and source currents are the same, so the rms currents are equal.
6. The fundamental frequency of the output voltage is 2ω, where ω is the frequency of the ac input since two periods of the output occur for every period of the input. The Fourier series of the output consists of a dc term and the even harmonics of the source frequency.

The Center-Tapped Transformer Rectifier

The voltage waveforms for a resistive load for the rectifier using the center-tapped transformer are shown in Fig. 4-2. Some basic observations for this circuit are as follows:

1. Kirchhoff's voltage law shows that only one diode can conduct at a time. Load current can be positive or zero but never negative.
2. The output voltage is $+v_{s_1}$ when D_1 conducts and is $-v_{s_2}$ when D_2 conducts. The transformer secondary voltages are related to the source voltage by $v_{s_1} = v_{s_2} = v_s(N_2/2N_1)$.
3. Kirchhoff's voltage law around the transformer secondary windings, D_1, and D_2 shows that the maximum voltage across a reverse-biased diode is *twice* the peak value of the load voltage.
4. Current in each half of the transformer secondary is reflected to the primary, resulting in an average source current of zero.
5. The transformer provides electrical isolation between the source and the load.
6. The fundamental frequency of the output voltage is 2ω since two periods of the output occur for every period of the input.

The lower peak diode voltage in the bridge rectifier makes it more suitable for high-voltage applications. The center-tapped transformer rectifier, in addition to including electrical isolation, has only one diode voltage drop between the source and load, making it desirable for low-voltage, high-current applications.

The following discussion focuses on the full-wave bridge rectifier but generally applies to the center-tapped circuit as well.

Resistive Load

The voltage across a resistive load for the bridge rectifier of Fig. 4-1 is expressed as

$$v_o(\omega t) = \begin{cases} V_m \sin \omega t & \text{for } 0 \leq \omega t \leq \pi \\ -V_m \sin \omega t & \text{for } \pi \leq \omega t \leq 2\pi \end{cases} \quad (4\text{-}1)$$

The dc component of the output voltage is the average value, and load current is simply the resistor voltage divided by resistance.

$$V_o = \frac{1}{\pi} \int_0^{\pi} V_m \sin \omega t \, d(\omega t) = \frac{2V_m}{\pi}$$

$$I_o = \frac{V_o}{R} = \frac{2V_m}{\pi R} \quad (4\text{-}2)$$

Power absorbed by the load resistor can be determined from $I_{rms}^2 R$, where I_{rms} for the full-wave rectified current waveform is the same as for an unrectified sine wave,

$$I_{rms} = \frac{I_m}{\sqrt{2}} \quad (4\text{-}3)$$

The source current for the full-wave rectifier with a resistive load is a sinusoid that is in phase with the voltage, so the power factor is 1.

RL Load

For an *RL* series-connected load (Fig. 4-3a), the method of analysis is similar to that for the half-wave rectifier with the freewheeling diode discussed in Chap. 3. After a transient that occurs during start-up, the load current i_o reaches a periodic steady-state condition similar to that in Fig. 4-3b.

For the bridge circuit, current is transferred from one pair of diodes to the other pair when the source changes polarity. The voltage across the *RL* load is a full-wave rectified sinusoid, as it was for the resistive load. The full-wave rectified sinusoidal voltage across the load can be expressed as a Fourier series consisting of a dc term and the even harmonics

$$v_o(t) = V_o + \sum_{n=2,4,\ldots}^{\infty} V_n \cos(n\omega_0 t + \pi)$$

where $(4\text{-}4)$

$$V_o = \frac{2V_m}{\pi} \quad \text{and} \quad V_n = \frac{2V_m}{\pi}\left(\frac{1}{n-1} - \frac{1}{n+1}\right)$$

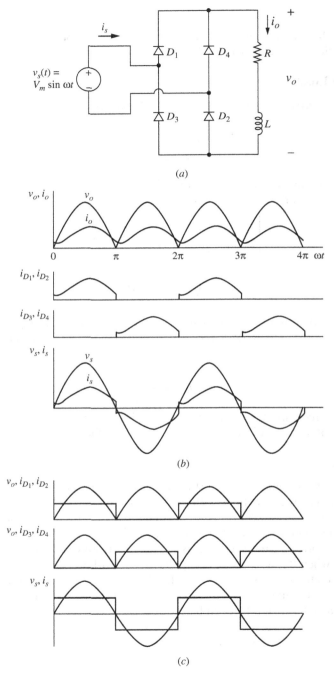

Figure 4-3 (*a*) Bridge rectifier with an *RL* load; (*b*) Voltages and currents; (*c*) Diode and source currents when the inductance is large and the current is nearly constant.

The current in the RL load is then computed using superposition, taking each frequency separately and combining the results. The dc current and current amplitude at each frequency are computed from

$$I_0 = \frac{V_0}{R}$$

$$I_n = \frac{V_n}{Z_n} = \frac{V_n}{|R + jn\omega L|}$$

(4-5)

Note that as the harmonic number n increases in Eq.(4-4), the voltage amplitude decreases. For an RL load, the impedance Z_n increases as n increases. The combination of decreasing V_n and increasing Z_n makes I_n decrease rapidly for increasing harmonic number. Therefore, the dc term and only a few, if any, of the ac terms are usually necessary to describe current in an RL load.

EXAMPLE 4-1

Full-Wave Rectifier with RL Load

The bridge rectifier circuit of Fig. 4-3a has an ac source with $V_m = 100$ V at 60 Hz and a series RL load with $R = 10\ \Omega$ and $L = 10$ mH. (a) Determine the average current in the load. (b) Estimate the peak-to-peak variation in load current based on the first ac term in the Fourier series. (c) Determine the power absorbed by the load and the power factor of the circuit. (d) Determine the average and rms currents in the diodes.

■ Solution

(a) The average load current is determined from the dc term in the Fourier series. The voltage across the load is a full-wave rectified sine wave that has the Fourier series determined from Eq. (4-4). Average output voltage is

$$V_0 = \frac{2V_m}{\pi} = \frac{2(200)}{\pi} = 63.7 \text{ V}$$

and average load current is

$$I_0 = \frac{V_0}{R} = \frac{63.7 \text{ V}}{10\ \Omega} = 6.37 \text{ A}$$

(b) Amplitudes of the ac voltage terms are determined from Eq. (4-4). For $n = 2$ and 4,

$$V_2 = \frac{2(100)}{\pi}\left(\frac{1}{1} - \frac{1}{3}\right) = 42.4 \text{ V}$$

$$V_4 = \frac{2(100)}{\pi}\left(\frac{1}{3} - \frac{1}{5}\right) = 8.49 \text{ V}$$

The amplitudes of first two ac current terms in the current Fourier series are computed from Eq. (4-5).

$$I_2 = \frac{42.4}{|10 + j(2)(377)(0.01)|} = \frac{42.4 \text{ V}}{12.5\ \Omega} = 3.39 \text{ A}$$

$$I_4 = \frac{8.49}{|10 + j(4)(377)(0.01)|} = \frac{8.49 \text{ V}}{18.1\ \Omega} = 0.47 \text{ A}$$

The current I_2 is much larger than I_4 and higher-order harmonics, so I_2 can be used to estimate the peak-to-peak variation in load current $\Delta i_o \approx 2(3.39) = 6.78$ A. Actual variation in i_o will be larger because of the higher-order terms.

(c) The power absorbed by the load is determined from I_{rms}^2. The rms current is then determined from Eq. (2-43) as

$$I_{\text{rms}} = \sqrt{\Sigma I_{n,\text{rms}}^2}$$

$$= \sqrt{(6.37)^2 + \left(\frac{3.39}{\sqrt{2}}\right)^2 + \left(\frac{0.47}{\sqrt{2}}\right)^2 + \cdots} \approx 6.81 \text{ A}$$

Adding more terms in the series would not be useful because they are small and have little effect on the result. Power in the load is

$$P = I_{\text{rms}}^2 R = (6.81)^2(10) = 464 \text{ W}$$

The rms source current is the same as the rms load current. Power factor is

$$\text{pf} = \frac{P}{S} = \frac{P}{V_{s,\text{rms}} I_{s,\text{rms}}} = \frac{464}{\left(\frac{100}{\sqrt{2}}\right)(6.81)} = 0.964$$

(d) Each diode conducts for one-half of the time, so

$$I_{D,\text{avg}} = \frac{I_o}{2} \approx \frac{6.37}{2} = 3.19 \text{ A}$$

and

$$I_{D,\text{rms}} = \frac{I_{\text{rms}}}{\sqrt{2}} = \frac{6.81}{\sqrt{2}} = 4.82 \text{ A}$$

In some applications, the load inductance may be relatively large or made large by adding external inductance. If the inductive impedance for the ac terms in the Fourier series effectively eliminates the ac current terms in the load, the load current is essentially dc. If $\omega L \gg R$,

$$i(\omega t) \approx I_o = \frac{V_o}{R} = \frac{2V_m}{\pi R} \quad \text{for } \omega L \gg R \qquad (4\text{-}6)$$

$$I_{\text{rms}} \approx I_o$$

Load and source voltages and currents are shown in Fig. 4-3c.

Source Harmonics

Nonsinusoidal source current is a concern in power systems. Source currents like that of Fig. 4-3 have a fundamental frequency equal to that of the source but are rich in the odd-numbered harmonics. Measures such as total harmonic distortion (THD) and distortion factor (DF) as presented in Chap. 2 describe the nonsinusoidal property of the source current. Where harmonics are of concern, filters can be added to the input of the rectifier.

PSpice Simulation

A PSpice simulation will give the output voltage, current, and power for full-wave rectifier circuits. Fourier analysis from the FOUR command or from Probe will give the harmonic content of voltages and currents in the load and source. The default diode model will give results that differ from the analytical results that assume an ideal diode. For the full-wave rectifier, two diodes will conduct at a time, resulting in two diode voltage drops. In some applications, the reduced voltage at the output may be significant. Since voltage drops across the diodes exist in real circuits, PSpice results are a better indicator of circuit performance than results that assume ideal diodes. (To simulate an ideal circuit in PSpice, a diode model with $n = 0.001$ will produce forward voltage drops in the microvolt range, approximating an ideal diode.)

EXAMPLE 4-2

PSpice Simulation of a Full-Wave Rectifier

For the full-wave bridge rectifier in Example 4-1, obtain the rms current and power absorbed by the load from a PSpice simulation.

■ **Solution**

The PSpice circuit for Fig. 4-3 is created using VSIN for the source, Dbreak for the diodes, and R and L for the load. A transient analysis is performed using a run time of 50 ms and data saved after 33.33 ms to obtain steady-state current.

The Probe output is used to determine the operating characteristics of the rectifier using the same techniques as presented in Chaps. 2 and 3. To obtain the average value of the load current, enter AVG(I(R1)). Using the cursor to identify the point at the end of the resulting trace, the average current is approximately 6.07 A. The Probe output is shown in Fig. 4-4.

Entering RMS(I(R1)) shows that the rms current is approximately 6.52 A. Power absorbed by the resistor can be computed from $I_{rms}^2 R$, or average power in the load can be computed directly from Probe by entering AVG(W(R1)), which yields 425.4 W. This is significantly less than the 464 W obtained in Example 4-1 when assuming ideal diodes.

The power supplied by the ac source is computed from AVG(W(V1)) as 444.6 W. When ideal diodes were assumed, power supplied by the ac source was identical to the power absorbed by the load, but this analysis reveals that power absorbed by the diodes in the bridge is $444.6 - 425.4 = 19.2$ W. Another way to determine power absorbed by the bridge is to enter AVG(W(D1)) to obtain the power absorbed by diode D_1, which is 4.8 W. Total power for the diodes is 4 times 4.8, or 19.2 W. Better models for power diodes would yield a more accurate estimate of power dissipation in the diodes.

Comparing the results of the simulation to the results based on ideal diodes shows how more realistic diode models reduce the current and power in the load.

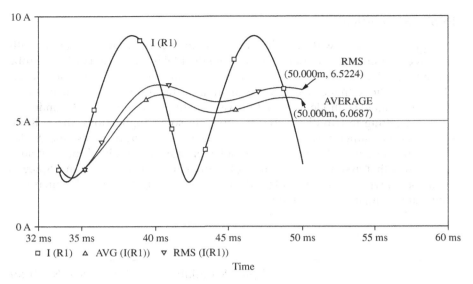

Figure 4-4 PSpice output for Example 4-2.

RL-Source Load

Another general industrial load may be modeled as a series resistance, inductance, and a dc voltage source, as shown in Fig. 4-5a. A dc motor drive circuit and a battery charger are applications for this model. There are two possible modes of operation for this circuit, the continuous-current mode and the discontinuous-current mode. In the continuous-current mode, the load current is always positive for steady-state operation (Fig. 4-5b). Discontinuous load current is characterized by current returning to zero during every period (Fig. 4-5c).

For continuous-current operation, one pair of diodes is always conducting, and the voltage across the load is a full-wave rectified sine wave. The only modification to the analysis that was done for an *RL* load is in the dc term of the Fourier series. The dc (average) component of current in this circuit is

$$I_o = \frac{V_o - V_{dc}}{R} = \frac{\frac{2V_m}{\pi} - V_{dc}}{R} \qquad (4\text{-}7)$$

The sinusoidal terms in the Fourier analysis are unchanged by the dc source provided that the current is continuous.

Discontinuous current is analyzed like the half-wave rectifier of Sec. 3.5. The load voltage is not a full-wave rectified sine wave for this case, so the Fourier series of Eq. (4-4) does not apply.

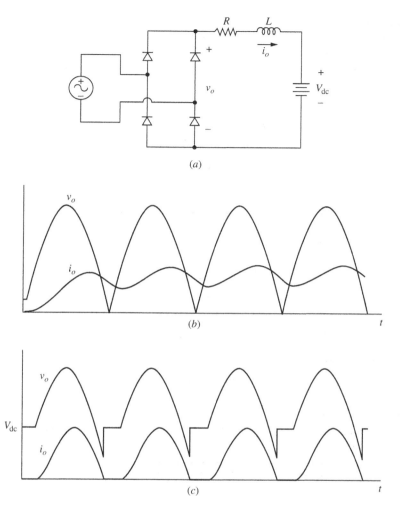

Figure 4-5 (*a*) Rectifier with *RL*-source load; (*b*) Continuous current: when the circuit is energized, the load current reaches the steady-state after a few periods; (*c*) Discontinuous current: the load current returns to zero during every period.

EXAMPLE 4-3

Full-Wave Rectifier with *RL*-Source Load—Continuous Current

For the full-wave bridge rectifier circuit of Fig. 4-5*a*, the ac source is 120 V rms at 60 Hz, $R = 2\ \Omega$, $L = 10$ mH, and $V_{dc} = 80$ V. Determine the power absorbed by the dc voltage source and the power absorbed by the load resistor.

Solution

For continuous current, the voltage across the load is a full-wave rectified sine wave which has the Fourier series given by Eq. (4-4). Equation (4-7) is used to compute the average current, which is used to compute power absorbed by the dc source,

$$I_0 = \frac{\frac{2V_m}{\pi} - V_{dc}}{R} = \frac{\frac{2\sqrt{2}(120)}{\pi} - 80}{2} = 14.0 \text{ A}$$

$$P_{dc} = I_0 V_{dc} = (14)(80) = 1120 \text{ W}$$

The first few terms of the Fourier series using Eqs. (4-4) and (4-5) are shown in Table 4-1.

Table 4-1 Fourier series components

n	V_n	Z_n	I_n
0	108	2.0	14.0
2	72.0	7.80	9.23
4	14.4	15.2	0.90

The rms current is computed from Eq. (2-43).

$$I_{rms} = \sqrt{14^2 + \left(\frac{9.23}{\sqrt{2}}\right)^2 + \left(\frac{0.90}{\sqrt{2}}\right)^2 + \cdots} \approx 15.46 \text{ A}$$

Power absorbed by the resistor is

$$P_R = I_{rms}^2 R = (15.46)^2(2) = 478 \text{ W}$$

PSpice Solution

PSpice simulation of the circuit of Fig 4-5a using the default diode model yields these results from Probe:

Quantity	Expression Entered	Result
I_o	AVG(I(R1))	11.9 A
I_{rms}	RMS(I(R1))	13.6 A
P_{ac}	AVG(W(Vs))	1383 W
P_{D1}	AVG(W(D1))	14.6 W
P_{dc}	AVG(W(VDC))	955 W
P_R	AVG(W(R))	370 W

Note that the simulation verifies the assumption of continuous load current.

Capacitance Output Filter

Placing a large capacitor in parallel with a resistive load can produce an output voltage that is essentially dc (Fig. 4-6). The analysis is very much like that of the half-wave rectifier with a capacitance filter in Chap. 3. In the full-wave circuit, the time that the capacitor discharges is smaller than that for the half-wave circuit

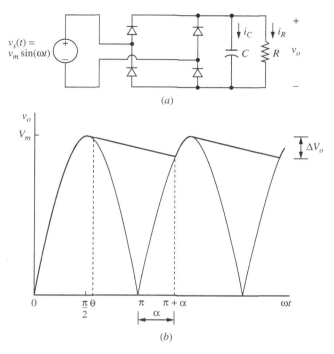

Figure 4-6 (a) Full-wave rectifier with capacitance filter; (b) Source and output voltage.

because of the rectified sine wave in the second half of each period. The output voltage ripple for the full-wave rectifier is approximately one-half that of the half-wave rectifier. The peak output voltage will be less in the full-wave circuit because there are two diode voltage drops rather than one.

The analysis proceeds exactly as for the half-wave rectifier. The output voltage is a positive sine function when one of the diode pairs is conducting and is a decaying exponential otherwise. Assuming ideal diodes,

$$v_o(\omega t) = \begin{cases} |V_m \sin \omega t| & \text{one diode pair on} \\ (V_m \sin \theta) e^{-(\omega t - \theta)/\omega RC} & \text{diodes off} \end{cases} \quad (4\text{-}8)$$

where θ is the angle where the diodes become reverse biased, which is the same as that for the half-wave rectifier and is found using Eq. (3-41).

$$\theta = \tan^{-1}(-\omega RC) = -\tan^{-1}(\omega RC) + \pi \quad (4\text{-}9)$$

The maximum output voltage is V_m, and the minimum output voltage is determined by evaluating v_o at the angle at which the second pair of diodes turns on, which is at $\omega t = \pi + \alpha$. At that boundary point,

$$(V_m \sin \theta) e^{-(\pi + \alpha - \theta)/\omega RC} = -V_m \sin(\pi + \alpha)$$

or

$$(\sin \theta)e^{-(\pi+\alpha-\theta)/\omega RC} - \sin \alpha = 0 \qquad (4\text{-}10)$$

which must be solved numerically for α.

The peak-to-peak voltage variation, or ripple, is the difference between maximum and minimum voltages.

$$\boxed{\Delta V_o = V_m - |V_m \sin(\pi + \alpha)| = V_m(1 - \sin \alpha)} \qquad (4\text{-}11)$$

This is the same as Eq. (3-49) for voltage variation in the half-wave rectifier, but α is larger for the full-wave rectifier and the ripple is smaller for a given load. Capacitor current is described by the same equations as for the half-wave rectifier.

In practical circuits where $\omega RC \gg \pi$.

$$\theta \approx \pi/2 \qquad \alpha \approx \pi/2 \qquad (4\text{-}12)$$

The minimum output voltage is then approximated from Eq. (4-9) for the diodes off evaluated at $\omega t = \pi$.

$$v_o(\pi + \alpha) = V_m e^{-(\pi + \pi/2 - \pi/2)/\omega RC} = V_m e^{-\pi/\omega RC}$$

The ripple voltage for the full-wave rectifier with a capacitor filter can then be approximated as

$$\Delta V_o \approx V_m(1 - e^{-\pi/\omega RC})$$

Furthermore, the exponential in the above equation can be approximated by the series expansion

$$e^{-\pi/\omega RC} \approx 1 - \frac{\pi}{\omega RC}$$

Substituting for the exponential in the approximation, the peak-to-peak ripple is

$$\boxed{\Delta V_o \approx \frac{V_m \pi}{\omega RC} = \frac{V_m}{2fRC}} \qquad (4\text{-}13)$$

Note that the approximate peak-to-peak ripple voltage for the full-wave rectifier is one-half that of the half-wave rectifier from Eq. (3-51). As for the half-wave rectifier, the peak diode current is much larger than the average diode current and Eq. (3-48) applies. The average source current is zero.

EXAMPLE 4-4

Full-Wave Rectifier with Capacitance Filter

The full-wave rectifier of Fig. 4-6a has a 120 V source at 60 Hz, $R = 500\,\Omega$, and $C = 100\,\mu\text{F}$. (a) Determine the peak-to-peak voltage variation of the output. (b) Determine the value of capacitance that would reduce the output voltage ripple to 1 percent of the dc value.

■ Solution

From the parameters given,

$$V_m = 120\sqrt{2} = 169.7 \text{ V}$$

$$\omega RC = (2\pi 60)(500)(10)^{-6} = 18.85$$

The angle θ is determined from Eq. (4-9).

$$\theta = -\tan^{-1}(18.85) + \pi = 1.62 \text{ rad} = 93°$$

$$V_m \sin \theta = 169.5 \text{ V}$$

The angle α is determined by the numerical solution of Eq. (4-10).

$$\sin(1.62)e^{-(\pi + \alpha - 1.62)/18.85} - \sin \alpha = 0$$

$$\alpha = 1.06 \text{ rad} = 60.6°$$

(a) Peak-to-peak output voltage is described by Eq. (4-11).

$$\Delta V_o = V_m(1 - \sin \alpha) = 169.7[1 - \sin(1.06)] = 22 \text{ V}$$

Note that this is the same load and source as for the half-wave rectifier of Example 3-9 where $\Delta V_o = 43$ V.

(b) With the ripple limited to 1 percent, the output voltage will be held close to V_m and the approximation of Eq. (4-13) applies.

$$\frac{\Delta V_o}{V_m} = 0.01 \approx \frac{1}{2fRC}$$

Solving for C,

$$C \approx \frac{1}{2fR(\Delta V_o/V_m)} = \frac{1}{(2)(60)(500)(0.01)} = 1670 \text{ μF}$$

Voltage Doublers

The rectifier circuit of Fig. 4-7a serves as a simple voltage doubler, having an output of twice the peak value of the source. For ideal diodes, C_1 charges to V_m through D_1 when the source is positive; C_2 charges to V_m through D_2 when the source is negative. The voltage across the load resistor is the sum of the capacitor voltages $2V_m$. This circuit is useful when the output voltage of a rectifier must be larger than the peak input voltage. Voltage doubler circuits avoid using a transformer to step up the voltage, saving expense, volume, and weight.

The full-wave rectifier with a capacitive output filter can be combined with the voltage doubler, as shown in Fig. 4-7b. When the switch is open, the circuit is similar to the full-wave rectifier of Fig. 4-6a, with output at approximately V_m when the capacitors are large. When the switch is closed, the circuit acts as the voltage doubler of Fig. 4-7a. Capacitor C_1 charges to V_m through D_1 when

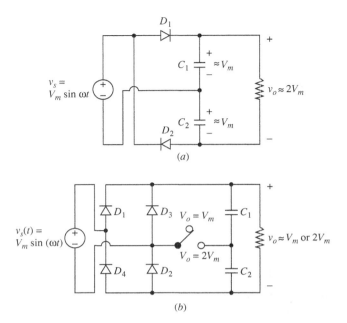

Figure 4-7 (*a*) Voltage doubler. (*b*) Dual-voltage rectifier.

the source is positive, and C_2 charges to V_m through D_4 when the source is negative. The output voltage is then $2V_m$. Diodes D_2 and D_3 remain reverse-biased in this mode.

This voltage doubler circuit is useful when equipment must be used on systems with different voltage standards. For example, a circuit could be designed to operate properly in both the United States, where the line voltage is 120 V, and places abroad where the line voltage is 240 V.

LC Filtered Output

Another full-wave rectifier configuration has an *LC* filter on the output, as shown in Fig. 4-8*a*. The purpose of the filter is to produce an output voltage that is close to purely dc. The capacitor holds the output voltage at a constant level, and the inductor smooths the current from the rectifier and reduces the peak current in the diodes from that of the current of Fig. 4-6*a*.

The circuit can operate in the continuous- or discontinuous-current mode. For continuous current, the inductor current is always positive, as illustrated in Fig. 4-8*b*. Discontinuous current is characterized by the inductor current returning to zero in each cycle, as illustrated in Fig. 4-8*c*. The continuous-current case is easier to analyze and is considered first.

Continuous Current for *LC* Filtered Output For continuous current, the voltage v_x in Fig. 4-8*a* is a full-wave rectified sine wave, which has an average

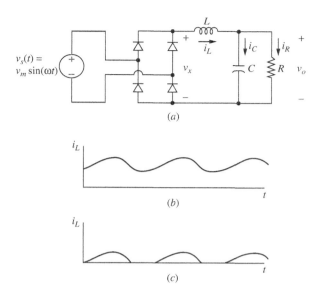

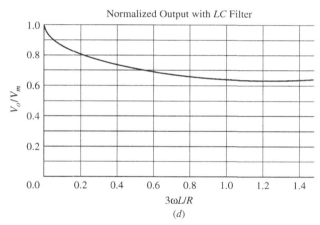

Figure 4-8 (*a*) Rectifier with *LC* filtered output; (*b*) Continuous inductor current; (*c*) Discontinuous inductor current; (*d*) Normalized output.

value of $2V_m/\pi$. Since the average voltage across the inductor in the steady state is zero, the average output voltage for continuous inductor current is

$$V_o = \frac{2V_m}{\pi} \qquad (4\text{-}14)$$

Average inductor current must equal the average resistor current because the average capacitor current is zero.

$$I_L = I_R = \frac{V_o}{R} = \frac{2V_m}{\pi R} \qquad (4\text{-}15)$$

The variation in inductor current can be estimated from the first ac term in the Fourier series. The first ac voltage term is obtained from Eq. (4-4) with $n = 2$. Assuming the capacitor to be a short circuit to ac terms, the harmonic voltage v_2 exists across the inductor. The amplitude of the inductor current for $n = 2$ is

$$I_2 = \frac{V_2}{Z_2} \approx \frac{V_2}{2\omega L} = \frac{4V_m/3\pi}{2\omega L} = \frac{2V_m}{3\pi \omega L} \quad (4\text{-}16)$$

For the current to always be positive, the amplitude of the ac term must be less than the dc term (average value). Using the above equations and solving for L,

$$I_2 < I_L$$

$$\frac{2V_m}{3\pi\omega L} < \frac{2V_m}{\pi R}$$

$$L > \frac{R}{3\omega}$$

or

$$\frac{3\omega L}{R} > 1 \quad \text{for continuous current} \quad (4\text{-}17)$$

If $3\omega L/R > 1$, the current is continuous and the output voltage is $2V_m/\pi$. Otherwise, the output voltage must be determined from analysis for discontinuous current, discussed as follows.

Discontinuous Current for *LC* Filtered Output For discontinuous inductor current, the current reaches zero during each period of the current waveform (Fig. 4-8c). Current becomes positive again when the bridge output voltage reaches the level of the capacitor voltage, which is at $\omega t = \alpha$.

$$\alpha = \sin^{-1}\left(\frac{V_o}{V_m}\right) \quad (4\text{-}18)$$

While current is positive, the voltage across the inductor is

$$v_L = V_m \sin(\omega t) - V_o \quad (4\text{-}19)$$

where the output voltage V_o is yet to be determined. Inductor current is expressed as

$$i_L(\omega t) = \frac{1}{\omega L} \int_\alpha^{\omega t} \left[V_m \sin(\omega t) - V_o\right] d(\omega t)$$

$$= \frac{1}{\omega L} \left[V_m\left(\cos \alpha - \cos \omega t\right)\right] - V_o(\omega t - \alpha) \quad (4\text{-}20)$$

$$\text{for } \alpha \leq \omega t \leq \beta \quad \text{when } \beta < \pi$$

which is valid until the current reaches zero, at $\omega t = \beta$.

The solution for the load voltage V_o is based on the fact that the average inductor current must equal the current in the load resistor. Unfortunately, a closed-form solution is not available, and an iterative technique is required.

A procedure for determining V_o is as follows:

1. Estimate a value for V_o slightly below V_m and solve for α in Eq. (4-18).
2. Solve for β numerically in Eq. (4-20) for inductor current,

$$i_L(\beta) = 0 = V_m(\cos\alpha - \cos\beta) - V_o(\beta - \alpha)$$

3. Solve for average inductor current I_L.

$$i_L = \frac{1}{\pi}\int_\alpha^\beta i_L(\omega t)\,d(\omega t) \tag{4-21}$$

$$= \frac{1}{\pi}\int_\alpha^\beta \frac{1}{\omega L}\Big[V_m(\cos\alpha - \cos\omega t) - V_o(\omega t - \alpha)\Big]d(\omega t)$$

4. Solve for load voltage V_o based upon the average inductor current from step 3.

$$I_R = I_L = \frac{V_o}{R}$$

or

$$V_o = I_L R \tag{4-22}$$

5. Repeat steps 1 to 4 until the computed value of V_o in step 4 equals the estimated V_o in step 1.

Output voltage for discontinuous current is larger than for continuous current. If there is no load, the capacitor charges to the peak value of the source so the maximum output is V_m. Figure 4-8d shows normalized output V_o/V_m as a function of $3\omega L/R$.

EXAMPLE 4-5

Full-Wave Rectifier with LC Filter

A full-wave rectifier has a source of $v_s(t) = 100\sin(377t)$ V. An LC filter as in Fig. 4-8a is used, with $L = 5$ mH and $C = 10{,}000$ μF. The load resistance is (a) 5 Ω and (b) 50 Ω. Determine the output voltage for each case.

■ **Solution**

Using Eq. (4-17), continuous inductor current exists when

$$R < 3\omega L = 3(377)(0.005) = 5.7\ \Omega$$

which indicates continuous current for 5 Ω and discontinuous current for 50 Ω.

(a) For $R = 5\ \Omega$ with continuous current, output voltage is determined from Eq. (4-14).

$$V_o = \frac{2V_m}{\pi} = \frac{2(100)}{\pi} = 63.7\text{ V}$$

(b) For $R = 50\ \Omega$ with discontinuous current, the iteration method is used to determine V_o. Initially, V_o is estimated to be 90 V. The results of the iteration are as follows:

Estimated V_o	α	β	Calculated V_o	
90	1.12	2.48	38.8	(Estimate is too high)
80	0.93	2.89	159	(Estimate is too low)
85	1.12	2.70	88.2	(Estimate is slightly low)
86	1.04	2.66	76.6	(Estimate is too high)
85.3	1.02	2.69	84.6	(Approximate solution)

Therefore, V_o is approximately 85.3 V. As a practical matter, three significant figures for the load voltage may not be justified when predicting performance of a real circuit. Knowing that the output voltage is slightly above 85 V after the third iteration is probably sufficient. Output could also be estimated from the graph of Fig. 4-8d.

PSpice Solution

The circuit is created using VSIN for the source and Dbreak for the diodes, with the diode model modified to represent an ideal diode by using $n = 0.01$. The voltage of the filter capacitor is initialized at 90 V, and small capacitors are placed across the diodes to avoid convergence problems. Both values of R are tested in one simulation by using a parametric sweep. The transient analysis must be sufficiently long to allow a steady-state periodic output to be observed. The Probe output for both load resistors is shown in Fig. 4-9. Average output voltage for each case is obtained from Probe by entering AVG(V(out+)−V(out−)) after restricting the data to represent steady-state output (after about 250 ms), resulting in $V_o = 63.6$ V for $R = 5\ \Omega$ (continuous current) and $V_o = 84.1$ V for $R = 50\ \Omega$ (discontinuous current). These values match very well with those of the analytical solution.

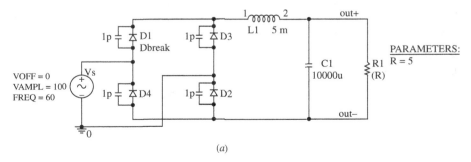

Figure 4-9 PSpice output for Example 4-6. (a) Full-wave rectifier with an *LC* filter. The small capacitors across the diodes help with convergence; (b) The output voltage for continuous and discontinuous inductor current.

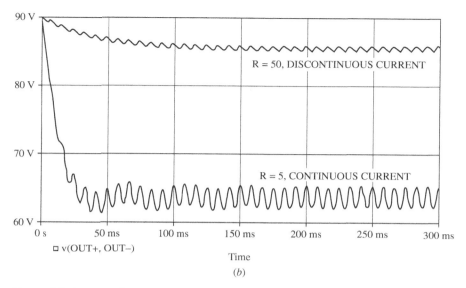

Figure 4-9 (*continued*)

4.3 CONTROLLED FULL-WAVE RECTIFIERS

A versatile method of controlling the output of a full-wave rectifier is to substitute controlled switches such as thyristors (SCRs) for the diodes. Output is controlled by adjusting the delay angle of each SCR, resulting in an output voltage that is adjustable over a limited range.

Controlled full-wave rectifiers are shown in Fig. 4-10. For the bridge rectifier, SCRs S_1 and S_2 will become forward-biased when the source becomes positive but will not conduct until gate signals are applied. Similarly, S_3 and S_4 will become forward-biased when the source becomes negative but will not conduct until they receive gate signals. For the center-tapped transformer rectifier, S_1 is forward-biased when v_s is positive, and S_2 is forward-biased when v_s is negative, but each will not conduct until it receives a gate signal.

The delay angle α is the angle interval between the forward biasing of the SCR and the gate signal application. If the delay angle is zero, the rectifiers behave exactly as uncontrolled rectifiers with diodes. The discussion that follows generally applies to both bridge and center-tapped rectifiers.

Resistive Load

The output voltage waveform for a controlled full-wave rectifier with a resistive load is shown in Fig. 4-10c. The average component of this waveform is determined from

$$V_o = \frac{1}{\pi} \int_{\alpha}^{\pi} V_m \sin(\omega t)\, d(\omega t) = \frac{V_m}{\pi}(1 + \cos \alpha) \qquad (4\text{-}23)$$

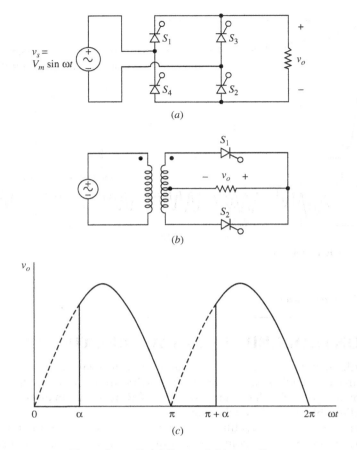

Figure 4-10 (a) Controlled full-wave bridge rectifier;
(b) Controlled full-wave center-tapped transformer rectifier;
(c) Output for a resistive load.

Average output current is then

$$I_o = \frac{V_o}{R} = \frac{V_m}{\pi R}(1 + \cos \alpha) \quad (4\text{-}24)$$

The power delivered to the load is a function of the input voltage, the delay angle, and the load components; $P = I_{rms}^2 R$ is used to determine the power in a resistive load, where

$$I_{rms} = \sqrt{\frac{1}{\pi}\int_\alpha^\pi \left(\frac{V_m}{R}\sin \omega t\right)^2 d(\omega t)}$$

$$= \frac{V_m}{R}\sqrt{\frac{1}{2} - \frac{\alpha}{2\pi} + \frac{\sin(2\alpha)}{4\pi}} \quad (4\text{-}25)$$

The rms current in the source is the same as the rms current in the load.

EXAMPLE 4-6

Controlled Full-Wave Rectifier with Resistive Load

The full-wave controlled bridge rectifier of Fig. 4-10a has an ac input of 120 V rms at 60 Hz and a 20-Ω load resistor. The delay angle is 40°. Determine the average current in the load, the power absorbed by the load, and the source voltamperes.

■ **Solution**

The average output voltage is determined from Eq. (4-23).

$$V_o = \frac{V_m}{\pi}(1 + \cos \alpha) = \frac{\sqrt{2}(120)}{\pi}(1 + \cos 40°) = 95.4 \text{ V}$$

Average load current is

$$I_o = \frac{V_o}{R} = \frac{95.4}{20} = 4.77 \text{ A}$$

Power absorbed by the load is determined from the rms current from Eq. (4-24), remembering to use α in radians.

$$I_{rms} = \frac{\sqrt{2}(120)}{20}\sqrt{\frac{1}{2} - \frac{0.698}{2\pi} + \frac{\sin[2(0.698)]}{4\pi}} = 5.80 \text{ A}$$

$$P = I_{rms}^2 R = (5.80)^2(20) = 673 \text{ W}$$

The rms current in the source is also 5.80 A, and the apparent power of the source is

$$S = V_{rms} I_{rms} = (120)(5.80) = 696 \text{ VA}$$

Power factor is

$$pf = \frac{P}{S} = \frac{672}{696} = 0.967$$

RL Load, Discontinuous Current

Load current for a controlled full-wave rectifier with an RL load (Fig. 4-11a) can be either continuous or discontinuous, and a separate analysis is required for each. Starting the analysis at $\omega t = 0$ with zero load current, SCRs S_1 and S_2 in the bridge rectifier will be forward-biased and S_3 and S_4 will be reverse-biased as the source voltage becomes positive. Gate signals are applied to S_1 and S_2 at $\omega t = \alpha$, turning S_1 and S_2 on. With S_1 and S_2 on, the load voltage is equal to the source voltage. For this condition, the circuit is identical to that of the controlled half-wave rectifier of Chap. 3, having a current function

$$i_o(\omega t) = \frac{V_m}{Z}\left[\sin(\omega t - \theta) - \sin(\alpha - \theta)e^{-(\omega t - \alpha)/\omega \tau}\right] \quad \text{for } \alpha \leq \omega t \leq \beta$$

where (4-26)

$$Z = \sqrt{R^2 + (\omega L)^2} \quad \theta = \tan^{-1}\left(\frac{\omega L}{R}\right) \quad \text{and} \quad \tau = \frac{L}{R}$$

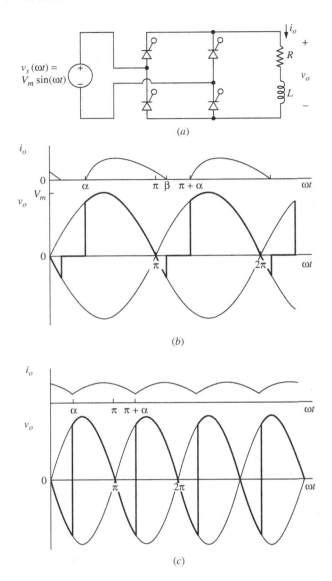

Figure 4-11 (a) Controlled rectifier with *RL* load; (b) Discontinuous current; (c) Continuous current.

The above current function becomes zero at $\omega t = \beta$. If $\beta < \pi + \alpha$, the current remains at zero until $\omega t = \pi + \alpha$ when gate signals are applied to S_3 and S_4 which are then forward-biased and begin to conduct. This mode of operation is called *discontinuous current*, which is illustrated in Fig. 4-11*b*.

$$\beta < \alpha + \pi \quad \rightarrow \quad \text{discontinuous current} \tag{4-27}$$

4.3 Controlled Full-Wave Rectifiers

Analysis of the controlled full-wave rectifier operating in the discontinuous-current mode is identical to that of the controlled half-wave rectifier except that the period for the output current is π rather than 2π rad.

EXAMPLE 4-7

Controlled Full-Wave Rectifier, Discontinuous Current

A controlled full-wave bridge rectifier of Fig. 4-11a has a source of 120 V rms at 60 Hz, $R = 10\,\Omega$, $L = 20$ mH, and $\alpha = 60°$. Determine (a) an expression for load current, (b) the average load current, and (c) the power absorbed by the load.

■ **Solution**
From the parameters given,

$$V_m = \frac{120}{\sqrt{2}} = 169.7 \text{ V}$$

$$Z = \sqrt{R^2 + (\omega L)^2} = \sqrt{10^2 + [(377)(0.02)]^2} = 12.5\,\Omega$$

$$\theta = \tan^{-1}\left(\frac{\omega L}{R}\right) = \tan^{-1}\left[\frac{(377)(0.02)}{10}\right] = 0.646 \text{ rad}$$

$$\omega\tau = \frac{\omega L}{R} = \frac{(377)(0.02)}{10} = 0.754 \text{ rad}$$

$$\alpha = 60° = 1.047 \text{ rad}$$

(a) Substituting into Eq. (4-26),

$$i_o(\omega t) = 13.6 \sin(\omega t - 0.646) - 21.2 e^{-\omega t/0.754} \text{ A} \quad \text{for } \alpha \le \omega t \le \beta$$

Solving $i_o(\beta) = 0$ numerically for β, $\beta = 3.78$ rad (216°). Since $\pi + \alpha = 4.19 > \beta$, the current is discontinuous, and the above expression for current is valid.

(b) Average load current is determined from the numerical integration of

$$I_o = \frac{1}{\pi}\int_\alpha^\beta i_o(\omega t)\,d(\omega t) = 7.05 \text{ A}$$

(c) Power absorbed by the load occurs in the resistor and is computed from $I_{rms}^2 R$, where

$$I_{rms} = \sqrt{\frac{1}{\pi}\int_\alpha^\beta i_o(\omega t)\,d(\omega t)} = 8.35 \text{ A}$$

$$P = (8.35)^2(10) = 697 \text{ W}$$

RL Load, Continuous Current

If the load current is still positive at $\omega t = \pi + \alpha$ when gate signals are applied to S_3 and S_4 in the above analysis, S_3 and S_4 are turned on and S_1 and S_2 are forced off. Since the initial condition for current in the second half-cycle is not zero, the current function does not repeat. Equation (4-26) is not valid in the steady state for continuous current. For an *RL* load with continuous current, the steady-state current and voltage waveforms are generally as shown in Fig. 4-11c.

The boundary between continuous and discontinuous current occurs when β for Eq. (4-26) is $\pi + \alpha$. The current at $\omega t = \pi + \alpha$ must be greater than zero for continuous-current operation.

$$i(\pi + \alpha) \geq 0$$

$$\sin(\pi + \alpha - \theta) - \sin(\pi + \alpha - \theta)\, e^{-(\pi + \alpha - \alpha)/\omega\tau} \geq 0$$

Using

$$\sin(\pi + \alpha - \theta) = \sin(\theta - \alpha)$$

$$\sin(\theta - \alpha)\left(1 - e^{-(\pi/\omega\tau)}\right) \geq 0$$

Solving for α,

$$\alpha \leq \theta$$

Using

$$\theta = \tan^{-1}\left(\frac{\omega L}{R}\right)$$

$$\alpha \leq \tan^{-1}\left(\frac{\omega L}{R}\right) \qquad \text{for continuous current} \qquad (4\text{-}28)$$

Either Eq. (4-27) or Eq. (4-28) can be used to check whether the load current is continuous or discontinuous.

A method for determining the output voltage and current for the continuous-current case is to use the Fourier series. The Fourier series for the voltage waveform for continuous-current case shown in Fig. 4-11c is expressed in general form as

$$v_o(\omega t) = V_o + \sum_{n=1}^{\infty} V_n \cos(n\omega_0 t + \theta_n) \qquad (4\text{-}29)$$

The dc (average) value is

$$\boxed{V_o = \frac{1}{\pi} \int_{\alpha}^{\alpha + \pi} V_m \sin(\omega t)\, d(\omega t) = \frac{2V_m}{\pi} \cos \alpha} \qquad (4\text{-}30)$$

The amplitudes of the ac terms are calculated from

$$V_n = \sqrt{a_n^2 + b_n^2} \qquad (4\text{-}31)$$

where

$$a_n = \frac{2V_m}{\pi}\left[\frac{\cos(n+1)\alpha}{n+1} - \frac{\cos(n-1)\alpha}{n-1}\right]$$

$$b_n = \frac{2V_m}{\pi}\left[\frac{\sin(n+1)\alpha}{n+1} - \frac{\sin(n-1)\alpha}{n-1}\right] \quad (4\text{-}32)$$

$$n = 2, 4, 6, \ldots$$

Figure 4-12 shows the relationship between normalized harmonic content of the output voltage and delay angle.

The Fourier series for current is determined by superposition as was done for the uncontrolled rectifier earlier in this chapter. The current amplitude at each frequency is determined from Eq. (4-5). The rms current is determined by combining the rms currents at each frequency. From Eq. (2-43),

$$I_{\text{rms}} = \sqrt{I_o^2 + \sum_{n=2,4,6\ldots}^{\infty}\left(\frac{I_n}{\sqrt{2}}\right)^2}$$

where

$$I_o = \frac{V_o}{R} \quad \text{and} \quad I_n = \frac{V_n}{Z_n} = \frac{V_n}{|R + jn\omega_0 L|} \quad (4\text{-}33)$$

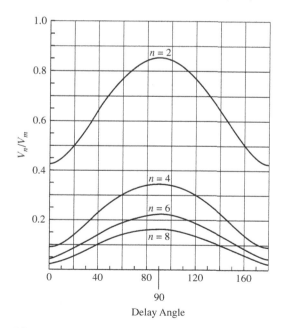

Figure 4-12 Output harmonic voltages as a function of delay angle for a single-phase controlled rectifier.

As the harmonic number increases, the impedance for the inductance increases. Therefore, it may be necessary to solve for only a few terms of the series to be able to calculate the rms current. If the inductor is large, the ac terms will become small, and the current is essentially dc.

EXAMPLE 4-8

Controlled Full-Wave Rectifier with *RL* Load, Continuous Current

A controlled full-wave bridge rectifier of Fig. 4-11a has a source of 120 V rms at 60 Hz, an *RL* load where $R = 10\ \Omega$ and $L = 100$ mH. The delay angle $\alpha = 60°$ (same as Example 4-7 except *L* is larger). (*a*) Verify that the load current is continuous. (*b*) Determine the dc (average) component of the current. (*c*) Determine the power absorbed by the load.

■ **Solution**

(*a*) Equation (4-28) is used to verify that the current is continuous.

$$\tan^{-1}\left(\frac{\omega L}{R}\right) = \tan^{-1}\left[\frac{(377)(0.1)}{10}\right] = 75°$$

$$\alpha = 60° < 75° \quad \therefore \text{ continuous current}$$

(*b*) The voltage across the load is expressed in terms of the Fourier series of Eq. (4-29). The dc term is computed from Eq. (4-30).

$$V_0 = \frac{2V_m}{\pi}\cos\alpha = \frac{2\sqrt{2}(120)}{\pi}\cos(60°) = 54.0\text{ V}$$

(*c*) The amplitudes of the ac terms are computed from Eqs. (4-31) and (4-32) and are summarized in the following table where, $Z_n = |R + j\omega L|$ and $I_n = V_n/Z_n$.

n	a_n	b_n	V_n	Z_n	I_n
0 (dc)	—	—	54.0	10	5.40
2	−90	−93.5	129.8	76.0	1.71
4	46.8	−18.7	50.4	151.1	0.33
6	−3.19	32.0	32.2	226.4	0.14

The rms current is computed from Eq. (4-33).

$$I_{\text{rms}} = \sqrt{(5.40)^2 + \left(\frac{1.71}{\sqrt{2}}\right)^2 + \left(\frac{0.33}{\sqrt{2}}\right)^2 + \left(\frac{0.14}{\sqrt{2}}\right)^2 + \cdots} \approx 5.54\text{ A}$$

Power is computed from $I_{\text{rms}}^2 R$.

$$P = (5.54)^2(10) = 307\text{ W}$$

Note that the rms current could be approximated accurately from the dc term and one ac term ($n = 2$). Higher-frequency terms are very small and contribute little to the power in the load.

PSpice Simulation of Controlled Full-Wave Rectifiers

To simulate the controlled full-wave rectifier in PSpice, a suitable SCR model must be chosen. As with the controlled half-wave rectifier of Chap. 3, a simple switch and diode can be used to represent the SCR, as shown in Fig. 4-13a. This circuit requires the full version of PSpice.

EXAMPLE 4-9

PSpice Simulation of a Controlled Full-Wave Rectifier

Use PSpice to determine the solution of the controlled full-wave rectifier in Example 4-8.

■ **Solution**

A PSpice circuit that uses the controlled-switch model for the SCRs is shown in Fig. 4-13a. (This circuit is too large for the demo version and requires the full production version of PSpice.)

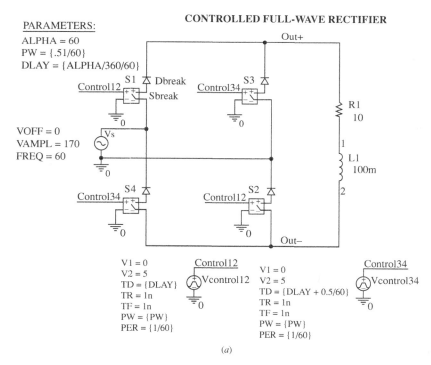

Figure 4-13 (a) PSpice circuit for a controlled full-wave rectifier of Example 4-8; (b) Probe output showing load voltage and current.

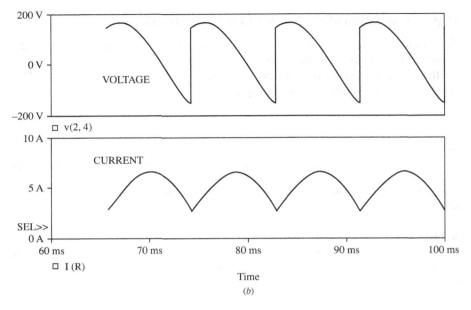

Figure 4-13 (*continued*)

Controlled Rectifier with *RL*-Source Load

The controlled rectifier with a load that is a series resistance, inductance, and dc voltage (Fig. 4-14) is analyzed much like the uncontrolled rectifier of Fig. 4-5*a* discussed earlier in this chapter. For the controlled rectifier, the SCRs may be turned on at any time that they are forward-biased, which is at an angle

$$\alpha \geq \sin^{-1}\left(\frac{V_{dc}}{V_m}\right) \quad (4\text{-}34)$$

For the continuous-current case, the bridge output voltage is the same as in Fig. 4-11*c*. The average bridge output voltage is

$$V_o = \frac{2V_m}{\pi} \cos \alpha \quad (4\text{-}35)$$

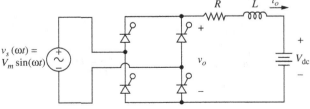

Figure 4-14 Controlled rectifier with *RL*-source load.

The average load current is

$$I_o = \frac{V_o - V_{dc}}{R} \quad (4\text{-}36)$$

The ac voltage terms are unchanged from the controlled rectifier with an RL load in Fig. 4-11a and are described by Eqs. (4-29) to (4-32). The ac current terms are determined from the circuit of Fig. 4-14c. Power absorbed by the dc voltage is

$$P_{dc} = I_o V_{dc} \quad (4\text{-}37)$$

Power absorbed by the resistor in the load is $I_{rms}^2 R$. If the inductance is large and the load current has little ripple, power absorbed by the resistor is approximately $I_o^2 R$.

EXAMPLE 4-10

Controlled Rectifier with RL-Source Load

The controlled rectifier of Fig. 4-14 has an ac source of 240 V rms at 60 Hz, $V_{dc} = 100$ V, $R = 5\ \Omega$, and an inductor large enough to cause continuous current. (a) Determine the delay angle α such that the power absorbed by the dc source is 1000 W. (b) Determine the value of inductance that will limit the peak-to-peak load current variation to 2 A.

■ **Solution**

(a) For the power in the 100-V dc source to be 1000 W, the current in it must be 10 A. The required output voltage is determined from Eq. (4-36) as

$$V_o = V_{dc} + I_o R = 100 + (10)(5) = 150\ \text{V}$$

The delay angle which will produce a 150 V dc output from the rectifier is determined from Eq. (4-35).

$$\alpha = \cos^{-1}\left(\frac{V_o \pi}{2 V_m}\right) = \cos^{-1}\left[\frac{(150)(\pi)}{2\sqrt{2}(240)}\right] = 46°$$

(b) Variation in load current is due to the ac terms in the Fourier series. The load current amplitude for each of the ac terms is

$$I_n = \frac{V_n}{Z_n}$$

where V_n is described by Eqs. (4-31) and (4-32) or can be estimated from the graph of Fig. 4-12. The impedance for the ac terms is

$$Z_n = |R + jn\omega_0 L|$$

Since the decreasing amplitude of the voltage terms and the increasing magnitude of the impedance both contribute to diminishing ac currents as n increases, the peak-to-peak current variation will be estimated from the first ac term. For $n = 2$, V_n/V_m is estimated from Fig. 4-12 as 0.68 for $\alpha = 46°$, making $V_2 = 0.68 V_m = 0.68\ (240\sqrt{2}) = 230$ V. The peak-to-peak variation of 2 A corresponds to a 1-A zero-to-peak amplitude. The required load impedance for $n = 2$ is then

$$Z_2 = \frac{V_2}{I_2} = \frac{230\ \text{V}}{1\ \text{A}} = 230\ \Omega$$

The 5-Ω resistor is insignificant compared to the total 230-Ω required impedance, so $Z_n \approx n\omega L$. Solving for L,

$$L \approx \frac{Z_2}{2\omega} = \frac{230}{2(377)} = 0.31 \text{ H}$$

A slightly larger inductance should be chosen to allow for the effect of higher-order ac terms.

Controlled Single-Phase Converter Operating as an Inverter

The above discussion focused on circuits operating as rectifiers, which means that the power flow is from the ac source to the load. It is also possible for power to flow from the load to the ac source, which classifies the circuit as an inverter.

For inverter operation of the converter in Fig. 4-14, power is supplied by the dc source, and power is absorbed by the bridge and is transferred to the ac system. The load current must be in the direction shown because of the SCRs in the bridge. For power to be supplied by the dc source, V_{dc} must be negative. For power to be absorbed by the bridge and transferred to the ac system, the bridge output voltage V_o must also be negative. Equation (4-35) applies, so a delay angle larger than 90° will result in a negative output voltage.

$$\begin{aligned} 0 < \alpha < 90° &\rightarrow V_o > 0 \quad \text{rectifier operation} \\ 90° < \alpha < 180° &\rightarrow V_o < 0 \quad \text{inverter operation} \end{aligned} \quad (4\text{-}38)$$

The voltage waveform for $\alpha = 150°$ and continuous inductor current is shown in Fig. 4-15. Equations (4-36) to (4-38) apply. If the inductor is large enough to

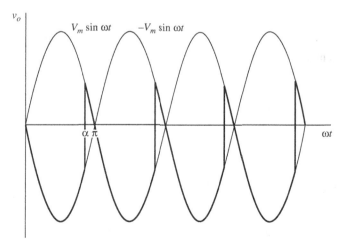

Figure 4-15 Output voltage for the controlled single-phase converter of Fig. 4-14 operating as an inverter, $\alpha = 150°$ and $V_{dc} < 0$.

effectively eliminate the ac current terms and the bridge is lossless, the power absorbed by the bridge and transferred to the ac system is

$$P_{\text{bridge}} = P_{\text{ac}} = -I_o V_o \qquad (4\text{-}39)$$

EXAMPLE 4-11

Single-Phase Bridge Operating as an Inverter

The dc voltage in Fig. 4-14 represents the voltage generated by an array of solar cells and has a value of 110 V, connected such that $V_{\text{dc}} = -110$ V. The solar cells are capable of producing 1000 W. The ac source is 120 V rms, $R = 0.5\ \Omega$, and L is large enough to cause the load current to be essentially dc. Determine the delay angle α such that 1000 W is supplied by the solar cell array. Determine the power transferred to the ac system and the losses in the resistance. Assume ideal SCRs.

■ Solution

For the solar cell array to supply 1000 W, the average current must be

$$I_o = \frac{P_{\text{dc}}}{V_{\text{dc}}} = \frac{1000}{110} = 9.09\ \text{A}$$

The average output voltage of the bridge is determined from Eq. (4-36).

$$V_o = I_o R + V_{\text{dc}} = (9.09)(0.5) + (-110) = -105.5\ \text{V}$$

The required delay angle is determined from Eq. (4-35).

$$\alpha = \cos^{-1}\left(\frac{V_o \pi}{2 V_m}\right) = \cos^{-1}\left[\frac{-105.5\pi}{2\sqrt{2}(120)}\right] = 165.5°$$

Power absorbed by the bridge and transferred to the ac system is determined from Eq. (4-39).

$$P_{\text{ac}} = -V_o I_o = (-9.09)(-105.5) = 959\ \text{W}$$

Power absorbed by the resistor is

$$P_R = I_{\text{rms}}^2 R \approx I_o^2 R = (9.09)^2 (0.5) = 41\ \text{W}$$

Note that the load current and power will be sensitive to the delay angle and the voltage drops across the SCRs because bridge output voltage is close to the dc source voltage. For example, assume that the voltage across a conducting SCR is 1 V. Two SCRs conduct at all times, so the average bridge output voltage is reduced to

$$V_o = -105.5 - 2 = -107.5\ \text{V}$$

Average load current is then

$$I_o = \frac{-107.5 - (-110)}{0.5} = 5.0\ \text{A}$$

Power delivered to the bridge is then reduced to

$$P_{\text{bridge}} = (107.5)(5.0) = 537.5 \text{ W}$$

Average current in each SCR is one-half the average load current. Power absorbed by each SCR is approximately

$$P_{\text{SCR}} = I_{\text{SCR}} V_{\text{SCR}} = \frac{1}{2} I_o V_{\text{SCR}} = \frac{1}{2}(5)(1) = 2.5 \text{ W}$$

Total power loss in the bridge is then $4(2.5) = 10$ W, and power delivered to the ac source is $537.5 - 10 = 527.5$ W.

4.4 THREE-PHASE RECTIFIERS

Three-phase rectifiers are commonly used in industry to produce a dc voltage and current for large loads. The three-phase full-bridge rectifier is shown in Fig. 4-16a. The three-phase voltage source is balanced and has phase sequence *a-b-c*. The source and the diodes are assumed to be ideal in the initial analysis of the circuit.

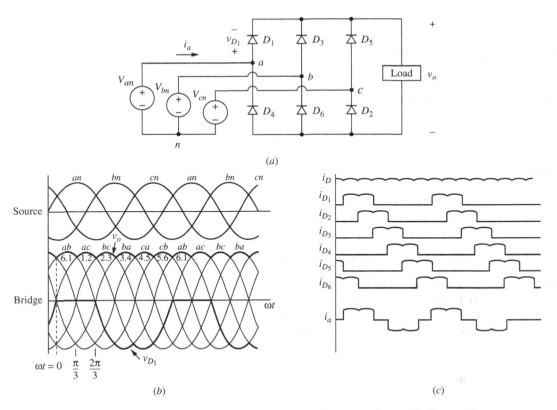

Figure 4-16 (*a*) Three-phase full-bridge rectifier; (*b*) Source and output voltages; (*c*) Currents for a resistive load.

4.4 Three-Phase Rectifiers

Some basic observations about the circuit are as follows:

1. Kirchhoff's voltage law around any path shows that only one diode in the top half of the bridge may conduct at one time (D_1, D_3, or D_5). The diode that is conducting will have its anode connected to the phase voltage that is highest at that instant.
2. Kirchhoff's voltage law also shows that only one diode in the bottom half of the bridge may conduct at one time (D_2, D_4, or D_6). The diode that is conducting will have its cathode connected to the phase voltage that is lowest at that instant.
3. As a consequence of items 1 and 2 above, D_1 and D_4 cannot conduct at the same time. Similarly, D_3 and D_6 cannot conduct simultaneously, nor can D_5 and D_2.
4. The output voltage across the load is one of the line-to-line voltages of the source. For example, when D_1 and D_2 are on, the output voltage is v_{ac}. Furthermore, the diodes that are on are determined by which line-to-line voltage is the highest at that instant. For example, when v_{ac} is the highest line-to-line voltage, the output is v_{ac}.
5. There are six combinations of line-to-line voltages (three phases taken two at a time). Considering one period of the source to be 360°, a transition of the highest line-to-line voltage must take place every 360°/6 = 60°. Because of the six transitions that occur for each period of the source voltage, the circuit is called a *six-pulse rectifier*.
6. The fundamental frequency of the output voltage is 6ω, where ω is the frequency of the three-phase source.

Figure 4-16b shows the phase voltages and the resulting combinations of line-to-line voltages from a balanced three-phase source. The current in each of the bridge diodes for a resistive load is shown in Fig. 4-16c. The diodes conduct in pairs (6,1), (1,2), (2,3), (3,4), (4,5), (5,6), (6,1), ... Diodes turn on in the sequence 1, 2, 3, 4, 5, 6, 1, ...

The current in a conducting diode is the same as the load current. To determine the current in each phase of the source, Kirchhoff's current law is applied at nodes a, b, and c,

$$i_a = i_{D_1} - i_{D_4}$$
$$i_b = i_{D_3} - i_{D_6} \qquad (4\text{-}40)$$
$$i_c = i_{D_5} - i_{D_2}$$

Since each diode conducts one-third of the time, resulting in

$$\boxed{\begin{aligned} I_{D,\text{avg}} &= \frac{1}{3} I_{o,\text{avg}} \\ I_{D,\text{rms}} &= \frac{1}{\sqrt{3}} I_{o,\text{rms}} \\ I_{s,\text{rms}} &= \sqrt{\frac{2}{3}} I_{o,\text{rms}} \end{aligned}} \qquad (4\text{-}41)$$

The apparent power from the three-phase source is

$$S = \sqrt{3}\, V_{L-L,\text{rms}}\, I_{S,\text{rms}} \qquad (4\text{-}42)$$

The maximum reverse voltage across a diode is the peak line-to-line voltage. The voltage waveform across diode D_1 is shown in Fig. 4-16b. When D_1 conducts, the voltage across it is zero. When D_1 is off, the output voltage is v_{ab} when D_3 is on and is v_{ac} when D_5 is on.

The periodic output voltage is defined as $v_o(\omega t) = V_{m,L-L} \sin(\omega t)$ for $\pi/3 \leq \omega t \leq 2\pi/3$ with period $\pi/3$ for the purpose of determining the Fourier series coefficients. The coefficients for the sine terms are zero from symmetry, enabling the Fourier series for the output voltage to be expressed as

$$v_o(t) = V_o + \sum_{n=6,12,18\ldots}^{\infty} V_n \cos(n\omega_0 t + \pi) \qquad (4\text{-}43)$$

The average or dc value of the output voltage is

$$\boxed{V_0 = \frac{1}{\pi/3} \int_{\pi/3}^{2\pi/3} V_{m,L-L} \sin(\omega t)\, d(\omega t) = \frac{3 V_{m,L-L}}{\pi} = 0.955\, V_{m,L-L}} \qquad (4\text{-}44)$$

where $V_{m,L-L}$ is the peak line-to-line voltage of the three-phase source, which is $\sqrt{2} V_{L-L,\text{rms}}$. The amplitudes of the ac voltage terms are

$$V_n = \frac{6 V_{m,L-L}}{\pi(n^2 - 1)} \qquad n = 6, 12, 18, \ldots \qquad (4\text{-}45)$$

Since the output voltage is periodic with period one-sixth of the ac supply voltage, the harmonics in the output are of order $6k\omega$, $k = 1, 2, 3 \ldots$ An advantage of the three-phase rectifier over the single-phase rectifier is that the output is inherently like a dc voltage, and the high-frequency low-amplitude harmonics enable filters to be effective.

In many applications, a load with series inductance results in a load current that is essentially dc. For a dc load current, the diode and ac line currents are shown in Fig. 4-17. The Fourier series of the currents in phase a of the ac line is

$$i_a(t) = \frac{2\sqrt{3}}{\pi} I_o \left(\cos \omega_0 t - \frac{1}{5} \cos 5\omega_0 t + \frac{1}{7} \cos 7\omega_0 t - \frac{1}{11} \cos 11\omega_0 t + \frac{1}{13} \cos 13\omega_0 t - \cdots \right) \qquad (4\text{-}46)$$

which consists of terms at the fundamental frequency of the ac system and harmonics of order $6k \pm 1$, $k = 1, 2, 3, \ldots$.

Because these harmonic currents may present problems in the ac system, filters are frequently necessary to prevent these harmonics from entering the ac system. A typical filtering scheme is shown in Fig. 4-18. Resonant filters are used to provide a path to ground for the fifth and seventh harmonics, which are the two lowest and are the strongest in amplitude. Higher-order harmonics are reduced with the high-pass filter. These filters prevent the harmonic currents from propagating through the ac power system. Filter components are chosen such that the impedance to the power system frequency is large.

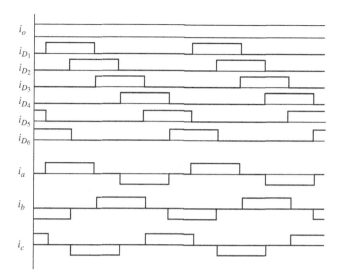

Figure 4-17 Three-phase rectifier currents when the output is filtered.

EXAMPLE 4-12

Three-Phase Rectifier

The three-phase rectifier of Fig. 4-16a has a three-phase source of 480 V rms line-to-line, and the load is a 25-Ω resistance in series with a 50-mH inductance. Determine (a) the dc level of the output voltage, (b) the dc and first ac term of the load current, (c) the average and rms current in the diodes, (d) the rms current in the source, and (e) the apparent power from the source.

■ Solution

(a) The dc output voltage of the bridge is obtained from Eq. (4-44).

$$V_o = \frac{3 V_{m,L-L}}{\pi} = \frac{3\sqrt{2}\,(480)}{\pi} = 648 \text{ V}$$

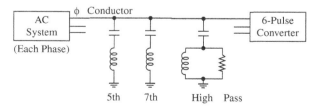

Figure 4-18 Filters for ac line harmonics.

(b) The average load current is

$$I_o = \frac{V_o}{R} = \frac{648}{25} = 25.9 \text{ A}$$

The first ac voltage term is obtained from Eq. (4-45) with $n = 6$, and current is

$$I_6 = \frac{V_6}{Z_6} = \frac{0.0546 V_m}{\sqrt{R^2 + (6\omega L)^2}} = \frac{0.0546\sqrt{2}(480)}{\sqrt{25^2 + [6(377)(0.05)]^2}} = \frac{37.0 \text{ V}}{115.8 \text{ }\Omega} = 0.32 \text{ A}$$

$$I_{6,\text{rms}} = \frac{0.32}{\sqrt{2}} = 0.23 \text{ A}$$

This and other ac terms are much smaller than the dc term and can be neglected.

(c) Average and rms diode currents are obtained from Eq. (4-41). The rms load current is approximately the same as average current since the ac terms are small.

$$I_{D,\text{avg}} = \frac{I_o}{3} = \frac{25.9}{3} = 8.63 \text{ A}$$

$$I_{D,\text{rms}} = \frac{I_{o,\text{rms}}}{\sqrt{3}} \approx \frac{25.9}{\sqrt{3}} = 15.0 \text{ A}$$

(d) The rms source current is also obtained from Eq. (4-41).

$$I_{s,\text{rms}} = \left(\sqrt{\frac{2}{3}}\right) I_{o,\text{rms}} \approx \left(\sqrt{\frac{2}{3}}\right) 25.9 = 21.2 \text{ A}$$

(e) The apparent power from the source is determined from Eq. (4-42).

$$S = \sqrt{3}(V_{L-L,\text{rms}})(I_{s,\text{rms}}) = \sqrt{3}\,(480)(21.2) = 17.6 \text{ kVA}$$

PSpice Solution

A circuit for this example is shown in Fig. 4-19a. VSIN is used for each of the sources. Dbreak, with the model changed to make $n = 0.01$, approximates an ideal diode. A transient analysis starting at 16.67 ms and ending at 50 ms represents steady-state currents.

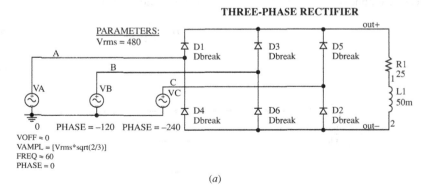

Figure 4-19 (a) PSpice circuit for a three-phase rectifier; (b) Probe output showing the current waveform and the Fourier analysis in one phase of the source.

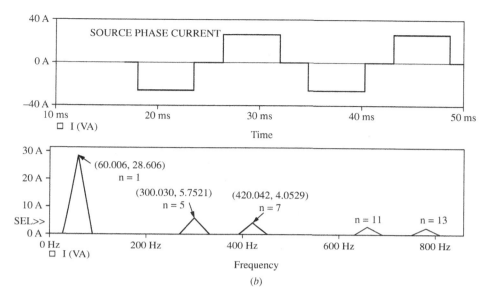

Figure 4-19 (*continued*)

All the circuit currents as calculated above can be verified. The Probe output in Fig. 4-19*b* shows the current and Fourier (FFT) components in one of the sources. Note that the harmonics correspond to those in Eq. (4-46).

4.5 CONTROLLED THREE-PHASE RECTIFIERS

The output of the three-phase rectifier can be controlled by substituting SCRs for diodes. Figure 4-20*a* shows a controlled six-pulse three-phase rectifier. With SCRs, conduction does not begin until a gate signal is applied while the SCR is forward-biased. Thus, the transition of the output voltage to the maximum instantaneous line-to-line source voltage can be delayed. The delay angle α is referenced from where the SCR would begin to conduct if it were a diode. The delay angle is the interval between when the SCR becomes forward-biased and when the gate signal is applied. Figure 4-20*b* shows the output of the controlled rectifier for a delay angle of 45°.

The average output voltage is

$$V_o = \frac{1}{\pi/3} \int_{\pi/3+\alpha}^{2\pi/3+\alpha} V_{m,L-L} \sin(\omega t)\, d(\omega t) = \frac{3V_{m,L-L}}{\pi} \cos \alpha \quad (4\text{-}47)$$

Equation (4-47) shows that the average output voltage is reduced as the delay angle α increases.

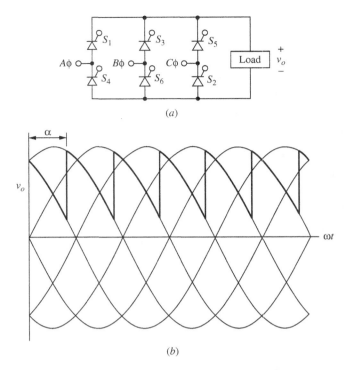

Figure 4-20 (*a*) A controlled three-phase rectifier; (*b*) Output voltage for $\alpha = 45°$.

Harmonics for the output voltage remain of order $6k$, but the amplitudes are functions of α. Figure 4-21 shows the first three normalized harmonic amplitudes.

EXAMPLE 4-13

A Controlled Three-Phase Rectifier

A three-phase controlled rectifier has an input voltage which is 480 V rms at 60 Hz. The load is modeled as a series resistance and inductance with $R = 10\ \Omega$ and $L = 50$ mH. (*a*) Determine the delay angle required to produce an average current of 50 A in the load. (*b*) Determine the amplitude of harmonics $n = 6$ and $n = 12$.

■ **Solution**

(*a*) The required dc component in the bridge output voltage is

$$V_o = I_o R = (50)(10) = 500\text{ V}$$

4.5 Controlled Three-Phase Rectifiers

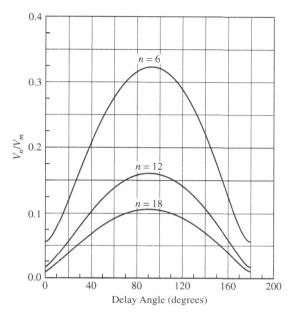

Figure 4-21 Normalized output voltage harmonics as a function of delay angle for a three-phase rectifier.

Equation (4-47) is used to determine the required delay angle:

$$\alpha = \cos^{-1}\left(\frac{V_o\pi}{3V_{m,L-L}}\right) = \cos^{-1}\left(\frac{500\pi}{3\sqrt{2}(480)}\right) = 39.5°$$

(b) Amplitudes of harmonic voltages are estimated from the graph in Fig. 4-21. For $\alpha = 39.5°$, normalized harmonic voltages are $V_6/V_m \approx 0.21$ and $V_{12}/V_m \approx 0.10$. Using $V_m = \sqrt{2}(480)$, $V_6 = 143$ V, and $V_{12} = 68$ V, harmonic currents are then

$$I_6 = \frac{V_6}{Z_6} = \frac{143}{\sqrt{10^2 + [6(377)(0.05)]^2}} = 1.26 \text{ A}$$

$$I_{12} = \frac{V_{12}}{Z_{12}} = \frac{68}{\sqrt{10^2 + [12(377)(0.05)]^2}} = 0.30 \text{ A}$$

Twelve-Pulse Rectifiers

The three-phase six-pulse bridge rectifier shows a marked improvement in the quality of the dc output over that of the single-phase rectifier. Harmonics of the output voltage are small and at frequencies that are multiples of 6 times the source frequency. Further reduction in output harmonics can be accomplished by

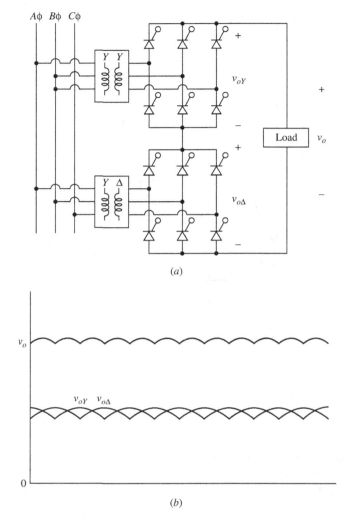

Figure 4-22 (a) A 12-pulse three-phase rectifier; (b) Output voltage for $\alpha = 0$.

using two six-pulse bridges as shown in Fig. 4-22a. This configuration is called a 12-pulse converter.

One of the bridges is supplied through a Y-Y connected transformer, and the other is supplied through a Y-Δ (or Δ-Y) transformer as shown. The purpose of the Y-Δ transformer connection is to introduce a 30° phase shift between the source and the bridge. This results in inputs to the two bridges which are 30° apart. The two bridge outputs are similar, but also shifted by 30°. The overall output voltage is the sum of the two bridge outputs. The delay angles for the bridges are typically the same. The dc output is the sum of the dc output of each bridge

$$V_o = V_{o,Y} + V_{o,\Delta} = \frac{3V_{m,L-L}}{\pi}\cos\alpha + \frac{3V_{m,L-L}}{\pi}\cos\alpha = \frac{6V_{m,L-L}}{\pi}\cos\alpha \quad (4\text{-}48)$$

The peak output of the 12-pulse converter occurs midway between alternate peaks of the 6-pulse converters. Adding the voltages at that point for $\alpha = 0$ gives

$$V_{o,\text{peak}} = 2V_{m,L-L}\cos(15°) = 1.932\, V_{m,L-L} \quad (4\text{-}49)$$

Figure 4-22b shows the voltages for $\alpha = 0$.

Since a transition between conducting thyristors occurs every 30°, there are a total of 12 such transitions for each period of the ac source. The output has harmonic frequencies that are multiples of 12 times the source frequency ($12k$, $k = 1, 2, 3, \ldots$). Filtering to produce a relatively pure dc output is less costly than that required for the 6-pulse rectifier.

Another advantage of using a 12-pulse converter rather than a 6-pulse converter is the reduced harmonics that occur in the ac system. The current in the ac lines supplying the Y-Y transformer is represented by the Fourier series

$$i_Y(t) = \frac{2\sqrt{3}}{\pi}I_o\left(\cos\omega_0 t - \frac{1}{5}\cos 5\omega_0 t + \frac{1}{7}\cos 7\omega_0 t\right.$$
$$\left. - \frac{1}{11}\cos 11\omega_0 t + \frac{1}{13}\cos 13\omega_0 t - \cdots\right) \quad (4\text{-}50)$$

The current in the ac lines supplying the Y-Δ transformer is represented by the Fourier series

$$i_\Delta(t) = \frac{2\sqrt{3}}{\pi}I_o\left(\cos\omega_0 t + \frac{1}{5}\cos 5\omega_0 t - \frac{1}{7}\cos 7\omega_0 t\right.$$
$$\left. - \frac{1}{11}\cos 11\omega_0 t + \frac{1}{13}\cos 13\omega_0 t + \cdots\right) \quad (4\text{-}51)$$

The Fourier series for the two currents are similar, but some terms have opposite algebraic signs. The ac system current, which is the sum of those transformer currents, has the Fourier series

$$i_{\text{ac}}(t) = i_Y(t) + i_\Delta(t)$$
$$= \frac{4\sqrt{3}}{\pi}I_o\left(\cos\omega_0 t - \frac{1}{11}\cos 11\omega_0 t + \frac{1}{13}\cos 13\omega_0 t \cdots\right) \quad (4\text{-}52)$$

Thus, some of the harmonics on the ac side are canceled by using the 12-pulse scheme rather than the 6-pulse scheme. The harmonics that remain in the ac system are of order $12k \pm 1$. Cancellation of harmonics $6(2n-1) \pm 1$ has resulted from this transformer and converter configuration.

This principle can be expanded to arrangements of higher pulse numbers by incorporating increased numbers of 6-pulse converters with transformers that have the appropriate phase shifts. The characteristic ac harmonics of a p-pulse converter will be $pk \pm 1$, $k = 1, 2, 3, \ldots$. Power system converters have a practical limitation of 12 pulses because of the large expense of producing high-voltage transformers with the appropriate phase shifts. However, lower-voltage industrial systems commonly have converters with up to 48 pulses.

The Three-Phase Converter Operating as an Inverter

The above discussion focused on circuits operating as rectifiers, meaning that the power flow is from the ac side of the converter to the dc side. It is also possible for the three-phase bridge to operate as an inverter, having power flow from the dc side to the ac side. A circuit that enables the converter to operate as an inverter is shown in Fig. 4-23a. Power is supplied by the dc source, and power is

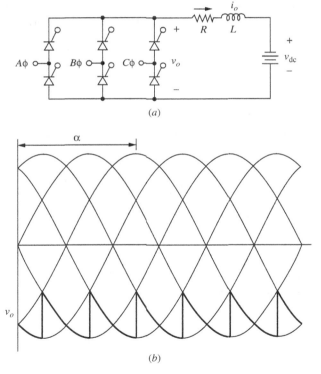

Figure 4-23 (a) Six-pulse three-phase converter operating as an inverter; (b) Bridge output voltage for $\alpha = 150°$.

4.5 Controlled Three-Phase Rectifiers

absorbed by the converter and transferred to the ac system. The analysis of the three-phase inverter is similar to that of the single-phase case.

The dc current must be in the direction shown because of the SCRs in the bridge. For power to be absorbed by the bridge and transferred to the ac system, the bridge output voltage must be negative. Equation (4-47) applies, so a delay angle larger than 90° results in a negative bridge output voltage.

$$
\begin{aligned}
0 < \alpha < 90° \quad & V_o > 0 \;\rightarrow\; \text{rectifier operation} \\
90° < \alpha < 180° \quad & V_o < 0 \;\rightarrow\; \text{inverter operation}
\end{aligned}
\quad (4\text{-}53)
$$

The output voltage waveform for $\alpha = 150°$ and continuous load current is shown in Fig. 4-23b.

EXAMPLE 4-14

Three-Phase Bridge Operating as an Inverter

The six-pulse converter of Fig. 4-23a has a delay angle $\alpha = 120°$. The three-phase ac system is 4160 V rms line-to-line. The dc source is 3000 V, $R = 2\,\Omega$, and L is large enough to consider the current to be purely dc. (a) Determine the power transferred to the ac source from the dc source. (b) Determine the value of L such that the peak-to-peak variation in load current is 10 percent of the average load current.

■ Solution

(a) The dc output voltage of the bridge is computed from Eq. (4-47) as

$$V_o = \frac{3V_{m,L-L}}{\pi}\cos\alpha = \frac{3\sqrt{2}(4160)}{\pi}\cos(120°) = -2809 \text{ V}$$

The average output current is

$$I_o = \frac{V_o + V_{dc}}{R} = \frac{-2809 + 3000}{2} = 95.5 \text{ A}$$

The power absorbed by the bridge and transferred back to the ac system is

$$P_{ac} = -I_o V_o = (-95.5)(-2809) = 268.3 \text{ kW}$$

Power supplied by the dc source is

$$P_{dc} = I_o V_{dc} = (95.5)(3000) = 286.5 \text{ kW}$$

Power absorbed by the resistance is

$$P_R = I_{rms}^2 R \approx I_o^2 R = (95.5)^2(2) = 18.2 \text{ kW}$$

(b) Variation in load current is due to the ac terms in the Fourier series. The load current amplitudes for each of the ac terms is

$$I_n = \frac{V_n}{Z_n}$$

where V_n can be estimated from the graph of Fig. 4-21 and

$$Z_n = |R + jn\omega_0 L|$$

Since the decreasing amplitude of the voltage terms and the increasing magnitude of the impedance both contribute to diminishing ac currents as n increases, the peak-to-peak current variation will be estimated from the first ac term. For $n = 6$, V_n/V_m is estimated from Fig. 4-21 as 0.28, making $V_6 = 0.28(4160\sqrt{2}) = 1650$ V. The peak-to-peak variation of 10 percent corresponds to a zero-to-peak amplitude of $(0.05)(95.5) = 4.8$ A. The required load impedance for $n = 6$ is then

$$Z_6 = \frac{V_6}{I_6} = \frac{1650 \text{ V}}{4.8 \text{ A}} = 343 \text{ }\Omega$$

The 2-Ω resistor is insignificant compared to the total 343-Ω required impedance, so $Z_6 \approx 6\omega_0 L$. Solving for L,

$$L \approx \frac{Z_6}{6\omega_0} = \frac{343}{6(377)} = 0.15 \text{ H}$$

4.6 DC POWER TRANSMISSION

The controlled 12-pulse converter of Fig. 4-22a is the basic element for dc power transmission. DC transmission lines are commonly used for transmission of electric power over very long distances. Examples include the Pacific Intertie; the Square Butte Project from Center, North Dakota, to Duluth, Minnesota; and the Cross Channel Link under the English Channel between England and France. Modern dc lines use SCRs in the converters, while very old converters used mercury-arc rectifiers.

Advantages of dc power transmission include the following:

1. The inductance of the transmission line has zero impedance to dc, whereas the inductive impedance for lines in an ac system is relatively large.
2. The capacitance that exists between conductors is an open circuit for dc. For ac transmission lines, the capacitive reactance provides a path for current, resulting in additional I^2R losses in the line. In applications where the conductors are close together, the capacitive reactance can be a significant problem for ac transmission lines, whereas it has no effect on dc lines.
3. There are two conductors required for dc transmission rather than three for conventional three-phase power transmission. (There will likely be an additional ground conductor in both dc and ac systems.)
4. Transmission towers are smaller for dc than ac because of only two conductors, and right-of-way requirements are less.
5. Power flow in a dc transmission line is controllable by adjustment of the delay angles at the terminals. In an ac system, power flow over a given transmission line is not controllable, being a function of system generation and load.

6. Power flow can be modulated during disturbances on one of the ac systems, resulting in increased system stability.
7. The two ac systems that are connected by the dc line do not need to be in synchronization. Furthermore, the two ac systems do not need to be of the same frequency. A 50-Hz system can be connected to a 60-Hz system via a dc link.

The disadvantage of dc power transmission is that a costly ac-dc converter, filters, and control system are required at each end of the line to interface with the ac system.

Figure 4-24a shows a simplified scheme for dc power transmission using six-pulse converters at each terminal. The two ac systems each have their own generators, and the purpose of the dc line is to enable power to be interchanged between the ac systems. The directions of the SCRs are such that current i_o will be positive as shown in the line.

In this scheme, one converter operates as a rectifier (power flow from ac to dc), and the other terminal operates as an inverter (power flow from dc to ac). Either terminal can operate as a rectifier or inverter, with the delay angle determining the mode of operation. By adjusting the delay angle at each terminal, power flow is controlled between the two ac systems via the dc link.

The inductance in the dc line is the line inductance plus an extra series inductor to filter harmonic currents. The resistance is that of the dc line conductors. For analysis purposes, the current in the dc line may be considered to be a ripple-free dc current.

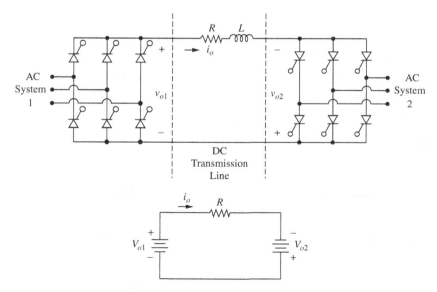

Figure 4-24 (*a*) An elementary dc transmission system; (*b*) Equivalent circuit.

Voltages at the terminals of the converters V_{o_1} and V_{o_2} are positive as shown for α between 0 and 90° and negative for α between 90 and 180°. The converter supplying power will operate with a positive voltage while the converter absorbing power will have a negative voltage.

With converter 1 in Fig. 4-24a operating as a rectifier and converter 2 operating as an inverter, the equivalent circuit for power computations is shown in Fig. 4-24b. The current is assumed to be ripple-free, enabling only the dc component of the Fourier series to be relevant. The dc current is

$$I_o = \frac{V_{o1} + V_{o2}}{R} \tag{4-54}$$

where

$$V_{o1} = \frac{3V_{m1,L-L}}{\pi} \cos \alpha_1$$
$$V_{o2} = \frac{3V_{m2,L-L}}{\pi} \cos \alpha_2 \tag{4-55}$$

Power supplied by the converter at terminal 1 is

$$P_1 = V_{o1} I_o \tag{4-56}$$

Power supplied by the converter at terminal 2 is

$$P_2 = V_{o2} I_o \tag{4-57}$$

EXAMPLE 4-15

DC Power Transmission

For the elementary dc transmission line represented in Fig. 4-24a, the ac voltage to each of the bridges is 230 kV rms line to line. The total line resistance is 10 Ω, and the inductance is large enough to consider the dc current to be ripple-free. The objective is to transmit 100 MW to ac system 2 from ac system 1 over the dc line. Design a set of operating parameters to accomplish this objective. Determine the required current-carrying capacity of the dc line, and compute the power loss in the line.

■ **Solution**

The relationships that are required are from Eqs. (4-54) to (4-57), where

$$P_2 = I_o V_{o2} = -100 \text{ MW} \qquad (100 \text{ MW absorbed})$$

The maximum dc voltage that is obtainable from each converter is, for $\alpha = 0$ in Eq. (4-47),

$$V_{o,\max} = \frac{3V_{m,L-L}}{\pi} = \frac{3\sqrt{2}\,(230 \text{ kV})}{\pi} = 310.6 \text{ kV}$$

The dc output voltages of the converters must have magnitudes less than 310.6 kV, so a voltage of -200 kV is arbitrarily selected for converter 2. This voltage must be negative because power must be absorbed at converter 2. The delay angle at converter 2 is then computed from Eq. (4-47).

$$V_{o2} = \frac{3V_{m,L-L}}{\pi} \cos \alpha_2 = (310.6 \text{ kV}) \cos \alpha_2 = -200 \text{ kV}$$

Solving for α_2,

$$\alpha_2 = \cos^{-1}\left(\frac{-200 \text{ kV}}{310.6 \text{ kV}}\right) = 130°$$

The dc current required to deliver 100 MW to converter 2 is then

$$I_o = \frac{100 \text{ MW}}{200 \text{ kV}} = 500 \text{ A}$$

which is the required current-carrying capacity of the line.

The required dc output voltage at converter 1 is computed as

$$V_{o1} = -V_{o2} + I_o R = 200 \text{ kV} + (500)(10) = 205 \text{ kV}$$

The required delay angle at converter 1 is computed from Eq. (4-47).

$$\alpha_1 = \cos^{-1} \frac{205 \text{ kV}}{310.6 \text{ kV}} = 48.7°$$

Power loss in the line is $I_{rms}^2 R$, where $I_{rms} \approx I_o$ because the ac components of line current are filtered by the inductor. Line loss is

$$P_{loss} = I_{rms}^2 R \approx (500)^2 (10) = 2.5 \text{ MW}$$

Note that the power supplied at converter 1 is

$$P_1 = V_{dc1} I_o = (205 \text{ kV})(500 \text{ A}) = 102.5 \text{ MW}$$

which is the total power absorbed by the other converter and the line resistance.

Certainly other combinations of voltages and current will meet the design objectives, as long as the dc voltages are less than the maximum possible output voltage and the line and converter equipment can carry the current. A better design might have higher voltages and a lower current to reduce power loss in the line. That is one reason for using 12-pulse converters and bipolar operation, as discussed next.

A more common dc transmission line has a 12-pulse converter at each terminal. This suppresses some of the harmonics and reduces filtering requirements. Moreover, a pair of 12-pulse converters at each terminal provides bipolar operation. One of the lines is energized at $+V_{dc}$ and the other is energized at $-V_{dc}$. In emergency situations, one pole of the line can operate without the other pole, with current returning through the ground path. Figure 4-25 shows a bipolar scheme for dc power transmission.

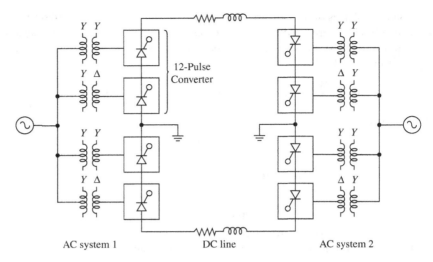

Figure 4-25 A dc transmission system with two 12-pulse converters at each terminal.

4.7 COMMUTATION: THE EFFECT OF SOURCE INDUCTANCE

Single-Phase Bridge Rectifier

An uncontrolled single-phase bridge rectifier with a source inductance L_s and an inductive load is shown in Fig. 4-26a. When the source changes polarity, source current cannot change instantaneously, and current must be transferred gradually from one diode pair to the other over a commutation interval u, as shown in Fig. 4-26b. Recall from Chap. 3 that commutation is the process of transferring the load current from one diode to another or, in this case, one diode pair to the other. (See Sec. 3.11.) During commutation, all four diodes are on, and the voltage across L_s is the source voltage $V_m \sin(\omega t)$.

Assume that the load current is a constant I_o. The current in L_s and the source during the commutation from D_1-D_2 to D_3-D_4 starts at $+I_o$ and goes to $-I_o$. This commutation interval starts when the source changes polarity at $\omega t = \pi$ as is expressed in

$$i_s(\omega t) = \frac{1}{\omega L_s} \int_{\pi}^{\omega t} V_m \sin(\omega t)\, d(\omega t) + I_o$$

Evaluating,

$$i_s(\omega t) = -\frac{V_m}{\omega L_s}(1 + \cos \omega t) + I_o \tag{4-58}$$

4.7 Commutation: The Effect of Source Inductance

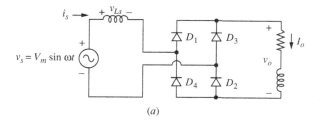

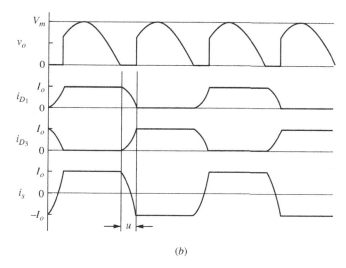

Figure 4-26 Commutation for the single-phase rectifier (a) circuit with source inductance L_s; (b) voltage and current waveforms.

When commutation is complete at $\omega t = \pi + u$,

$$i(\pi + u) = -I_o = -\frac{V_m}{\omega L_S}[1 + \cos(\pi + u)] + I_o \quad (4\text{-}59)$$

Solving for the commutation angle u,

$$u = \cos^{-1}\left(1 - \frac{2I_o \omega L_S}{V_m}\right) = \cos^{-1}\left(1 - \frac{2I_o X_S}{V_m}\right) \quad (4\text{-}60)$$

where $X_s = \omega L_s$ is the reactance of the source. Figure 4-26b shows the effect of the source reactance on the load current and voltage.

Average load voltage is

$$V_o = \frac{1}{\pi} \int_u^\pi V_m \sin(\omega t) \, d(\omega t) = \frac{V_m}{\pi}(1 + \cos u)$$

Using u from Eq. (4-60),

$$V_o = \frac{2V_m}{\pi}\left(1 - \frac{I_o X_s}{V_m}\right) \qquad (4\text{-}61)$$

Thus, source inductance lowers the average output voltage of full-wave rectifiers.

Three-Phase Rectifier

For the uncontrolled three-phase bridge rectifier with source reactance (Fig. 4-27a), assume that diodes D_1 and D_2 are conducting and the load current is a constant I_o. The next transition has load current transferred from D_1 to D_3 in the top half of the bridge. The equivalent circuit during commutation from D_1 to D_3 is shown in Fig. 4-27b. The voltage across L_a is

$$v_{La} = \frac{v_{AB}}{2} = \frac{V_{m,L-L}}{2}\sin(\omega t) \qquad (4\text{-}62)$$

Current in L_a starts at I_o and decreases to zero in the commutation interval,

$$i_{La}(\pi + u) = 0 = \frac{1}{\omega L_a}\int_\pi^{\pi+u}\frac{V_{m,L-L}}{2}\sin(\omega t)\,d(\omega t) + I_o \qquad (4\text{-}63)$$

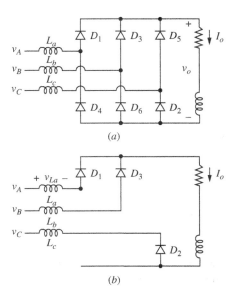

Figure 4-27 Commutation for the three-phase rectifier. (*a*) Circuit; (*b*) Circuit during commutation from D_1 to D_3; (*c*) Output voltage and diode currents.

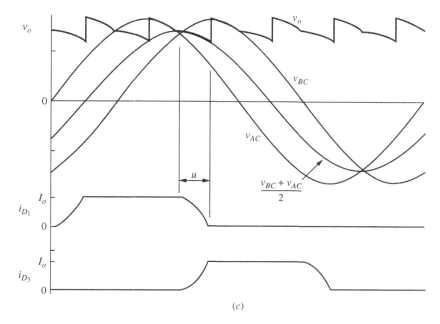

Figure 4-27 (*continued*)

Solving for u,

$$u = \cos^{-1}\left(1 - \frac{2\omega L_a I_o}{V_{m,L-L}}\right) = \cos^{-1}\left(1 - \frac{2X_s I_o}{V_{m,L-L}}\right) \quad (4\text{-}64)$$

During the commutation interval from D_1 to D_3, the converter output voltage is

$$v_o = \frac{v_{bc} + v_{ac}}{2} \quad (4\text{-}65)$$

Output voltage and diode currents are shown in Fig. 4-27c. Average output voltage for the three-phase converter with a nonideal source is

$$\boxed{V_o = \frac{3V_{m,L-L}}{\pi}\left(1 - \frac{X_s I_o}{V_{m,L-L}}\right)} \quad (4\text{-}66)$$

Therefore, *source inductance lowers the average output voltage of three-phase rectifiers*.

4.8 Summary

- Single-phase full-wave rectifiers can be of the bridge or center-tapped transformer types.
- The average source current for single-phase full-wave rectifiers is zero.
- The Fourier series method can be used to analyze load currents.

- A large inductor in series with a load resistor produces a load current that is essentially dc.
- A filter capacitor on the output of a rectifier can produce an output voltage that is nearly dc. An *LC* output filter can further improve the quality of the dc output and reduce the peak current in the diodes.
- Switches such as SCRs can be used to control the output of a single-phase or three-phase rectifier.
- Under certain circumstances, controlled converters can be operated as inverters.
- The 6-pulse three-phase rectifiers have 6 diodes or SCRs, and 12-pulse rectifiers have 12 diodes or SCRs.
- Three-phase bridge rectifiers produce an output that is inherently like dc.
- DC power transmission has a three-phase converter at each end of a dc line. One converter is operated as a rectifier and the other is operated as a converter.
- Source inductance reduces the dc output of a single-phase or three-phase rectifier.

4.9 Bibliography

S. B. Dewan and A. Straughen, *Power Semiconductor Circuits*, Wiley, New York, 1975.

J. Dixon, *Power Electronics Handbook*, edited by M. H. Rashid, Academic Press, San Diego, 2001, Chapter 12.

E. W. Kimbark, *Direct Current Transmission*, Wiley-Interscience, New York, 1971.

P. T. Krein, *Elements of Power Electronics*, Oxford University Press, 1998.

Y.-S. Lee and M. H. L. Chow, *Power Electronics Handbook*, edited by M. H. Rashid, Academic Press, San Diego, 2001, Chapter 10.

N. Mohan, T. M. Undeland, and W. P. Robbins, *Power Electronics: Converters, Applications, and Design*, 3d ed., Wiley, New York, 2003.

M. H. Rashid, *Power Electronics: Circuits, Devices, and Systems*, 3d ed., Prentice-Hall, Upper Saddle River, N.J., 2004.

B. Wu, *High-Power Converters and AC Drives*, Wiley, New York, 2006.

Problems

Uncontrolled Single-Phase Rectifiers

4-1. A single-phase full-wave bridge rectifier has a resistive load of 18 Ω and an ac source of 120-V rms. Determine the average, peak, and rms currents in the load and in each diode.

4-2. A single-phase rectifier has a resistive load of 25 Ω. Determine the average current and peak reverse voltage across each of the diodes for (*a*) a bridge rectifier with an ac source of 120 V rms and 60 Hz and (*b*) a center-tapped transformer rectifier with 120 V rms on each half of the secondary winding.

4-3. A single-phase bridge rectifier has an *RL* load with $R = 15\ \Omega$ and $L = 60$ mH. The ac source is $v_s = 100 \sin(377t)$ V. Determine the average and rms currents in the load and in each diode.

4-4. A single-phase bridge rectifier has an *RL* load with $R = 10\ \Omega$ and $L = 25$ mH. The ac source is $v_s = 170 \sin(377t)$ V. Determine the average and rms currents in the load and in each diode.

4-5. A single-phase bridge rectifier has an RL load with $R = 15\ \Omega$ and $L = 30$ mH. The ac source is 120 V rms, 60 Hz. Determine (a) the average load current, (b) the power absorbed by the load, and (c) the power factor.

4-6. A single-phase bridge rectifier has an RL load with $R = 12\ \Omega$ and $L = 20$ mH. The ac source is 120 V rms and 60 Hz. Determine (a) the average load current, (b) the power absorbed by the load, and (c) the power factor.

4-7. A single-phase center-tapped transformer rectifier has an ac source of 240 V rms and 60 Hz. The overall transformer turns ratio is 3:1 (80 V between the extreme ends of the secondary and 40 V on each tap). The load is a resistance of 4 Ω. Determine (a) the average load current, (b) the rms load current, (c) the average source current, and (d) the rms source current. Sketch the current waveforms of the load and the source.

4-8. Design a center-tapped transformer rectifier to produce an average current of 10.0 A in a 15-Ω resistive load. Both 120- and 240-V rms 60-Hz sources are available. Specify which source to use and specify the turns ratio of the transformer.

4-9. Design a center-tapped transformer rectifier to produce an average current of 5.0 A in an RL load with $R = 10\ \Omega$ and $L = 50$ mH. Both 120- and 240-V rms 60-Hz sources are available. Specify which source to use and specify the turns ratio of the transformer.

4-10. An electromagnet is modeled as a 200-mH inductance in series with a 4-Ω resistance. The average current in the inductance must be 10 A to establish the required magnetic field. Determine the amount of additional series resistance required to produce the required average current from a bridge rectifier supplied from a single-phase 120-V, 60-Hz source.

4-11. The full-wave rectifier of Fig. 4-3a has $v_s(\omega t) = 170 \sin \omega t$ V, $R = 3\ \Omega$, $L = 15$ mH, $V_{dc} = 48$ V, and $\omega = 2\pi(60)$ rad/s. Determine (a) the power absorbed by the dc source, (b) the power absorbed by the resistor, and (c) the power factor. (d) Estimate the peak-to-peak variation in the load current by considering only the first ac term in the Fourier series for current.

4-12. The full-wave rectifier of Fig. 4-3a has $v_s(\omega t) = 340 \sin \omega t$ V, $R = 5\ \Omega$, $L = 40$ mH, $V_{dc} = 96$ V, and $\omega = 2\pi(60)$ rad/s. Determine (a) the power absorbed by the dc source, (b) the power absorbed by the resistor, and (c) the power factor. (d) Estimate the peak-to-peak variation in the load current by considering only the first ac term in the Fourier series for current.

4-13. The peak-to-peak variation in load current in Example 4-1 based on I_2 was estimated to be 6.79 A. Compare this estimate with that obtained from a PSpice simulation. (a) Use the default diode model Dbreak. (b) Modify the diode model to make $n = 0.01$ to approximate an ideal diode.

4-14. (a) In Example 4-3, the inductance is changed to 8 mH. Simulate the circuit in PSpice and determine whether the inductor current is continuous or discontinuous. Determine the power absorbed by the dc voltage using PSpice. (b) Repeat part (a), using $L = 4$ mH.

4-15. The single-phase full-wave bridge rectifier of Fig. 4-5a has an RL-source load with $R = 4\ \Omega$, $L = 40$ mH, and $V_{dc} = 24$ V. The ac source is 120 V rms at 60 Hz. Determine (a) the power absorbed by the dc source, (b) the power absorbed by the resistor, and (c) the power factor.

4-16. The single-phase full-wave bridge rectifier of Fig. 4-5a has an *RL*-source load with $R = 5\ \Omega$, $L = 60$ mH, and $V_{dc} = 36$ V. The ac source is 120 V rms at 60 Hz. Determine (*a*) the power absorbed by the dc source, (*b*) the power absorbed by the resistor, and (*c*) the power factor.

4-17. Simulate the circuit of Prob. 4-16 using $L = 40$ mH and again with $L = 100$ μH. Discuss the differences in the behavior of the circuits for the two inductors. Observe steady-state conditions. Use the PSpice default diode model.

4-18. The full-wave rectifier of Fig. 4-6 has a 120-V rms 60 Hz source and a load resistance of 200 Ω. Determine the filter capacitance required to limit the peak-to-peak output voltage ripple to 1 percent of the dc output. Determine the peak and average diode currents.

4-19. The full-wave rectifier of Fig. 4-6 has a 60-Hz ac source with $V_m = 100$ V. It is to supply a load that requires a dc voltage of 100 V and will draw 0.5 A. Determine the filter capacitance required to limit the peak-to-peak output voltage ripple to 1 percent of the dc output. Determine the peak and average diode currents.

4-20. In Example 3-9, the half-wave rectifier of Fig. 3-11a has a 120 V rms source at 60 Hz, $R = 500\ \Omega$. The capacitance required for a 1 percent ripple in output voltage was determined to be 3333 μF. Determine the capacitance required for a 1 percent ripple if a full-wave rectifier is used instead. Determine the peak diode currents for each circuit. Discuss the advantages and disadvantages of each circuit.

4-21. Determine the output voltage for the full-wave rectifier with an *LC* filter of Fig. 4-8a if $L = 10$ mH and (*a*) $R = 7\ \Omega$ and (*b*) $R = 20\ \Omega$. The source is 120 V rms at 60 Hz. Assume the capacitor is sufficiently large to produce a ripple-free output voltage. (*c*) Modify the PSpice circuit in Example 4-5 to determine V_o for each case. Use the default diode model.

4-22. For the full-wave rectifier with an *LC* filter in Example 4-5, the inductor has a series resistance of 0.5 Ω. Use PSpice to determine the effect on the output voltage for each load resistance.

Controlled Single-phase Rectifiers

4-23. The controlled single-phase bridge rectifier of Fig. 4-10a has a 20-Ω resistive load and has a 120-V rms, 60-Hz ac source. The delay angle is 45°. Determine (*a*) the average load current, (*b*) the rms load current, (*c*) the rms source current, and (*d*) the power factor.

4-24. Show that the power factor for the controlled full-wave rectifier with a resistive load is

$$\text{pf} = \sqrt{1 - \frac{\alpha}{\pi} + \frac{\sin(2\alpha)}{2\pi}}$$

4-25. The controlled single-phase full-wave bridge rectifier of Fig. 4-11a has an *RL* load with $R = 25\ \Omega$ and $L = 50$ mH. The source is 240 V rms at 60 Hz. Determine the average load current for (*a*) $\alpha = 15°$ and (*b*) $\alpha = 75°$.

4-26. The controlled single-phase full-wave bridge rectifier of Fig. 4-11a has an *RL* load with $R = 30\ \Omega$ and $L = 75$ mH. The source is 120 V rms at 60 Hz. Determine the average load current for (*a*) $\alpha = 20°$ and (*b*) $\alpha = 80°$.

4-27. Show that the power factor for the full-wave rectifier with RL load where L is large and the load current is considered dc is $2\sqrt{2}/\pi$.

4-28. A 20-Ω resistive load requires an average current that varies from 4.5 to 8.0 A. An isolation transformer is placed between a 120-V rms 60-Hz ac source and a controlled single-phase full-wave rectifier. Design a circuit to meet the current requirements. Specify the transformer turns ratio and the range of delay angle.

4-29. An electromagnet is modeled as a 100-mH inductance in series with a 5-Ω resistance. The average current in the inductance must be 10 A to establish the required magnetic field. Determine the delay angle required for a controlled single-phase rectifier to produce the required average current from a single-phase 120-V, 60-Hz source. Determine if the current is continuous or discontinuous. Estimate the peak-to-peak variation in current based on the first ac term in the Fourier series.

4-30. The full-wave converter used as an inverter in Fig. 4-14 has an ac source of 240 V rms at 60 Hz, $R = 10\ \Omega$, $L = 0.8$ H, and $V_{dc} = -100$ V. The delay angle for the converter is 105°. Determine the power supplied to the ac system from the dc source. Estimate the peak-to-peak ripple in load current from the first ac term in the Fourier series.

4-31. An array of solar cells produces 100 V dc. A single-phase ac power system is 120 V rms at 60 Hz. (a) Determine the delay angle for the controlled converter in the arrangement of Fig. 4-14 ($V_{dc} = -100$) such that 2000 W is transmitted to the ac system. Assume L is large enough to produce a current that is nearly ripple-free. The equivalent resistance is 0.8 Ω. Assume that the converter is lossless. (b) Determine the power supplied by the solar cells. (c) Estimate the value of inductance such that the peak-to-peak variation in solar cell current is less than 2.5 A.

4-32. An array of solar panels produces a dc voltage. Power produced by the solar panels is to be delivered to an ac power system. The method of interfacing the solar panels with the power system is via a full-wave SCR bridge as shown in Fig. 4-14 except with the dc source having the opposite polarity. Individual solar panels produce a voltage of 12 V. Therefore, the voltage from the solar panel array can be established at any multiple of 12 by connecting the panels in appropriate combinations. The ac source is $\sqrt{2}(120) \sin(377t)$ V. The resistance is 1 Ω. Determine values of V_{dc}, delay angle α, and inductance L such that the power delivered to the ac system is 1000 W and the maximum variation in solar panel current is no more than 10 percent of the average current. There are several solutions to this problem.

4-33. A full-wave converter operating as an inverter is used to transfer power from a wind generator to a single-phase 240-V rms 60-Hz ac system. The generator produces a dc output of 150 V and is rated at 5000 W. The equivalent resistance in the generator circuit is 0.6 Ω. Determine (a) the converter delay angle for rated generator output power, (b) the power absorbed by the ac system, and (c) the inductance required to limit the current peak-to-peak ripple to 10 percent of the average current.

Three-phase Uncontrolled Rectifiers

4-34. A three-phase rectifier is supplied by a 480-V rms line-to-line 60-Hz source. The load is a 50-Ω resistor. Determine (a) the average load current, (b) the rms load current, (c) the rms source current, and (d) the power factor.

168 CHAPTER 4 Full-Wave Rectifiers

4-35. A three-phase rectifier is supplied by a 240-V rms line-to-line 60-Hz source. The load is an 80-Ω resistor. Determine (a) the average load current, (b) the rms load current, (c) the rms source current, and (d) the power factor.

4-36. A three-phase rectifier is supplied by a 480-V rms line-to-line 60-Hz source. The RL load is a 100-Ω resistor in series with a 15-mH inductor. Determine (a) the average and rms load currents, (b) the average and rms diode currents, (c) the rms source current, and (d) the power factor.

4-37. Use PSpice to simulate the three-phase rectifier of Prob. 4-31. Use the default diode model Dbreak. Determine the average and rms values of load current, diode current, and source current. Compare your results to Eq. (4-41). How much power is absorbed by the diodes?

4-38. Using the PSpice circuit of Example 4-12, determine the harmonic content of the line current in the ac source. Compare the results with Eq. (4-46). Determine the total harmonic distortion of the source current.

Three-phase Controlled Rectifiers

4-39. The three-phase controlled rectifier of Fig. 4-20a is supplied from a 4160-V rms line-to-line 60-Hz source. The load is a 120-Ω resistor. (a) Determine the delay angle required to produce an average load current of 25 A. (b) Estimate the amplitudes of the voltage harmonics V_6, V_{12}, and V_{18}. (c) Sketch the currents in the load, S_1, S_4, and phase A of the ac source.

4-40. The three-phase controlled rectifier of Fig. 4-20a is supplied from a 480-V rms line-to-line 60-Hz source. The load is a 50-Ω resistor. (a) Determine the delay angle required to produce an average load current of 10 A. (b) Estimate the amplitudes of the voltage harmonics V_6, V_{12}, and V_{18}. (c) Sketch the currents in the load, S_1, S_4, and phase A of the ac source.

4-41. The six-pulse controlled three-phase converter of Fig. 4-20a is supplied from a 480-V rms line-to-line 60-Hz three-phase source. The delay angle is 35°, and the load is a series RL combination with $R = 50 \, \Omega$ and $L = 50$ mH. Determine (a) the average current in the load, (b) the amplitude of the sixth harmonic current, and (c) the rms current in each line from the ac source.

4-42. The six-pulse controlled three-phase converter of Fig. 4-20a is supplied from a 480-V rms line-to-line 60-Hz three-phase source. The delay angle is 50°, and the load is a series RL combination with $R = 10 \, \Omega$ and $L = 10$ mH. Determine (a) the average current in the load, (b) the amplitude of the sixth harmonic current, and (c) the rms current in each line from the ac source.

4-43. The six-pulse controlled three-phase converter of Fig. 4-20a is supplied form a 480-V rms line-to-line 60-Hz three-phase source. The load is a series RL combination with $R = 20 \, \Omega$. (a) Determine the delay angle required for an average load current of 20 A. (b) Determine the value of L such that the first ac current term ($n = 6$) is less than 2 percent of the average current. (c) Verify your results with a PSpice simulation.

4-44. A three-phase converter is operating as an inverter and is connected to a 300-V dc source as shown in Fig. 4-23a. The ac source is 240 V rms line to line at 60 Hz. The resistance is 0.5 Ω, and the inductor is large enough to consider the load current to be ripple-free. (a) Determine the delay angle α such that the output voltage of the converter is $V_o = -280$ V. (*b*) Determine the power supplied or absorbed by each component in the circuit. The SCRs are assumed to be ideal.

4-45. An inductor having superconducting windings is used to store energy. The controlled six-pulse three-phase converter of Fig. 4-20*a* is used to recover the stored energy and transfer it to a three-phase ac system. Model the inductor as a 1000-A current source load, and determine the required delay angle such that 1.5 MW is transferred to the ac system which is 4160 V line-to-line rms at 60 Hz. What is the rms current in each phase of the ac system?

4-46. A power company has installed an array of solar cells to be used as an energy source. The array produces a dc voltage of 1000 V and has an equivalent series resistance of 0.1 Ω. The peak-to-peak variation in solar cell current should not exceed 5 percent of the average current. The interface between the solar cell array and the ac system is the controlled six-pulse three-phase converter of Fig. 4-23*a*. A three-phase transformer is placed between the converter and a 12.5-kV line-to-line rms 60-Hz ac line. Design a system to transfer 100 kW to the ac power system from the solar cell array. (The ac system must absorb 100 kW.) Specify the transformer turns ratio, converter delay angle, and the values of any other circuit components. Determine the power loss in the resistance.

Dc Power Transmission

4-47. For the elementary dc transmission line represented in Fig. 4-24*a*, the ac voltage to each of the bridges is 345 kV rms line to line. The total line resistance is 15 Ω, and the inductance is large enough to consider the dc current to be ripple-free. AC system 1 is operated with $\alpha = 45.0°$, and ac system 2 has $\alpha = 134.4°$. (*a*) Determine the power absorbed or supplied by each ac system. (*b*) Determine the power loss in the line.

4-48. For the elementary dc transmission line represented in Fig. 4-24*a*, the ac voltage to each of the bridges is 230 kV rms line to line. The total line resistance is 12 Ω, and the inductance is large enough to consider the dc current to be ripple-free. The objective is to transmit 80 MW to ac system 2 from ac system 1 over the dc line. Design a set of operating parameters to accomplish this objective. Determine the required current-carrying capacity of the dc line, and compute the power loss in the line.

4-49. For the elementary dc transmission line represented in Fig. 4-24*a*, the ac voltage to each of the bridges is 345 kV rms line-to-line. The total line resistance is 20 Ω, and the inductance is large enough to consider the dc current to be ripple-free. The objective is to transmit 300 MW to ac system 2 from ac system 1 over the dc line. Design a set of operating parameters to accomplish this objective. Determine the required current-carrying capacity of the dc line, and compute the power loss in the line.

Design Problems

4-50. Design a circuit that will produce an average current that is to vary from 8 to 12 A in an 8-Ω resistor. Single-phase ac sources of 120 and 240 V rms at 60 Hz are available. The current must have a peak-to-peak variation of no more than 2.5 A. Determine the average and rms currents and maximum voltage for each circuit element. Simulate your circuit in PSpice to verify that it meets the specifications. Give alternative circuits that could be used to satisfy the design specifications, and give reasons for your selection.

4-51. Design a circuit that will produce a current which has an average value of 15 A in a resistive load of 20 Ω. The peak-to-peak variation in load current must be no more than 10 percent of the dc current. Voltage sources available are a single-phase 480 V rms, 60 Hz source and a three-phase 480 V rms line-to-line 60 Hz source. You may include additional elements in the circuit. Determine the average, rms, and peak currents in each circuit element. Simulate your circuit in PSpice to verify that it meets the specifications. Give alternative circuits that could be used to satisfy the design specifications, and give reasons for your selection.

CHAPTER 5

AC Voltage Controllers
AC to ac Converters

5.1 INTRODUCTION

An ac voltage controller is a converter that controls the voltage, current, and average power delivered to an ac load from an ac source. Electronic switches connect and disconnect the source and the load at regular intervals. In a switching scheme called phase control, switching takes place during every cycle of the source, in effect removing some of the source waveform before it reaches the load. Another type of control is integral-cycle control, whereby the source is connected and disconnected for several cycles at a time.

The phase-controlled ac voltage controller has several practical uses including light-dimmer circuits and speed control of induction motors. The input voltage source is ac, and the output is ac (although not sinusoidal), so the circuit is classified as an ac-ac converter.

5.2 THE SINGLE-PHASE AC VOLTAGE CONTROLLER

Basic Operation

A basic single-phase voltage controller is shown in Fig. 5-1a. The electronic switches are shown as parallel thyristors (SCRs). This SCR arrangement makes it possible to have current in either direction in the load. This SCR connection is called antiparallel or inverse parallel because the SCRs carry current in opposite directions. A triac is equivalent to the antiparallel SCRs. Other controlled switching devices can be used instead of SCRs.

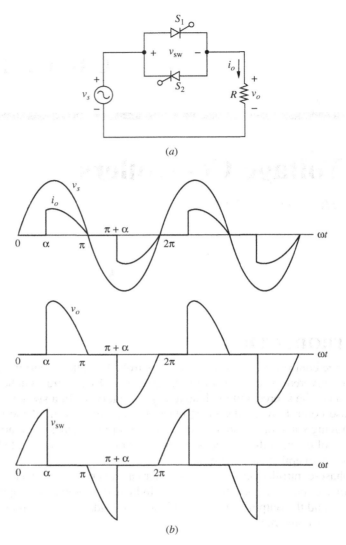

Figure 5-1 (*a*) Single-phase ac voltage controller with a resistive load; (*b*) Waveforms.

The principle of operation for a single-phase ac voltage controller using phase control is quite similar to that of the controlled half-wave rectifier of Sec. 3.9. Here, load current contains both positive and negative half-cycles. An analysis identical to that done for the controlled half-wave rectifier can be done on a half-cycle for the voltage controller. Then, by symmetry, the result can be extrapolated to describe the operation for the entire period.

Some basic observations about the circuit of Fig. 5-1a are as follows:

1. The SCRs cannot conduct simultaneously.
2. The load voltage is the same as the source voltage when either SCR is on. The load voltage is zero when both SCRs are off.
3. The switch voltage v_{sw} is zero when either SCR is on and is equal to the source voltage when neither is on.
4. The average current in the source and load is zero if the SCRs are on for equal time intervals. The average current in each SCR is not zero because of unidirectional SCR current.
5. The rms current in each SCR is $1/\sqrt{2}$ times the rms load current if the SCRs are on for equal time intervals. (Refer to Chap. 2.)

For the circuit of Fig. 5-1a, S_1 conducts if a gate signal is applied during the positive half-cycle of the source. Just as in the case of the SCR in the controlled half-wave rectifier, S_1 conducts until the current in it reaches zero. Where this circuit differs from the controlled half-wave rectifier is when the source is in its negative half-cycle. A gate signal is applied to S_2 during the negative half-cycle of the source, providing a path for negative load current. If the gate signal for S_2 is a half period later than that of S_1, analysis for the negative half-cycle is identical to that for the positive half, except for algebraic sign for the voltage and current.

Single-Phase Controller with a Resistive Load

Figure 5-1b shows the voltage waveforms for a single-phase phase-controlled voltage controller with a resistive load. These are the types of waveforms that exist in a common incandescent light-dimmer circuit. Let the source voltage be

$$v_s(\omega t) = V_m \sin \omega t \qquad (5\text{-}1)$$

Output voltage is

$$v_o(\omega t) = \begin{cases} V_m \sin \omega t & \text{for } \alpha < \omega t < \pi \text{ and } \alpha + \pi < \omega t < 2\pi \\ 0 & \text{otherwise} \end{cases} \qquad (5\text{-}2)$$

The rms load voltage is determined by taking advantage of positive and negative symmetry of the voltage waveform, necessitating evaluation of only a half-period of the waveform:

$$\boxed{V_{o,\text{rms}} = \sqrt{\frac{1}{\pi} \int_\alpha^\pi [V_m \sin (\omega t)]^2 d(\omega t)} = \frac{V_m}{\sqrt{2}} \sqrt{1 - \frac{\alpha}{\pi} + \frac{\sin (2\alpha)}{2\pi}}} \qquad (5\text{-}3)$$

Note that for $\alpha = 0$, the load voltage is a sinusoid that has the same rms value as the source. Normalized rms load voltage is plotted as a function of α in Fig. 5-2.

The rms current in the load and the source is

$$I_{o,\text{rms}} = \frac{V_{o,\text{rms}}}{R} \qquad (5\text{-}4)$$

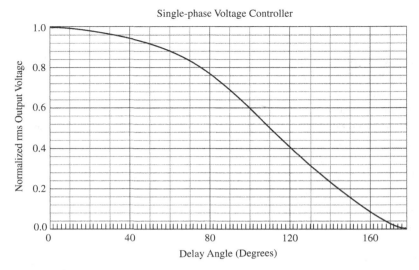

Figure 5-2 Normalized rms load voltage vs. delay angle for a single-phase ac voltage controller with a resistive load.

and the power factor of the load is

$$\text{pf} = \frac{P}{S} = \frac{P}{V_{s,\text{rms}} I_{s,\text{rms}}} = \frac{V_{o,\text{rms}}^2/R}{V_{s,\text{rms}}(V_{o,\text{rms}}/R)} = \frac{V_{o,\text{rms}}}{V_{s,\text{rms}}}$$

$$= \frac{\frac{V_m}{\sqrt{2}}\sqrt{1 - \frac{\alpha}{\pi} + \frac{(\sin 2\alpha)}{2\pi}}}{V_m/\sqrt{2}}$$

$$\boxed{\text{pf} = \sqrt{1 - \frac{\alpha}{\pi} + \frac{\sin(2\alpha)}{2\pi}}} \quad (5\text{-}5)$$

Note that pf = 1 for $\alpha = 0$, which is the same as for an uncontrolled resistive load, and the power factor for $\alpha > 0$ is less than 1.

The average source current is zero because of half-wave symmetry. The average SCR current is

$$I_{\text{SCR,avg}} = \frac{1}{2\pi}\int_{\alpha}^{\pi} \frac{V_m \sin(\omega t)}{R} d(\omega t) = \frac{V_m}{2\pi R}(1 + \cos \alpha) \quad (5\text{-}6)$$

Since each SCR carries one-half of the line current, the rms current in each SCR is

$$I_{\text{SCR,rms}} = \frac{I_{o,\text{rms}}}{\sqrt{2}} \quad (5\text{-}7)$$

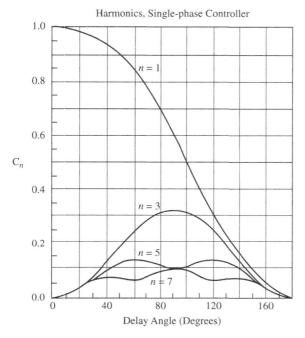

Figure 5-3 Normalized harmonic content vs. delay angle for a single-phase ac voltage controller with a resistive load; C_n is the normalized amplitude. (See Chap. 2.)

Since the source and load current is nonsinusoidal, harmonic distortion is a consideration when designing and applying ac voltage controllers. Only odd harmonics exist in the line current because the waveform has half-wave symmetry. Harmonic currents are derived from the defining Fourier equations in Chap. 2. Normalized harmonic content of the line currents vs. α is shown in Fig. 5-3. Base current is source voltage divided by resistance, which is the current for $\alpha = 0$.

EXAMPLE 5-1

Single-Phase Controller with a Resistive Load

The single-phase ac voltage controller of Fig. 5-1a has a 120-V rms 60-Hz source. The load resistance is 15 Ω. Determine (*a*) the delay angle required to deliver 500 W to the load, (*b*) the rms source current, (*c*) the rms and average currents in the SCRs, (*d*) the power factor, and (*e*) the total harmonic distortion (THD) of the source current.

■ **Solution**

(*a*) The required rms voltage to deliver 500 W to a 15-Ω load is

$$P = \frac{V_{o,\text{rms}}^2}{R}$$

$$V_{o,\text{rms}} = \sqrt{PR} = \sqrt{(500)(15)} = 86.6 \text{ V}$$

The relationship between output voltage and delay angle is described by Eq. (5-3) and Fig. 5-2. From Fig. 5-2, the delay angle required to obtain a normalized output of $86.6/120 = 0.72$ is approximately 90°. A more precise solution is obtained from the numerical solution for α in Eq. (5-3), expressed as

$$86.6 - 120\sqrt{1 - \frac{\alpha}{\pi} + \frac{\sin(2\alpha)}{2\pi}} = 0$$

which yields

$$\alpha = 1.54 \text{ rad} = 88.1°$$

(b) Source rms current is

$$I_{o,\text{rms}} = \frac{V_{o,\text{rms}}}{R} = \frac{86.6}{15} = 5.77 \text{ A}$$

(c) SCR currents are determined from Eqs. (5-6) and (5-7),

$$I_{\text{SCR,rms}} = \frac{I_{\text{rms}}}{\sqrt{2}} = \frac{5.77}{\sqrt{2}} = 4.08 \text{ A}$$

$$I_{\text{SCR,avg}} = \frac{\sqrt{2}(120)}{2\pi(15)}\left[1 + \cos(88.1°)\right] = 1.86 \text{ A}$$

(d) The power factor is

$$\text{pf} = \frac{P}{S} = \frac{500}{(120)(5.77)} = 0.72$$

which could also be computed from Eq. (5-5).

(e) Base rms current is

$$I_{\text{base}} = \frac{V_{s,\text{rms}}}{R} = \frac{120}{15} = 8.0 \text{ A}$$

The rms value of the current's fundamental frequency is determined from C_1 in the graph of Fig. 5-3.

$$C_1 \approx 0.61 \Rightarrow I_{1,\text{rms}} = C_1 I_{\text{base}} = (0.61)(8.0) = 4.9 \text{ A}$$

The THD is computed from Eq. (2-68),

$$\text{THD} = \frac{\sqrt{I_{\text{rms}}^2 - I_{1,\text{rms}}^2}}{I_{1,\text{rms}}} = \frac{\sqrt{5.77^2 - 4.9^2}}{4.9} = 0.63 = 63\%$$

Single-Phase Controller with an *RL* Load

Figure 5-4*a* shows a single-phase ac voltage controller with an *RL* load. When a gate signal is applied to S_1 at $\omega t = \alpha$, Kirchhoff's voltage law for the circuit is expressed as

$$V_m \sin(\omega t) = Ri_o(t) + L\frac{di_o(t)}{dt} \tag{5-8}$$

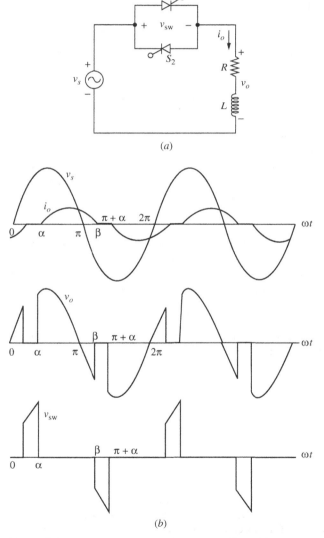

Figure 5-4 (*a*) Single-phase ac voltage controller with an *RL* load; (*b*) Typical waveforms.

The solution for current in this equation, outlined in Sec. 3.9, is

$$i_o(\omega t) = \begin{cases} \dfrac{V_m}{Z}\left[\sin(\omega t - \theta) - \sin(\alpha - \theta)e^{(\alpha - \omega t)/\omega \tau}\right] & \text{for } \alpha \leq \omega t \leq \beta \\ 0 & \text{otherwise} \end{cases}$$

where (5-9)

$$Z = \sqrt{R^2 + (\omega L)^2}, \quad \text{and} \quad \theta = \tan^{-1}\left(\dfrac{\omega L}{R}\right)$$

The extinction angle β is the angle at which the current returns to zero, when $\omega t = \beta$,

$$i_o(\beta) = 0 = \dfrac{V_m}{Z}\left[\sin(\beta - \theta) - \sin(\alpha - \theta)e^{(\alpha - \beta)/\omega \tau}\right] \tag{5-10}$$

which must be solved numerically for β.

A gate signal is applied to S_2 at $\omega t = \pi + \alpha$, and the load current is negative but has a form identical to that of the positive half-cycle. Figure 5-4b shows typical waveforms for a single-phase ac voltage controller with an RL load.

The conduction angle γ is defined as

$$\gamma = \beta - \alpha \tag{5-11}$$

In the interval between π and β when the source voltage is negative and the load current is still positive, S_2 cannot be turned on because it is not forward-biased. The gate signal to S_2 must be delayed at least until the current in S_1 reaches zero, at $\omega t = \beta$. The delay angle is therefore at least $\beta - \pi$.

$$\alpha \geq \beta - \pi \tag{5-12}$$

The limiting condition when $\beta - \alpha = \pi$ is determined from an examination of Eq. (5-10). When $\alpha = \theta$, Eq. (5-10) becomes

$$\sin(\beta - \alpha) = 0$$

which has a solution

$$\beta - \alpha = \pi$$

Therefore,

$$\gamma = \pi \quad \text{when} \quad \alpha = \theta \tag{5-13}$$

If $\alpha < \theta$, $\gamma = \pi$, provided that the gate signal is maintained beyond $\omega t = \theta$.

In the limit, when $\gamma = \pi$, one SCR is always conducting, and the voltage across the load is the same as the voltage of the source. The load voltage and current are sinusoids for this case, and the circuit is analyzed using phasor analysis for ac circuits. *The power delivered to the load is continuously controllable between the two extremes corresponding to full source voltage and zero.*

5.2 The Single-Phase AC Voltage Controller

This SCR combination can act as a *solid-state relay*, connecting or disconnecting the load from the ac source by gate control of the SCRs. The load is disconnected from the source when no gate signal is applied, and the load has the same voltage as the source when a gate signal is continuously applied. In practice, the gate signal may be a high-frequency series of pulses rather than a continuous dc signal.

An expression for rms load current is determined by recognizing that the square of the current waveform repeats every π rad. Using the definition of rms,

$$I_{o,\text{rms}} = \sqrt{\frac{1}{\pi}\int_{\alpha}^{\beta} i_o^2(\omega t)\, d(\omega t)} \quad (5\text{-}14)$$

where $i_o(\omega t)$ is described in Eq. (5-9).

Power absorbed by the load is determined from

$$P = I_{o,\text{rms}}^2 R \quad (5\text{-}15)$$

The rms current in each SCR is

$$I_{\text{SCR,rms}} = \frac{I_{o,\text{rms}}}{\sqrt{2}} \quad (5\text{-}16)$$

The average load current is zero, but each SCR carries one-half of the current waveform, making the average SCR current

$$I_{\text{SCR,avg}} = \frac{1}{2\pi}\int_{\alpha}^{\beta} i_o(\omega t)\, d(\omega t) \quad (5\text{-}17)$$

EXAMPLE 5-2

Single-Phase Voltage Controller with RL Load

For the single-phase voltage controller of Fig. 5-4a, the source is 120 V rms at 60 Hz, and the load is a series RL combination with $R = 20\ \Omega$ and $L = 50$ mH. The delay angle α is 90°. Determine (*a*) an expression for load current for the first half-period, (*b*) the rms load current, (*c*) the rms SCR current, (*d*) the average SCR current, (*e*) the power delivered to the load, and (*f*) the power factor.

■ **Solution**

(*a*) The current is expressed as in Eq. (5-9). From the parameters given,

$$Z = \sqrt{R^2 + (\omega L)^2} = \sqrt{(20)^2 + [(377)(0.05)]^2} = 27.5\ \Omega$$

$$\theta = \tan^{-1}\left(\frac{\omega L}{R}\right) = \tan^{-1}\frac{(377)(0.05)}{20} = 0.756\ \text{rad}$$

$$\omega\tau = \omega\left(\frac{L}{R}\right) = 377\left(\frac{0.05}{20}\right) = 0.943\ \text{rad}$$

$$\frac{V_m}{Z} = \frac{120\sqrt{2}}{27.5} = 6.18\ \text{A}$$

$$\alpha = 90° = 1.57\ \text{rad}$$

$$\frac{V_m}{Z} \sin(\alpha - \theta) e^{\alpha/\omega\tau} = 23.8 \text{ A}$$

The current is then expressed in Eq. (5-9) as

$$i_o(\omega t) = 6.18 \sin(\omega t - 0.756) - 23.8 e^{-\omega t/0.943} \text{ A} \quad \text{for } \alpha \leq \omega t \leq \beta$$

The extinction angle β is determined from the numerical solution of $i(\beta) = 0$ in the above equation, yielding

$$\beta = 3.83 \text{ rad} = 220°$$

Note that the conduction angle $\gamma = \beta - \alpha = 2.26$ rad $= 130°$, which is less than the limit of $180°$.

(b) The rms load current is determined from Eq. (5-14).

$$I_{o,\text{rms}} = \sqrt{\frac{1}{\pi} \int_{1.57}^{3.83} \left[6.18 \sin(\omega t - 0.756) - 23.8 e^{-\omega t/0.943}\right] d(\omega t)} = 2.71 \text{ A}$$

(c) The rms current in each SCR is determined from Eq. (5-16).

$$I_{\text{SCR,rms}} = \frac{I_{o,\text{rms}}}{\sqrt{2}} = \frac{2.71}{\sqrt{2}} = 1.92 \text{ A}$$

(d) Average SCR current is obtained from Eq. (5-17).

$$I_{\text{SCR,avg}} = \frac{1}{2\pi} \int_{1.57}^{3.83} \left[6.18 \sin(\omega t - 0.756) - 23.8 e^{-\omega t/0.943}\right] d(\omega t) = 1.04 \text{ A}$$

(e) Power absorbed by the load is

$$P = I_{o,\text{rms}}^2 R = (2.71)^2 (20) = 147 \text{ W}$$

(f) Power factor is determined from P/S.

$$\text{pf} = \frac{P}{S} = \frac{P}{V_{s,\text{rms}} I_{s,\text{rms}}} = \frac{147}{(120)(2.71)} = 0.45 = 45\%$$

PSpice Simulation of Single-Phase AC Voltage Controllers

The PSpice simulation of single-phase voltage controllers is very similar to the simulation of the controlled half-wave rectifier. The SCR is modeled with a diode and voltage-controlled switch. The diodes limit the currents to positive values, thus duplicating SCR behavior. The two switches are complementary, each closed for one-half the period.

The Schematic Capture circuit requires the full version, whereas the text CIR file will run on the PSpice A/D Demo version.

EXAMPLE 5-3

PSpice Simulation of a Single-Phase Voltage Controller

Use PSpice to simulate the circuit of Example 5-2. Determine the rms load current, the rms and average SCR currents, load power, and total harmonic distortion in the source current. Use the default diode model in the SCR.

■ Solution

The circuit for the simulation is shown in Fig. 5-5. This requires the full version of Schematic Capture.

The PSpice circuit file for the A/D Demo version is as follows:

```
SINGLE-PHASE VOLTAGE CONTROLLER (voltcont.cir)
*** OUTPUT VOLTAGE IS V(3), OUTPUT CURRENT IS I(R) ***
*************** INPUT PARAMETERS *******************
.PARAM VS = 120                 ;source rms voltage
.PARAM ALPHA = 90               ;delay angle in degrees
.PARAM R = 20                   ;load resistance
.PARAM L = 50mH                 ;load inductance
.PARAM F = 60                   ;frequency
.PARAM TALPHA = {ALPHA/(360*F)} PW 5 {0.5/F} ;converts angle to time delay

**************** CIRCUIT DESCRIPTION ********************
VS 1 0 SIN(0 {VS*SQRT(2)} {F})
S1 1 2 11 0 SMOD
D1 2 3 DMOD                     ; FORWARD SCR
S2 3 5 0 11 SMOD
```

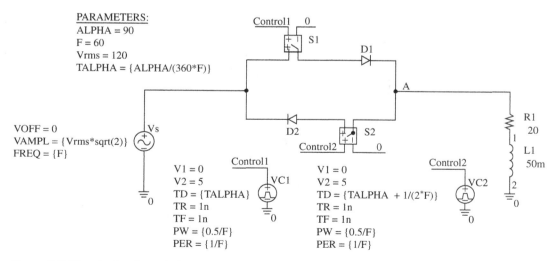

Figure 5-5 The circuit schematic for a single-phase ac voltage controller. The full version of Schematic Capture is required for this circuit.

```
D2 5 1 DMOD                         ; REVERSE SCR
R 3 4 {R}
L 4 0 {L}

*************** MODELS AND COMMANDS ********************
.MODEL DMOD D
.MODEL SMOD VSWITCH (RON=.01)
VCONTROL 11 0 PULSE(-10 10 {TALPHA} 0 0 {PW} {1/F})   ;control for both
switches
.TRAN .1MS 33.33MS 16.67MS .1MS UIC     ;one period of output
.FOUR 60 I(R)                           ;Fourier Analysis to get THD
.PROBE
.END
```

Using the PSpice A/D input file for the simulation, the Probe output of load current and related quantities is shown in Fig. 5-6. From Probe, the following quantities are obtained:

Quantity	Expression	Result
RMS load current	RMS(I(R))	2.59 A
RMS SCR current	RMS(I(S1))	1.87 A
Average SCR current	AVG(I(S1))	1.01 A
Load power	AVG(W(R))	134 W
Total harmonic distortion	(from the output file)	31.7%

Note that the nonideal SCRs (using the default diode) result in smaller currents and load power than for the analysis in Example 5-2 which assumed ideal SCRs. A model for the particular SCR that will be used to implement the circuit will give a more accurate prediction of actual circuit performance.

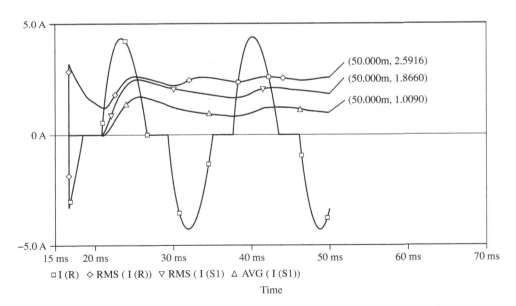

Figure 5-6 Probe output for Example 5-3.

5.3 THREE-PHASE VOLTAGE CONTROLLERS

Y-Connected Resistive Load

A three-phase voltage controller with a Y-connected resistive load is shown in Fig. 5-7a. The power delivered to the load is controlled by the delay angle α on each SCR. The six SCRs are turned on in the sequence 1-2-3-4-5-6, at 60° intervals. Gate signals are maintained throughout the possible conduction angle.

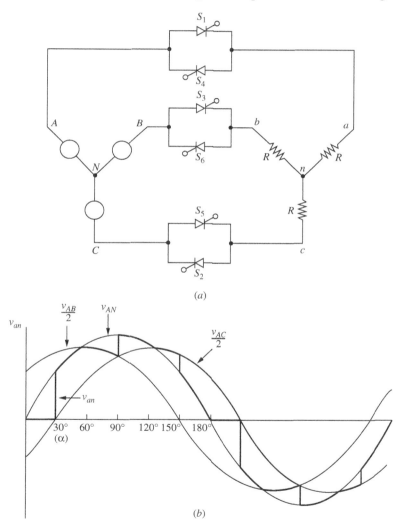

Figure 5-7 (a) Three-phase ac voltage controller with a Y-connected resistive load; (b) Load voltage v_{an} for α = 30°; (c) Load voltages and switch currents for a three-phase resistive load for α = 30°; (d) Load voltage v_{an} for α = 75°; (e) Load voltage v_{an} for α = 120°.

184 CHAPTER 5 AC Voltage Controllers

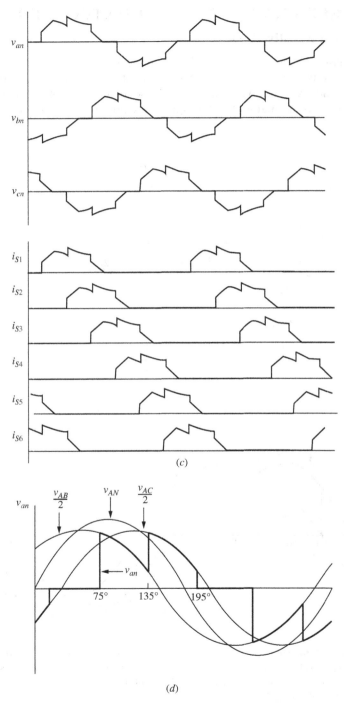

Figure 5-7 (continued)

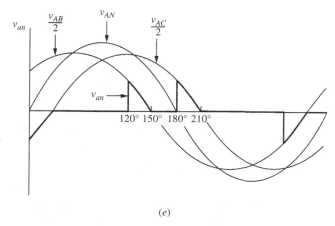

(e)

Figure 5-7 (*continued*)

The instantaneous voltage across each phase of the load is determined by which SCRs are conducting. At any instant, three SCRs, two SCRs, or no SCRs are on. The instantaneous load voltages are either a line-to-neutral voltage (three on), one-half of a line-to-line voltage (two on), or zero (none on).

When three SCRs are on (one in each phase), all three phase voltages are connected to the source, corresponding to a balanced three-phase source connected to a balanced three-phase load. The voltage across each phase of the load is the corresponding line-to-neutral voltage. For example, if S_1, S_2, and S_6 are on, $v_{an} = v_{AN}$, $v_{bn} = v_{BN}$, and $v_{cn} = v_{CN}$. When two SCRs are on, the line-to-line voltage of those two phases is equally divided between the two load resistors that are connected. For example, if only S_1 and S_2 are on, $v_{an} = v_{AC}/2$, $v_{cn} = v_{CA}/2$, and $v_{bn} = 0$.

Which SCRs are conducting depends on the delay angle α and on the source voltages at a particular instant. The following are the ranges of α that produce particular types of load voltages with an example for each:

For $0 < \alpha < 60°$:

Two or three SCRs conduct at any one time for this range of α. Figure 5-7b shows the load line-to-neutral voltage v_{an} for $\alpha = 30°$. At $\omega t = 0$, S_5 and S_6 are conducting and there is no current in R_a, making $v_{an} = 0$. At $\omega t = \pi/6$ (30°), S_1 receives a gate signal and begins to conduct; S_5 and S_6 remain on, and $v_{an} = v_{AN}$. The current in S_5 reaches zero at 60°, turning S_5 off. With S_1 and S_6 remaining on, $v_{an} = v_{AB}/2$. At 90°, S_2 is turned on; the three SCRs S_1, S_2, and S_6 are then on; and $v_{an} = v_{AN}$. At 120°, S_6 turns off, leaving S_1 and S_2 on, so $v_{an} = v_{AC}/2$. As the firing sequence for the SCRs proceeds, the number of SCRs on at a particular instant alternates between 2 and 3. All three phase-to-neutral load voltages and switch currents are shown in Fig. 5-7c. For intervals to exist when three SCRs are on, the delay angle must be less than 60°.

For $60° < \alpha < 90°$:

Only two SCRs conduct at any one time when the delay angle is between 60 and 90°. Load = voltage v_{an} for $\alpha = 75°$ is shown in Fig. 5-7d. Just prior to 75°, S_5 and S_6 are conducting, and $v_{an} = 0$. When S_1 is turned on at 75°, S_6 continues to conduct, but S_5 must turn off because v_{CN} is negative. Voltage v_{an} is then $v_{AB}/2$. When S_2 is turned on at 135°, S_6 is forced off, and $v_{an} = v_{AC}/2$. The next SCR to turn on is S_3, which forces S_1 off, and $v_{an} = 0$. One SCR is always forced off when an SCR is turned on for α in this range. Load voltages are one-half line-to-line voltages or zero.

For $90° < \alpha < 150°$:

Only two SCRs can conduct at any one time in this mode. Additionally, there are intervals when no SCRs conduct. Figure 5-7e shows the load voltage v_{an} for $\alpha = 120°$. In the interval just prior to 120°, no SCRs are on, and $v_{an} = 0$. At $\alpha = 120°$, S_1 is given a gate signal, and S_6 still has a gate signal applied. Since v_{AB} is positive, both S_1 and S_6 are forward-biased and begin to conduct, and $v_{an} = v_{AB}/2$. Both S_1 and S_6 turn off when v_{AB} becomes negative. When a gate signal is applied to S_2, it turns on, and S_1 turns on again.

For $\alpha > 150°$, there is no time interval when an SCR is forward-biased while a gate signal is applied. Output voltage is zero for this condition.

Normalized output voltage vs. delay angle is shown in Fig. 5-8. Note that a delay angle of zero corresponds to the load being connected directly to the

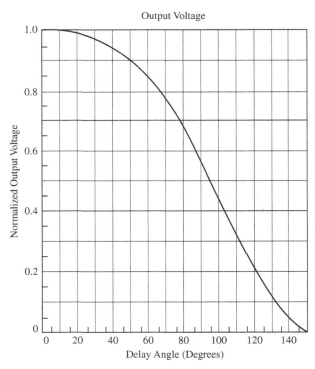

Figure 5-8 Normalized rms output voltage for a three-phase ac voltage controller with a resistive load.

three-phase source. The range of output voltage for the three-phase voltage controller is between full source voltage and zero.

Harmonic currents in the load and line for the three-phase ac voltage controller are the odd harmonics of order $6n \pm 1$, $n = 1, 2, 3, \ldots$ (that is, 5th, 7th, 11th, 13th).... Harmonic filters may be required in some applications to prevent harmonic currents from propagating into the ac system.

Since analysis of the three-phase ac voltage controller is cumbersome, simulation is a practical means of obtaining rms output voltages and power delivered to a load. PSpice simulation is presented in Example 5-4.

Y-Connected *RL* Load

The load voltages for a three-phase voltage controller with an *RL* load are again characterized by being a line-to-neutral voltage, one-half of a line-to-line voltage, or zero. The analysis is much more difficult for an *RL* load than for a resistive load, and simulation provides results that would be extremely difficult to obtain analytically. Example 5-4 illustrates the use of PSpice for a three-phase ac voltage controller.

EXAMPLE 5-4

PSpice Simulation of a Three-Phase Voltage Controller

Use PSpice to obtain the power delivered to a *Y*-connected three-phase load. Each phase of the load is a series *RL* combination with $R = 10 \, \Omega$ and $L = 30$ mH. The three-phase source is 480 V rms line-to-line at 60 Hz, and the delay angle α is 75°. Determine the rms value of the line currents, the power absorbed by the load, the power absorbed by the SCRs, and the total harmonic distortion (THD) of the source currents.

■ **Solution**

A PSpice A/D input file for the *Y*-connected three-phase voltage controller with an *RL* load is as follows:

```
THREE-PHASE VOLTAGE CONTROLLER-R-L LOAD  (3phvc.cir)
*SOURCE AND LOAD ARE Y-CONNECTED (UNGROUNDED)
******************** INPUT PARAMETERS **************************
.PARAM Vs = 480                  ; rms line-to-line voltage
.PARAM ALPHA = 75                ; delay angle in degrees
.PARAM R = 10                    ; load resistance (y-connected)
.PARAM L = 30mH                  ; load inductance
.PARAM F = 60                    ; source frequency

******************** COMPUTED PARAMETERS **********************
.PARAM Vm = {Vs*SQRT(2)/SQRT(3)} ; convert to peak line-neutral volts
.PARAM DLAY = {1/(6*F)}          ; switching interval is 1/6 period
.PARAM PW = {.5/F} TALPHA={ALPHA/(F*360)}
.PARAM TRF = 10US                ; rise and fall time for pulse switch control
```

```
*********************** THREE-PHASE SOURCE ***********************
VAN 1 0 SIN(0 {VM} 60)
VBN 2 0 SIN(0 {VM} 60 0 0 -120)
VCN 3 0 SIN(0 {VM} 60 0 0 -240)

*************************** SWITCHES *****************************
S1 1 8 18 0 SMOD                         ; A-phase
D1 8 4 DMOD
S4 4 9 19 0 SMOD
D4 9 1 DMOD

S3 2 10 20 0 SMOD                        ; B-phase
D3 10 5 DMOD
S6 5 11 21 0 SMOD
D6 11 2 DMOD

S5 3 12 22 0 SMOD                        ; C-phase
D5 12 6 DMOD
S2 6 13 23 0 SMOD
D2 13 3 DMOD

*************************** LOAD *********************************
RA 4 4A {R}                              ; van = v(4,7)
LA 4A 7 {L}

RB 5 5A {R}                              ; vbn = v(5,7)
LB 5A 7 {L}

RC 6 6A {R}                              ; vcn = v(6,7)
LC 6A 7 {L}

*********************** SWITCH CONTROL ***************************
V1 18 0 PULSE(-10 10 {TALPHA}         {TRF} {TRF} {PW} {1/F})
V4 19 0 PULSE(-10 10 {TALPHA+3*DLAY}  {TRF} {TRF} {PW} {1/F})
V3 20 0 PULSE(-10 10 {TALPHA+2*DLAY}  {TRF} {TRF} {PW} {1/F})
V6 21 0 PULSE(-10 10 {TALPHA+5*DLAY}  {TRF} {TRF} {PW} {1/F})
V5 22 0 PULSE(-10 10 {TALPHA+4*DLAY}  {TRF} {TRF} {PW} {1/F})
V2 23 0 PULSE(-10 10 {TALPHA+DLAY}    {TRF} {TRF} {PW} {1/F})

********************* MODELS AND COMMANDS ************************
.MODEL SMOD VSWITCH(RON=0.01)
.MODEL DMOD D
.TRAN .1MS 50MS 16.67ms .05MS UIC
.FOUR 60 I(RA)                           ; Fourier analysis of line current
.PROBE
.OPTIONS NOPAGE ITL5=0
.END
```

Probe output of the steady-state current in one of the phases is shown in Fig. 5-9. The rms line current, load power, and power absorbed by the SCRs are obtained by entering the appropriate expression in Probe. The THD in the source current is determined from the Fourier analysis in the output file. The results are summarized in the following table.

Quantity	Expression	Result
RMS line current	RMS(I(RA))	12.86 A
Load power	3*AVG(V(4,7)*I(RA))	4960 W
Total SCR power absorbed	6*AVG(V(1,4)*I(S1))	35.1 W
THD of source current	(from the output file)	13.1%

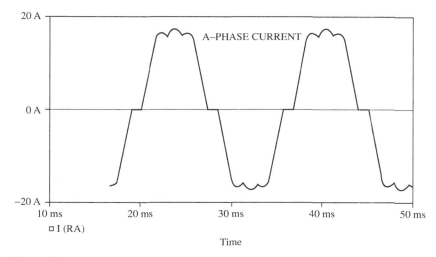

Figure 5-9 Probe output for Example 5-4.

Delta-Connected Resistive Load

A three-phase ac voltage controller with a delta-connected resistive load is shown in Fig. 5-10a. The voltage across a load resistor is the corresponding line-to-line voltage when a SCR in the phase is on. The delay angle is referenced to the zero crossing of the line-to-line voltage. SCRs are turned on in the sequence 1-2-3-4-5-6.

The line current in each phase is the sum of two of the delta currents:

$$\begin{aligned} i_a &= i_{ab} - i_{ca} \\ i_b &= i_{bc} - i_{ab} \\ i_c &= i_{ca} - i_{bc} \end{aligned} \quad (5\text{-}18)$$

The relationship between rms line and delta currents depends on the conduction angle of the SCRs. For small conduction angles (large α), the delta currents do not overlap (Fig. 5-10b), and the rms line currents are

$$I_{L,\text{rms}} = \sqrt{2}\, I_{\Delta,\text{rms}} \quad (5\text{-}19)$$

For large conduction angles (small α), the delta currents overlap (Fig. 5-10c), and the rms line current is larger than $\sqrt{2}I_\Delta$. In the limit when $\gamma = \pi$ ($\alpha = 0$), the delta currents and line currents are sinusoids. The rms line current is determined from ordinary three-phase analysis.

$$I_{L,\text{rms}} = \sqrt{3} I_{\Delta,\text{rms}} \quad (5\text{-}20)$$

The range of rms line current is therefore

$$\sqrt{2}\, I_{\Delta,\text{rms}} \leq I_{L,\text{rms}} \leq \sqrt{3}\, I_{\Delta,\text{rms}} \quad (5\text{-}21)$$

depending on α.

Figure 5-10
(a) Three-phase ac voltage controller with a delta-connected resistive load;
(b) Current waveforms for α = 130°;
(c) Current waveforms for α = 90°.

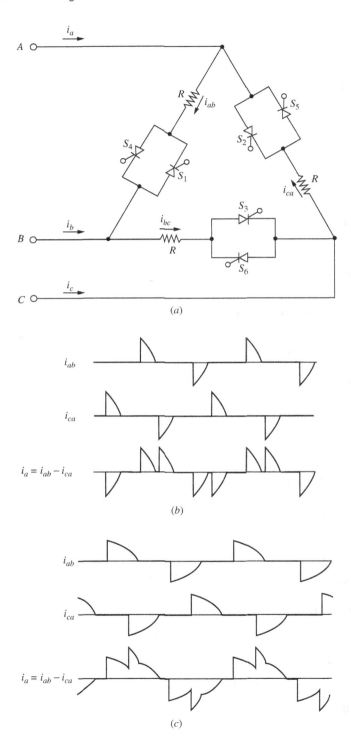

Use of the delta-connected three-phase voltage controller requires the load to be broken to allow thyristors to be inserted in each phase, which is often not feasible.

5.4 INDUCTION MOTOR SPEED CONTROL

Squirrel-cage induction motor speed can be controlled by varying the voltage and/or frequency. The ac voltage controller is suitable for some speed control applications. The torque produced by an induction motor is proportional to the square of the applied voltage. Typical torque-speed curves for an induction motor are shown in Fig. 5-11. If a load has a torque-speed characteristic like that also shown in Fig. 5-11, speed can be controlled by adjusting the motor voltage. Operating speed corresponds to the intersection of the torque-speed curves of the motor and the load. A fan or pump is a suitable load for this type of speed control, where the torque requirement is approximately proportional to the square of the speed.

Single-phase induction motors are controlled with the circuit of Fig. 5-4a, and three-phase motors are controlled with the circuit of Fig. 5-7a. Energy efficiency is poor when using this type of control, especially at low speeds. The large slip at low speeds results in large rotor losses. Typical applications exist where the load is small, such as single-phase fractional-horsepower motors, or where the time of low-speed operation is short. Motor speed control using a variable-frequency source from an inverter circuit (Chap. 8) is usually a preferred method.

5.5 STATIC VAR CONTROL

Capacitors are routinely placed in parallel with inductive loads for power factor improvement. If a load has a constant reactive voltampere (VAR) requirement, a fixed capacitor can be selected to correct the power factor to unity. However, if a

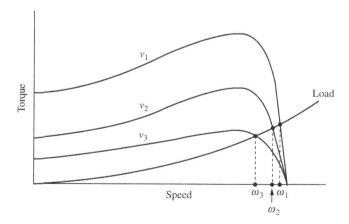

Figure 5-11 Torque-speed curves for an induction motor.

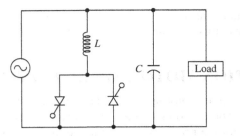

Figure 5-12 Static VAR control.

load has a varying VAR requirement, the fixed-capacitor arrangement results in a changing power factor.

The circuit of Fig. 5-12 represents an application of an ac voltage controller to maintain a unity power factor for varying load VAR requirements. The power factor correction capacitance supplies a fixed amount of reactive power, generally greater than required by the load. The parallel inductance absorbs a variable amount of reactive power, depending on the delay angle of the SCRs. The net reactive power supplied by the inductor-capacitor combination is controlled to match that absorbed by the load. As the VAR requirement of the load changes, the delay angle is adjusted to maintain unity power factor. This type of power factor correction is known as *static VAR control*.

The SCRs are placed in the inductor branch rather than in the capacitor branch because very high currents could result from switching a capacitor with a SCR.

Static VAR control has the advantage of being able to adjust to changing load requirements very quickly. Reactive power is continuously adjustable with static VAR control, rather than having discrete levels as with capacitor banks which are switched in and out with circuit breakers. Static VAR control is becoming increasingly prevalent in installations with rapidly varying reactive power requirements, such as electric arc furnaces. Filters are generally required to remove the harmonic currents generated by the switched inductance.

5.6 Summary

- Voltage controllers use electronic switches to connect and disconnect a load to an ac source at regular intervals. This type of circuit is classified as an ac-ac converter.
- Voltage controllers are used in applications such as single-phase light-dimmer circuits, single-phase or three-phase induction motor control, and static VAR control.
- The delay angle for the thyristors controls the time interval for the switch being on and thereby controls the effective value of voltage at the load. The range of control for load voltage is between full ac source voltage and zero.
- An ac voltage controller can be designed to function in either the fully on or fully off mode. This application is used as a solid-state relay.
- The load and source current and voltage in ac voltage controller circuits may contain significant harmonics. For equal delay angles in the positive and negative half-cycles, the average source current is zero, and only odd harmonics exist.

- Three-phase voltage controllers can have Y- or Δ-connected loads.
- Simulation of single-phase or three-phase voltage controllers provides an efficient analysis method.

5.7 Bibliography

B. K. Bose, *Power Electronics and Motor Drives: Advances and Trends*, Academic Press, New York, 2006.

A. K. Chattopadhyay, *Power Electronics Handbook*, edited by M. H. Rashid, Academic Press, New York, 2001, Chapter 16.

M. A. El-Sharkawi, *Fundamentals of Electric Drives*, Brooks/Cole, Pacific Grove, Calif., 2000.

B. M. Han and S. I. Moon, "Static Reactive-Power Compensator Using Soft-Switching Current-Source Inverter," *IEEE Transactions on Power Electronics*, vol. 48, no. 6, December 2001.

N. Mohan, T. M. Undeland, and W. P. Robbins, *Power Electronics: Converters, Applications, and Design*, 3d ed., Wiley, New York, 2003.

M. H. Rashid, *Power Electronics: Circuits, Devices, and Systems*, 3d ed., Prentice-Hall, Upper Saddle River, N. J., 2004.

R. Valentine, *Motor Control Electronics Handbook*, McGraw-Hill, New York, 1996.

B. Wu, *High-Power Converters and AC Drives*, Wiley, New York, 2006.

Problems

Single-phase Voltage Controllers

5-1. The single-phase ac voltage controller of Fig. 5-1a has a 480-V rms 60-Hz source and a load resistance of 50 Ω. The delay angle α is 60°. Determine (a) the rms load voltage, (b) the power absorbed by the load, (c) the power factor, (e) the average and rms currents in the SCRs, and (f) the THD of the source current.

5-2. The single-phase ac voltage controller of Fig. 5-1a has a 120-V rms 60-Hz source and a load resistance of 20 Ω. The delay angle α is 45°. Determine (a) the rms load voltage, (b) the power absorbed by the load, (c) the power factor, (d) the average and rms currents in the SCRs, and (e) the THD of the source current.

5-3. The single-phase ac voltage controller of Fig. 5-1a has a 240-V rms source and a load resistance of 35 Ω. (a) Determine the delay angle required to deliver 800 W to the load. (b) Determine the rms current in each SCR. (c) Determine the power factor.

5-4. A resistive load absorbs 200 W when connected to a 120-V rms 60-Hz ac voltage source. Design a circuit which will result in 200 W absorbed by the same resistance when the source is 240 V rms at 60 Hz. What is the peak load voltage in each case?

5-5. The single-phase ac voltage controller of Fig. 5-1a has a 120-V rms source at 60 Hz and a load resistance of 40 Ω. Determine the range of α so that the output power can be controlled from 200 to 400 W. Determine the range of power factor that will result.

5-6. Design a circuit to deliver power in the range of 750 to 1500 W to a 32-Ω resistor from a 240-V rms 60-Hz source. Determine the maximum rms and average currents in the switching devices, and determine the maximum voltage across the devices.

5-7. Design a circuit to deliver a constant 1200 W of power to a load that varies in resistance from 20 to 40 Ω. The ac source is 240 V rms, 60 Hz. Determine the maximum rms and average currents in the devices, and determine the maximum voltage across the devices.

5-8. Design a light-dimmer for a 120-V, 100-W incandescent lightbulb. The source is 120 V rms, 60 Hz. Specify the delay angle for the triac to produce an output power of (*a*) 75 W (*b*) 25 W. Assume that the bulb is a load of constant resistance.

5-9. A single-phase ac voltage controller is similar to Fig. 5-1*a* except that S_2 is replaced with a diode. S_1 operates at a delay angle α. Determine (*a*) an expression for rms load voltage as a function of α and V_m and (*b*) the range of rms voltage across a resistive load for this circuit.

5-10. The single-phase ac voltage controller of Fig. 5-1*a* is operated with unequal delays on the two SCRs ($\alpha_1 \neq \alpha_2$). Derive expressions for the rms load voltage and average load voltage in terms of V_m, α_1, and α_2.

5-11. The single-phase ac voltage controller of Fig. 5-4*a* has a 120-V rms 60-Hz source. The series RL load has $R = 18\ \Omega$ and $L = 30$ mH. The delay angle $\alpha = 60°$. Determine (*a*) an expression for current, (*b*) rms load current, (*c*) rms current in each of the SCRs, and (*d*) power absorbed by the load. (*e*) Sketch the waveforms of output voltage and voltage across the SCRs.

5-12. The single-phase ac voltage controller of Fig. 5-4*a* has a 120-V rms 60-Hz source. The RL load has $R = 22\ \Omega$ and $L = 40$ mH. The delay angle $\alpha = 50°$. Determine (*a*) an expression for current, (*b*) rms load current, (*c*) rms current in each of the SCRs, and (*d*) power absorbed by the load. (*e*) Sketch the waveforms of output voltage and voltage across the SCRs.

5-13. The single-phase ac voltage controller of Fig. 5-4*a* has a 120-V rms 60-Hz source. The RL load has $R = 12\ \Omega$ and $L = 24$ mH. The delay angle α is 115°. Determine the rms load current.

5-14. The single-phase ac voltage controller of Fig. 5-4*a* has a 120-V rms 60-Hz source. The RL load has $R = 12\ \Omega$ and $L = 20$ mH. The delay angle α is 70°. (*a*) Determine the power absorbed by the load for ideal SCRs. (*b*) Determine the power in the load from a PSpice simulation. Use the default diode and $R_{on} = 0.1\ \Omega$ in the SCR model. (*c*) Determine the THD of the source current from the PSpice output.

5-15. Use PSpice to determine the delay angle required in the voltage controller of Fig. 5-4*a* to deliver (*a*) 400 W, and (*b*) 700 W to an RL load with $R = 15\ \Omega$ and $L = 15$ mH from a 120-V rms 60-Hz source.

5-16. Use PSpice to determine the delay angle required in the voltage controller of Fig. 5-4*a* to deliver (*a*) 600 W, and (*b*) 1000 W to an RL load with $R = 25\ \Omega$ and $L = 60$ mH from a 240-V rms 60-Hz source.

5-17. Design a circuit to deliver 250 W to an RL series load, where $R = 24\ \Omega$ and $L = 35$ mH. The source is 120 V rms at 60 Hz. Specify the rms and average currents in the devices. Specify the maximum voltage across the devices.

Three-phase Voltage Controllers

5-18. The three-phase voltage controller of Fig. 5-7*a* has a 480-V rms line-to-line source and a resistive load with 35 Ω in each phase. Simulate the circuit in PSpice to determine the power absorbed by the load if the delay angle α is (*a*) 20°, (*b*) 80°, and (*c*) 115°.

5-19. The three-phase Y-connected voltage controller has a 240-V rms, 60-Hz line-to-line source. The load in each phase is a series RL combination with $R = 16\,\Omega$ and $L = 50$ mH. The delay angle α is 90°. Simulate the circuit in PSpice to determine the power absorbed by the load. On a graph of one period of A-phase current, indicate the intervals when each SCR conducts. Do your analysis for steady-state current.

5-20. For the delta-connected resistive load in the three-phase voltage controller of Fig. 5-10, determine the smallest delay angle such that the rms line current is described by $I_{\text{line rms}} = \sqrt{2} I_{\Delta\text{rms}}$.

5-21. Modify the PSpice circuit file for the three-phase controller for analysis of a delta-connected load. Determine the rms values of the delta currents and the line currents for a 480-V rms source, a resistive load of $R = 25\,\Omega$ in each phase, and a delay angle of 45°. Hand in a Probe output showing i_{ab} and i_a.

5-22. A three-phase ac voltage controller has a 480-V rms, 60-Hz source. The load is Y-connected, and each phase has series RLC combination with $R = 14\,\Omega$, $L = 10$ mH, and $C = 1\,\mu\text{F}$. The delay angle is 70°. Use PSpice to determine (a) the rms load current, (b) the power absorbed by the load, and (c) the THD of the line current. Also hand in a graph of one period of A-phase current, indicating which SCRs are conducting at each time. Do your analysis for steady-state current.

5-23. For a three-phase ac voltage controller with a Y-connected load, the voltage across the S_1-S_4 SCR pair is zero when either is on. In terms of the three-phase source voltages, what is the voltage across the S_1-S_4 pair when both are off?

CHAPTER 6

DC-DC Converters

Dc-dc converters are power electronic circuits that convert a dc voltage to a different dc voltage level, often providing a regulated output. The circuits described in this chapter are classified as switched-mode dc-dc converters, also called switching power supplies or switchers. This chapter describes some basic dc-dc converter circuits. Chapter 7 describes some common variations of these circuits that are used in many dc power supply designs.

6.1 LINEAR VOLTAGE REGULATORS

Before we discuss switched-mode converters, it is useful to review the motivation for an alternative to linear dc-dc converters that was introduced in Chapt. 1. One method of converting a dc voltage to a lower dc voltage is a simple circuit as shown in Fig. 6-1. The output voltage is

$$V_o = I_L R_L$$

where the load current is controlled by the transistor. By adjusting the transistor base current, the output voltage may be controlled over a range of 0 to roughly V_s. The base current can be adjusted to compensate for variations in the supply voltage or the load, thus regulating the output. This type of circuit is called a linear dc-dc converter or a linear regulator because the transistor operates in the linear region, rather than in the saturation or cutoff regions. The transistor in effect operates as a variable resistance.

While this may be a simple way of converting a dc supply voltage to a lower dc voltage and regulating the output, the low efficiency of this circuit is a serious drawback for power applications. The power absorbed by the load is $V_o I_L$, and

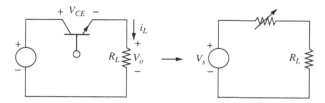

Figure 6-1 A basic linear regulator.

the power absorbed by the transistor is $V_{CE}I_L$, assuming a small base current. The power loss in the transistor makes this circuit inefficient. For example, if the output voltage is one-quarter of the input voltage, the load resistor absorbs one-quarter of the source power, which is an efficiency of 25 percent. The transistor absorbs the other 75 percent of the power supplied by the source. Lower output voltages result in even lower efficiencies. Therefore, the linear voltage regulator is suitable only for low-power applications.

6.2 A BASIC SWITCHING CONVERTER

An efficient alternative to the linear regulator is the switching converter. In a switching converter circuit, the transistor operates as an electronic switch by being completely on or completely off (saturation or cutoff for a BJT or the triode and cutoff regions of a MOSFET). This circuit is also known as a dc chopper.

Assuming the switch is ideal in Fig. 6-2, the output is the same as the input when the switch is closed, and the output is zero when the switch is open. Periodic

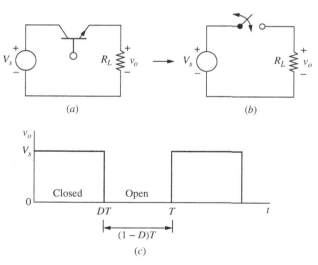

Figure 6-2 (a) A basic dc-dc switching converter; (b) Switching equivalent; (c) Output voltage.

opening and closing of the switch results in the pulse output shown in Fig. 6-2c. The average or dc component of the output voltage is

$$V_o = \frac{1}{T}\int_0^T v_o(t)\,dt = \frac{1}{T}\int_0^{DT} V_s\,dt = V_s D \qquad (6\text{-}1)$$

The dc component of the output voltage is controlled by adjusting the duty ratio D, which is the fraction of the switching period that the switch is closed

$$D \equiv \frac{t_{on}}{t_{on} + t_{off}} = \frac{t_{on}}{T} = t_{on}f \qquad (6\text{-}2)$$

where f is the switching frequency. The dc component of the output voltage will be less than or equal to the input voltage for this circuit.

The power absorbed by the ideal switch is zero. When the switch is open, there is no current in it; when the switch is closed, there is no voltage across it. Therefore, all power is absorbed by the load, and the energy efficiency is 100 percent. Losses will occur in a real switch because the voltage across it will not be zero when it is on, and the switch must pass through the linear region when making a transition from one state to the other.

6.3 THE BUCK (STEP-DOWN) CONVERTER

Controlling the dc component of a pulsed output voltage of the type in Fig. 6-2c may be sufficient for some applications, such as controlling the speed of a dc motor, but often the objective is to produce an output that is purely dc. One way of obtaining a dc output from the circuit of Fig. 6-2a is to insert a low-pass filter after the switch. Figure 6-3a shows an *LC* low-pass filter added to the basic converter. The diode provides a path for the inductor current when the switch is opened and is reverse-biased when the switch is closed. This circuit is called a *buck converter* or a *step-down converter* because the output voltage is less than the input.

Voltage and Current Relationships

If the low-pass filter is ideal, the output voltage is the average of the input voltage to the filter. The input to the filter, v_x in Fig. 6-3a, is V_s when the switch is closed and is zero when the switch is open, provided that the inductor current remains positive, keeping the diode on. If the switch is closed periodically at a duty ratio D, the average voltage at the filter input is $V_s D$, as in Eq. (6-1).

This analysis assumes that the diode remains forward-biased for the entire time when the switch is open, implying that the inductor current remains positive. An inductor current that remains positive throughout the switching period is known as *continuous current*. Conversely, discontinuous current is characterized by the inductor current's returning to zero during each period.

6.3 The Buck (Step-Down) Converter

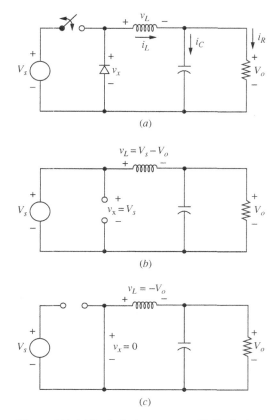

Figure 6-3 (*a*) Buck dc-dc converter; (*b*) Equivalent circuit for the switch closed; (*c*) Equivalent circuit for the switch open.

Another way of analyzing the operation of the buck converter of Fig. 6-3*a* is to examine the inductor voltage and current. This analysis method will prove useful for designing the filter and for analyzing circuits that are presented later in this chapter.

Buck converters and dc-dc converters in general, have the following properties when operating in the steady state:

1. The inductor current is periodic.

$$i_L(t+T) = i_L(t) \tag{6-3}$$

2. The average inductor voltage is zero (see Sec. 2.3).

$$V_L = \frac{1}{T}\int_t^{t+T} v_L(\lambda)d\lambda = 0 \tag{6-4}$$

3. The average capacitor current is zero (see Sec. 2.3).

$$I_C = \frac{1}{T}\int_{t}^{t+T} i_C(\lambda)\,d\lambda = 0 \qquad (6\text{-}5)$$

4. The power supplied by the source is the same as the power delivered to the load. For nonideal components, the source also supplies the losses.

$$\begin{aligned} P_s &= P_o & \text{ideal} \\ P_s &= P_o + \text{losses} & \text{nonideal} \end{aligned} \qquad (6\text{-}6)$$

Analysis of the buck converter of Fig. 6-3a begins by making these assumptions:

1. The circuit is operating in the steady state.
2. The inductor current is continuous (always positive).
3. The capacitor is very large, and the output voltage is held constant at voltage V_o. This restriction will be relaxed later to show the effects of finite capacitance.
4. The switching period is T; the switch is closed for time DT and open for time $(1-D)T$.
5. The components are ideal.

The key to the analysis for determining the output V_o is to examine the inductor current and inductor voltage first for the switch closed and then for the switch open. The net change in inductor current over one period must be zero for steady-state operation. The average inductor voltage is zero.

Analysis for the Switch Closed When the switch is closed in the buck converter circuit of Fig. 6-3a, the diode is reverse-biased and Fig. 6-3b is an equivalent circuit. The voltage across the inductor is

$$v_L = V_s - V_o = L\frac{di_L}{dt}$$

Rearranging,

$$\frac{di_L}{dt} = \frac{V_s - V_o}{L} \qquad \text{switch closed}$$

Since the derivative of the current is a positive constant, the current increases linearly as shown in Fig. 6-4b. The change in current while the switch is closed is computed by modifying the preceding equation.

$$\frac{di_L}{dt} = \frac{\Delta i_L}{\Delta t} = \frac{\Delta i_L}{DT} = \frac{V_s - V_o}{L}$$

$$(\Delta i_L)_{\text{closed}} = \left(\frac{V_s - V_o}{L}\right)DT \qquad (6\text{-}7)$$

6.3 The Buck (Step-Down) Converter

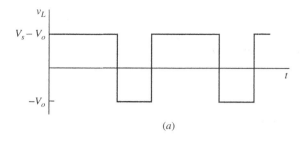

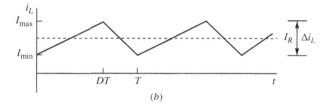

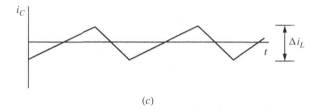

Figure 6-4 Buck converter waveforms: (*a*) Inductor voltage; (*b*) Inductor current; (*c*) Capacitor current.

Analysis for the Switch Open When the switch is open, the diode becomes forward-biased to carry the inductor current and the equivalent circuit of Fig. 6-3*c* applies. The voltage across the inductor when the switch is open is

$$v_L = -V_o = L\frac{di_L}{dt}$$

Rearranging,

$$\frac{di_L}{dt} = \frac{-V_o}{L} \quad \text{switch open}$$

The derivative of current in the inductor is a negative constant, and the current decreases linearly as shown in Fig. 6-4*b*. The change in inductor current when the switch is open is

$$\frac{\Delta i_L}{\Delta t} = \frac{\Delta i_L}{(1-D)T} = -\frac{V_o}{L}$$

$$(\Delta i_L)_{\text{open}} = -\left(\frac{V_o}{L}\right)(1-D)T \tag{6-8}$$

Steady-state operation requires that the inductor current at the end of the switching cycle be the same as that at the beginning, meaning that the net change in inductor current over one period is zero. This requires

$$(\Delta i_L)_{\text{closed}} + (\Delta i_L)_{\text{open}} = 0$$

Using Eqs. (6-7) and (6-8),

$$\left(\frac{V_s - V_o}{L}\right)(DT) - \left(\frac{V_o}{L}\right)(1-D)T = 0$$

Solving for V_o,

$$\boxed{V_o = V_s D} \tag{6-9}$$

which is the same result as Eq. (6-1). *The buck converter produces an output voltage that is less than or equal to the input.*

An alternative derivation of the output voltage is based on the inductor voltage, as shown in Fig. 6-4a. Since the average inductor voltage is zero for periodic operation,

$$V_L = (V_s - V_o)DT + (-V_o)(1-D)T = 0$$

Solving the preceding equation for V_o yields the same result as Eq. (6-9), $V_o = V_s D$.

Note that the output voltage depends on only the input and the duty ratio D. If the input voltage fluctuates, the output voltage can be regulated by adjusting the duty ratio appropriately. A feedback loop is required to sample the output voltage, compare it to a reference, and set the duty ratio of the switch accordingly. Regulation techniques are discussed in Chap. 7.

The average inductor current must be the same as the average current in the load resistor, since the average capacitor current must be zero for steady-state operation:

$$I_L = I_R = \frac{V_o}{R} \tag{6-10}$$

Since the change in inductor current is known from Eqs. (6-7) and (6-8), the maximum and minimum values of the inductor current are computed as

$$I_{\max} = I_L + \frac{\Delta i_L}{2}$$
$$= \frac{V_o}{R} + \frac{1}{2}\left[\frac{V_o}{L}(1-D)T\right] = V_o\left(\frac{1}{R} + \frac{1-D}{2Lf}\right) \tag{6-11}$$

$$I_{\min} = I_L - \frac{\Delta i_L}{2}$$
$$= \frac{V_o}{R} - \frac{1}{2}\left[\frac{V_o}{L}(1-D)T\right] = V_o\left(\frac{1}{R} - \frac{1-D}{2Lf}\right) \tag{6-12}$$

where $f = 1/T$ is the switching frequency.

For the preceding analysis to be valid, continuous current in the inductor must be verified. An easy check for continuous current is to calculate the minimum inductor current from Eq. (6-12). Since the minimum value of inductor current must be positive for continuous current, a negative minimum calculated from Eq. (6-12) is not allowed due to the diode and indicates discontinuous current. The circuit will operate for discontinuous inductor current, but the preceding analysis is not valid. Discontinuous-current operation is discussed later in this chapter.

Equation (6-12) can be used to determine the combination of L and f that will result in continuous current. Since $I_{min} = 0$ is the boundary between continuous and discontinuous current,

$$I_{min} = 0 = V_o\left(\frac{1}{R} - \frac{1-D}{2Lf}\right)$$

$$(Lf)_{min} = \frac{(1-D)R}{2} \tag{6-13}$$

If the desired switching frequency is established,

$$\boxed{L_{min} = \frac{(1-D)R}{2f} \quad \text{for continuous current}} \tag{6-14}$$

where L_{min} is the minimum inductance required for continuous current. In practice, a value of inductance greater than L_{min} is desirable to ensure continuous current.

In the design of a buck converter, the peak-to-peak variation in the inductor current is often used as a design criterion. Equation (6-7) can be combined with Eq. (6-9) to determine the value of inductance for a specified peak-to-peak inductor current for continuous-current operation:

$$\Delta i_L = \left(\frac{V_s - V_o}{L}\right)DT = \left(\frac{V_s - V_o}{Lf}\right)D = \frac{V_o(1-D)}{Lf} \tag{6-15}$$

or

$$\boxed{L = \left(\frac{V_s - V_o}{\Delta i_L f}\right)D = \frac{V_o(1-D)}{\Delta i_L f}} \tag{6-16}$$

Since the converter components are assumed to be ideal, the power supplied by the source must be the same as the power absorbed by the load resistor.

$$P_s = P_o$$
$$V_s I_s = V_o I_o \tag{6-17}$$

or

$$\frac{V_o}{V_s} = \frac{I_s}{I_o}$$

Note that the preceding relationship is similar to the voltage-current relationship for a transformer in ac applications. Therefore, the buck converter circuit is equivalent to a dc transformer.

Output Voltage Ripple

In the preceding analysis, the capacitor was assumed to be very large to keep the output voltage constant. In practice, the output voltage cannot be kept perfectly constant with a finite capacitance. The variation in output voltage, or ripple, is computed from the voltage-current relationship of the capacitor. The current in the capacitor is

$$i_C = i_L - i_R$$

shown in Fig. 6-5a.

While the capacitor current is positive, the capacitor is charging. From the definition of capacitance,

$$Q = CV_o$$
$$\Delta Q = C \Delta V_o$$
$$\Delta V_o = \frac{\Delta Q}{C}$$

The change in charge ΔQ is the area of the triangle above the time axis

$$\Delta Q = \frac{1}{2}\left(\frac{T}{2}\right)\left(\frac{\Delta i_L}{2}\right) = \frac{T \Delta i_L}{8}$$

resulting in

$$\Delta V_o = \frac{T \Delta i_L}{8C}$$

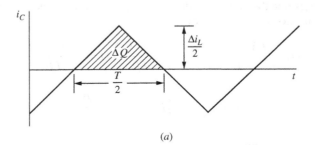

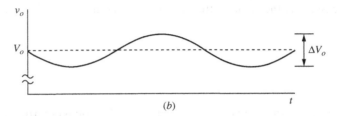

Figure 6-5 Buck converter waveforms. (*a*) Capacitor current; (*b*) Capacitor ripple voltage.

Using Eq. (6-8) for Δi_L,

$$\Delta V_o = \frac{TV_o}{8CL}(1-D)T = \frac{V_o(1-D)}{8LCf^2} \qquad (6\text{-}18)$$

In this equation, ΔV_o is the peak-to-peak ripple voltage at the output, as shown in Fig. 6-5b. It is also useful to express the ripple as a fraction of the output voltage,

$$\boxed{\frac{\Delta V_o}{V_o} = \frac{1-D}{8LCf^2}} \qquad (6\text{-}19)$$

In design, it is useful to rearrange the preceding equation to express required capacitance in terms of specified voltage ripple:

$$\boxed{C = \frac{1-D}{8L(\Delta V_o/V_o)f^2}} \qquad (6\text{-}20)$$

If the ripple is not large, the assumption of a constant output voltage is reasonable and the preceding analysis is essentially valid.

EXAMPLE 6-1

Buck Converter

The buck dc-dc converter of Fig. 6-3a has the following parameters:

$V_s = 50$ V
$D = 0.4$
$L = 400$ μH
$C = 100$ μF
$f = 20$ kHz
$R = 20$ Ω

Assuming ideal components, calculate (a) the output voltage V_o, (b) the maximum and minimum inductor current, and (c) the output voltage ripple.

■ **Solution**

(a) The inductor current is assumed to be continuous, and the output voltage is computed from Eq. (6-9),

$$V_o = V_s D = (50)(0.4) = 20 \text{ V}$$

(b) Maximum and minimum inductor currents are computed from Eqs. (6-11) and (6-12).

$$I_{max} = V_o\left(\frac{1}{R} + \frac{1-D}{2Lf}\right)$$

$$= 20\left[\frac{1}{20} + \frac{1-0.4}{2(400)(10)^{-6}(20)(10)^3}\right]$$

$$= 1 + \frac{1.5}{2} = 1.75 \text{ A}$$

206 CHAPTER 6 DC-DC Converters

$$I_{\min} = V_o\left(\frac{1}{R} - \frac{1-D}{2Lf}\right)$$

$$= 1 - \frac{1.5}{2} = 0.25 \text{ A}$$

The average inductor current is 1 A, and $\Delta i_L = 1.5$ A. Note that the minimum inductor current is positive, verifying that the assumption of continuous current was valid.

(c) The output voltage ripple is computed from Eq. (6-19).

$$\frac{\Delta V_o}{V_o} = \frac{1-D}{8LCf^2} = \frac{1-0.4}{8(400)(10)^{-6}(100)(10)^{-6}(20,000)^2}$$

$$= 0.00469 = 0.469\%$$

Since the output ripple is sufficiently small, the assumption of a constant output voltage was reasonable.

Capacitor Resistance—The Effect on Ripple Voltage

The output voltage ripple in Eqs. (6-18) and (6-19) is based on an ideal capacitor. A real capacitor can be modeled as a capacitance with an equivalent series resistance (ESR) and an equivalent series inductance (ESL). The ESR may have a significant effect on the output voltage ripple, often producing a ripple voltage greater than that of the ideal capacitance. The inductance in the capacitor is usually not a significant factor at typical switching frequencies. Figure 6.6 shows a capacitor model that is appropriate for most applications.

The ripple due to the ESR can be approximated by first determining the current in the capacitor, assuming the capacitor to be ideal. For the buck converter in the continuous-current mode, capacitor current is the triangular current waveform of Fig. 6-4c. The voltage variation across the capacitor resistance is

$$\Delta V_{o,\text{ESR}} = \Delta i_C r_C = \Delta i_L r_C \qquad (6\text{-}21)$$

To estimate a worst-case condition, one could assume that the peak-to-peak ripple voltage due to the ESR algebraically adds to the ripple due to the capacitance. However, the peaks of the capacitor and the ESR ripple voltages will not coincide, so

$$\Delta V_o < \Delta V_{o,C} + \Delta V_{o,\text{ESR}} \qquad (6\text{-}22)$$

where $\Delta V_{o,C}$ is ΔV_o in Eq. (6-18). The ripple voltage due to the ESR can be much larger than the ripple due to the pure capacitance. In that case, the output capacitor is chosen on the basis of the equivalent series resistance rather than capacitance only.

$$\Delta V_o \approx \Delta V_{o,\text{ESR}} = \Delta i_C r_C \qquad (6\text{-}23)$$

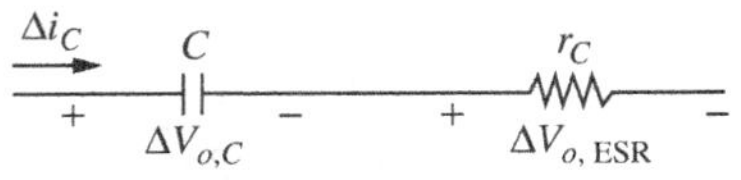

Figure 6-6 A model for the capacitor including the equivalent series resistance (ESR).

Capacitor ESR is inversely proportional to the capacitance value—a larger capacitance results in a lower ESR. Manufacturers provide what are known as *low-ESR capacitors* for power supply applications.

In Example 6-1, the 100-μF capacitor may have an ESR of $r_C = 0.1\ \Omega$. The ripple voltage due to the ESR is calculated as

$$\Delta V_{o,\text{ESR}} = \Delta i_C r_C = \Delta i_L r_C = (1.5\ \text{A})(0.1\ \Omega) = 0.15\ \text{V}$$

Expressed as a percent, $\Delta V_o/V_o$ is $0.15/20 = 0.75$ percent. The total ripple can then be approximated as 0.75 percent.

Synchronous Rectification for the Buck Converter

Many buck converters use a second MOSFET in place of the diode. When S_2 is on and S_1 is off, current flows upward out of the drain of S_2. The advantage of this configuration is that the second MOSFET will have a much lower voltage drop across it compared to a diode, resulting in higher circuit efficiency. This is especially important in low-voltage, high-current applications. A Shottky diode would have a voltage of 0.3 to 0.4 V across it while conducting, whereas a MOSFET will have an extremely low voltage drop due to an $R_{DS_{on}}$ as low as single-digit milliohms. This circuit has a control scheme known as *synchronous switching*, or *synchronous rectification*. The second MOSFET is known as a *synchronous rectifier*. The two MOSFETs must not be on at the same time to prevent a short circuit across the source, so a "dead time" is built into the switching control—one MOSFET is turned off before the other is turned on. A diode is placed in parallel with the second MOSFET to provide a conducting path for inductor current during the dead time when both MOSFETs are off. This diode may be the MOSFET body diode, or it may be an extra diode, most likely a Shottky diode, for improved switching. The synchronous buck converter should be operated in the continuous-current mode because the MOSFET would allow the inductor current to go negative.

Other converter topologies presented in this chapter and in Chap. 7 can utilize MOSFETs in place of diodes.

6.4 DESIGN CONSIDERATIONS

Most buck converters are designed for continuous-current operation. The choice of switching frequency and inductance to give continuous current is given by Eq. (6-13), and the output voltage ripple is described by Eqs. (6-16) and (6-21). Note that as the switching frequency increases, the minimum size of the inductor to produce continuous current and the minimum size of the capacitor to limit output ripple both decrease. Therefore, high switching frequencies are desirable to reduce the size of both the inductor and the capacitor.

The tradeoff for high switching frequencies is increased power loss in the switches, which is discussed later in this chapter and in Chap. 10. Increased power loss in the switches means that heat is produced. This decreases the converter's efficiency and may require a large heat sink, offsetting the reduction in size of the inductor and capacitor. Typical switching frequencies are above 20 kHz to avoid audio noise, and they extend well into the 100s of kilohertz and into the megahertz range. Some designers consider about 500 kHz to be the best compromise

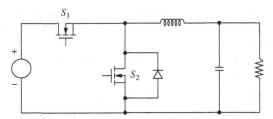

Figure 6-7 A synchronous buck converter. The MOSFET S_2 carries the inductor current when S_1 is off to provide a lower voltage drop than a diode.

between small component size and efficiency. Other designers prefer to use lower switching frequencies of about 50 kHz to keep switching losses small, while still others prefer frequencies larger than 1 MHz. As switching devices improve, switching frequencies will increase.

For low-voltage, high-current applications, the synchronous rectification scheme of Fig. 6-7 is preferred over using a diode for the second switch. The voltage across the conducting MOSFET will be much less than that across a diode, resulting in lower losses.

The inductor value should be larger than L_{min} in Eq. (6-14) to ensure continuous-current operation. Some designers select a value 25 percent larger than L_{min}. Other designers use different criteria, such as setting the inductor current variation, Δi_L in Eq. (6-15), to a desired value, such as 40 percent of the average inductor current. A smaller Δi_L results in lower peak and rms inductor currents and a lower rms capacitor current but requires a larger inductor.

The inductor wire must be rated at the rms current, and the core should not saturate for peak inductor current. The capacitor must be selected to limit the output ripple to the design specifications, to withstand peak output voltage, and to carry the required rms current.

The switch (usually a MOSFET with a low $R_{DS_{on}}$) and diode (or second MOSFET for synchronous rectification) must withstand maximum voltage stress when off and maximum current when on. The temperature ratings must not be exceeded, often requiring a heat sink.

Assuming ideal switches and an ideal inductor in the initial design is usually reasonable. However, the ESR of the capacitor should be included because it typically gives a more significant output voltage ripple than the ideal device and greatly influences the choice of capacitor size.

EXAMPLE 6-2

Buck Converter Design 1

Design a buck converter to produce an output voltage of 18 V across a 10-Ω load resistor. The output voltage ripple must not exceed 0.5 percent. The dc supply is 48 V. Design for continuous inductor current. Specify the duty ratio, the switching frequency, the values of the inductor and capacitor, the peak voltage rating of each device, and the rms current in the inductor and capacitor. Assume ideal components.

■ Solution

Using the buck converter circuit in Fig. 6-3a, the duty ratio for continuous-current operation is determined from Eq. (6-9):

$$D = \frac{V_o}{V_s} = \frac{18}{48} = 0.375$$

The switching frequency and inductor size must be selected for continuous-current operation. Let the switching frequency arbitrarily be 40 kHz, which is well above the audio range and is low enough to keep switching losses small. The minimum inductor size is determined from Eq. (6-14).

$$L_{min} = \frac{(1-D)(R)}{2f} = \frac{(1-0.375)(10)}{2(40,000)} = 78 \,\mu H$$

Let the inductor be 25 percent larger than the minimum to ensure that inductor current is continuous.

$$L = 1.25 L_{min} = (1.25)(78 \,\mu H) = 97.5 \,\mu H$$

Average inductor current and the change in current are determined from Eqs. (6-10) and (6-17).

$$I_L = \frac{V_o}{R} = \frac{18}{10} = 1.8 \text{ A}$$

$$\Delta i_L = \left(\frac{V_s - V_o}{L}\right) DT = \frac{48-18}{97.5(10)^{-6}}(0.375)\left(\frac{1}{40,000}\right) = 2.88 \text{ A}$$

The maximum and minimum inductor currents are determined from Eqs. (6-11) and (6-12).

$$I_{max} = I_L + \frac{\Delta i_L}{2} = 1.8 + 1.44 = 3.24 \text{ A}$$

$$I_{min} = I_L - \frac{\Delta i_L}{2} = 1.8 - 1.44 = 0.36 \text{ A}$$

The inductor must be rated for rms current, which is computed as in Chap. 2 (see Example 2-8). For the offset triangular wave,

$$I_{L,rms} = \sqrt{I_L^2 + \left(\frac{\Delta i_L/2}{\sqrt{3}}\right)^2} = \sqrt{(1.8)^2 + \left(\frac{1.44}{\sqrt{3}}\right)^2} = 1.98 \text{ A}$$

The capacitor is selected using Eq. (6-20).

$$C = \frac{1-D}{8L(\Delta V_o/V_o)f^2} = \frac{1-0.375}{8(97.5)(10)^{-6}(0.005)(40,000)^2} = 100 \,\mu F$$

Peak capacitor current is $\Delta i_L/2 = 1.44$ A, and rms capacitor current for the triangular waveform is $1.44/\sqrt{3} = 0.83$ A. The maximum voltage across the switch and diode is V_s, or 48 V. The inductor voltage when the switch is closed is $V_s - V_o = 48 - 18 = 30$ V. The inductor voltage when the switch is open is $V_o = 18$ V. Therefore, the inductor must withstand 30 V. The capacitor must be rated for the 18-V output.

EXAMPLE 6-3

Buck Converter Design 2

Power supplies for telecommunications applications may require high currents at low voltages. Design a buck converter that has an input voltage of 3.3 V and an output voltage of 1.2 V. The output current varies between 4 and 6 A. The output voltage ripple must not exceed 2 percent. Specify the inductor value such that the peak-to-peak variation in inductor current does not exceed 40 percent of the average value. Determine the required rms current rating of the inductor and of the capacitor. Determine the maximum equivalent series resistance of the capacitor.

■ Solution

Because of the low voltage and high output current in this application, the synchronous rectification buck converter of Fig. 6-7 is used. The duty ratio is determined from Eq. (6-9).

$$D = \frac{V_o}{V_s} = \frac{1.2}{3.3} = 0.364$$

The switching frequency and inductor size must be selected for continuous-current operation. Let the switching frequency arbitrarily be 500 kHz to give a good tradeoff between small component size and low switching losses.

The average inductor current is the same as the output current. Analyzing the circuit for an output current of 4 A,

$$I_L = I_o = 4 \text{ A}$$

$$\Delta i_L = (40\%)(4) = 1.6 \text{ A}$$

Using Eq. (6-16),

$$L = \left(\frac{V_s - V_o}{\Delta i_L f}\right)D = \frac{3.3 - 1.2}{(1.6)(500{,}000)}(0.364) = 0.955 \text{ μH}$$

Analyzing the circuit for an output current of 6 A,

$$I_L = I_o = 6 \text{ A}$$

$$\Delta i_L = (40\%)(6) = 2.4 \text{ A}$$

resulting in

$$L = \left(\frac{V_s - V_o}{\Delta i_L f}\right)D = \frac{3.3 - 1.2}{(2.4)(500{,}000)}(0.364) = 0.636 \text{ μH}$$

Since 0.636 μH would be too small for the 4-A output, use $L = 0.955$ μH, which would be rounded to 1 μH.

Inductor rms current is determined from

$$I_{L,\text{rms}} = \sqrt{I_L^2 + \left(\frac{\Delta i_L/2}{\sqrt{3}}\right)^2}$$

(see Chap. 2). From Eq. (6-15), the variation in inductor current is 1.6 A for each output current. Using the 6-A output current, the inductor must be rated for an rms current of

$$I_{L,\text{rms}} = \sqrt{6^2 + \left(\frac{0.8}{\sqrt{3}}\right)^2} = 6.02 \text{ A}$$

Note that the average inductor current would be a good approximation to the rms current since the variation is relatively small.

Using $L = 1$ µH in Eq. (6-20), the minimum capacitance is determined as

$$C = \frac{1 - D}{8L(\Delta V_o/V_o)f^2} = \frac{1 - 0.364}{8(1)(10)^{-6}(0.02)(500{,}000)^2} = 0.16 \text{ µF}$$

The allowable output voltage ripple of 2 percent is $(0.02)(1.2) = 24$ mV. The maximum ESR is computed from Eq. (6-23).

$$\Delta V_o \approx r_C \Delta i_C = r_C \Delta i_L$$

or

$$r_C = \frac{\Delta V_o}{\Delta i_C} = \frac{24 \text{ mV}}{1.6 \text{ A}} = 15 \text{ m}\Omega$$

At this point, the designer would search manufacturer's specifications for a capacitor having 15-mΩ ESR. The capacitor may have to be much larger than the calculated value of 0.16 µF to meet the ESR requirement. Peak capacitor current is $\Delta i_L/2 = 0.8$ A, and rms capacitor current for the triangular waveform is $0.8/\sqrt{3} = 0.46$ A.

6.5 THE BOOST CONVERTER

The boost converter is shown in Fig. 6-8. This is another switching converter that operates by periodically opening and closing an electronic switch. It is called a boost converter because the output voltage is larger than the input.

Voltage and Current Relationships

The analysis assumes the following:

1. Steady-state conditions exist.
2. The switching period is T, and the switch is closed for time DT and open for $(1-D)T$.
3. The inductor current is continuous (always positive).
4. The capacitor is very large, and the output voltage is held constant at voltage V_o.
5. The components are ideal.

The analysis proceeds by examining the inductor voltage and current for the switch closed and again for the switch open.

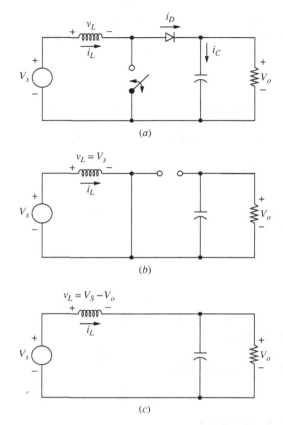

Figure 6-8 The boost converter. (a) Circuit; (b) Equivalent circuit for the switch closed; (c) Equivalent circuit for the switch open.

Analysis for the Switch Closed When the switch is closed, the diode is reverse-biased. Kirchhoff's voltage law around the path containing the source, inductor, and closed switch is

$$v_L = V_s = L\frac{di_L}{dt} \quad \text{or} \quad \frac{di_L}{dt} = \frac{V_s}{L} \qquad (6\text{-}24)$$

The rate of change of current is a constant, so the current increases linearly while the switch is closed, as shown in Fig. 6-9b. The change in inductor current is computed from

$$\frac{\Delta i_L}{\Delta t} = \frac{\Delta i_L}{DT} = \frac{V_s}{L}$$

Solving for Δi_L for the switch closed,

$$(\Delta i_L)_{\text{closed}} = \frac{V_s DT}{L} \qquad (6\text{-}25)$$

6.5 The Boost Converter

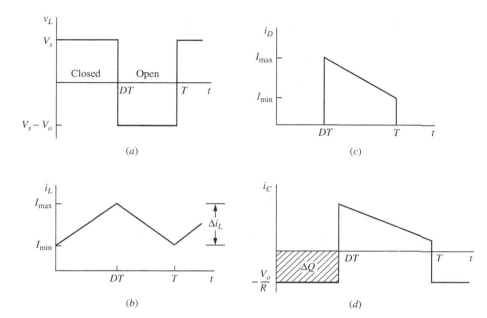

Figure 6-9 Boost converter waveforms. (*a*) Inductor voltage; (*b*) Inductor current; (*c*) Diode current; (*d*) Capacitor current.

Analysis for the Switch Open When the switch is opened, the inductor current cannot change instantaneously, so the diode becomes forward-biased to provide a path for inductor current. Assuming that the output voltage V_o is a constant, the voltage across the inductor is

$$v_L = V_s - V_o = L\frac{di_L}{dt}$$

$$\frac{di_L}{dt} = \frac{V_s - V_o}{L}$$

The rate of change of inductor current is a constant, so the current must change linearly while the switch is open. The change in inductor current while the switch is open is

$$\frac{\Delta i_L}{\Delta t} = \frac{\Delta i_L}{(1-D)T} = \frac{V_s - V_o}{L}$$

Solving for Δi_L,

$$(\Delta i_L)_{\text{open}} = \frac{(V_s - V_o)(1-D)T}{L} \qquad (6\text{-}26)$$

For steady-state operation, the net change in inductor current must be zero. Using Eqs. (6-25) and (6-26),

$$(\Delta i_L)_{\text{closed}} + (\Delta i_L)_{\text{open}} = 0$$

$$\frac{V_s D T}{L} + \frac{(V_s - V_o)(1 - D)T}{L} = 0$$

Solving for V_o,

$$V_s(D + 1 - D) - V_o(1 - D) = 0$$

$$\boxed{V_o = \frac{V_s}{1 - D}} \qquad (6\text{-}27)$$

Also, the average inductor voltage must be zero for periodic operation. Expressing the average inductor voltage over one switching period,

$$V_L = V_s D + (V_s - V_o)(1 - D) = 0$$

Solving for V_o yields the same result as in Eq. (6-27).

Equation (6-27) shows that if the switch is always open and D is zero, the output voltage is the same as the input. As the duty ratio is increased, the denominator of Eq. (6-27) becomes smaller, resulting in a larger output voltage. *The boost converter produces an output voltage that is greater than or equal to the input voltage.* However, the output voltage cannot be less than the input, as was the case with the buck converter.

As the duty ratio of the switch approaches 1, the output voltage goes to infinity according to Eq. (6-27). However, Eq. (6-27) is based on ideal components. Real components that have losses will prevent such an occurrence, as shown later in this section. Figure 6-9 shows the voltage and current waveforms for the boost converter.

The average current in the inductor is determined by recognizing that the average power supplied by the source must be the same as the average power absorbed by the load resistor. Output power is

$$P_o = \frac{V_o^2}{R} = V_o I_o$$

and input power is $V_s I_s = V_s I_L$. Equating input and output powers and using Eq. (6-27),

$$V_s I_L = \frac{V_o^2}{R} = \frac{[V_s/(1-D)]^2}{R} = \frac{V_s^2}{(1-D)^2 R}$$

By solving for average inductor current and making various substitutions, I_L can be expressed as

$$\boxed{I_L = \frac{V_s}{(1-D)^2 R} = \frac{V_o^2}{V_s R} = \frac{V_o I_o}{V_s}} \qquad (6\text{-}28)$$

Maximum and minimum inductor currents are determined by using the average value and the change in current from Eq. (6-25).

$$I_{max} = I_L + \frac{\Delta i_L}{2} = \frac{V_s}{(1-D)^2 R} + \frac{V_s DT}{2L} \qquad (6\text{-}29)$$

$$I_{min} = I_L - \frac{\Delta i_L}{2} = \frac{V_s}{(1-D)^2 R} - \frac{V_s DT}{2L} \qquad (6\text{-}30)$$

Equation (6-27) was developed with the assumption that the inductor current is continuous, meaning that it is always positive. A condition necessary for continuous inductor current is for I_{min} to be positive. Therefore, the boundary between continuous and discontinuous inductor current is determined from

$$I_{min} = 0 = \frac{V_s}{(1-D)^2 R} - \frac{V_s DT}{2L}$$

or

$$\frac{V_s}{(1-D)^2 R} = \frac{V_s DT}{2L} = \frac{V_s D}{2Lf}$$

The minimum combination of inductance and switching frequency for continuous current in the boost converter is therefore

$$(Lf)_{min} = \frac{D(1-D)^2 R}{2} \qquad (6\text{-}31)$$

or

$$\boxed{L_{min} = \frac{D(1-D)^2 R}{2f}} \qquad (6\text{-}32)$$

A boost converter designed for continuous-current operation will have an inductor value greater than L_{min}.

From a design perspective, it is useful to express L in terms of a desired Δi_L,

$$L = \frac{V_s DT}{\Delta i_L} = \frac{V_s D}{\Delta i_L f} \qquad (6\text{-}33)$$

Output Voltage Ripple

The preceding equations were developed on the assumption that the output voltage was a constant, implying an infinite capacitance. In practice, a finite capacitance will result in some fluctuation in output voltage, or ripple.

The peak-to-peak output voltage ripple can be calculated from the capacitor current waveform, shown in Fig. 6-9d. The change in capacitor charge can be calculated from

$$|\Delta Q| = \left(\frac{V_o}{R}\right)DT = C\Delta V_o$$

An expression for ripple voltage is then

$$\Delta V_o = \frac{V_o DT}{RC} = \frac{V_o D}{RCf}$$

or

$$\boxed{\frac{\Delta V_o}{V_o} = \frac{D}{RCf}} \quad (6\text{-}34)$$

where f is the switching frequency. Alternatively, expressing capacitance in terms of output voltage ripple yields

$$C = \frac{D}{R(\Delta V_o/V_o)f} \quad (6\text{-}35)$$

As with the buck converter, equivalent series resistance of the capacitor can contribute significantly to the output voltage ripple. The peak-to-peak variation in capacitor current (Fig. 6-9) is the same as the maximum current in the inductor. The voltage ripple due to the ESR is

$$\Delta V_{o,\text{ESR}} = \Delta i_C r_C = I_{L,\max} r_C \quad (6\text{-}36)$$

EXAMPLE 6-4

Boost Converter Design 1

Design a boost converter that will have an output of 30 V from a 12-V source. Design for continuous inductor current and an output ripple voltage of less than one percent. The load is a resistance of 50 Ω. Assume ideal components for this design.

■ **Solution**

First, determine the duty ratio from Eq. (6-27),

$$D = 1 - \frac{V_s}{V_o} = 1 - \frac{12}{30} = 0.6$$

If the switching frequency is selected at 25 kHz to be above the audio range, then the minimum inductance for continuous current is determined from Eq. (6-32).

$$L_{\min} = \frac{D(1-D)^2(R)}{2f} = \frac{0.6(1-0.6)^2(50)}{2(25{,}000)} = 96 \ \mu\text{H}$$

To provide a margin to ensure continuous current, let $L = 120 \ \mu\text{H}$. Note that L and f are selected somewhat arbitrarily and that other combinations will also give continuous current.

Using Eqs. (6-28) and (6-25),

$$I_L = \frac{V_s}{(1-D)^2(R)} = \frac{12}{(1-0.6)^2(50)} = 1.5 \text{ A}$$

$$\frac{\Delta i_L}{2} = \frac{V_s DT}{2L} = \frac{(12)(0.6)}{(2)(120)(10)^{-6}(25,000)} = 1.2 \text{ A}$$

$$I_{\max} = 1.5 + 1.2 = 2.7 \text{ A}$$

$$I_{\min} = 1.5 - 1.2 = 0.3 \text{ A}$$

The minimum capacitance required to limit the output ripple voltage to 1 percent is determined from Eq. (6-35).

$$C \geq \frac{D}{R(\Delta V_o/V_o)f} = \frac{0.6}{(50)(0.01)(25,000)} = 48 \text{ }\mu\text{F}$$

EXAMPLE 6-5

Boost Converter Design 2

A boost converter is required to have an output voltage of 8 V and supply a load current of 1 A. The input voltage varies from 2.7 to 4.2 V. A control circuit adjusts the duty ratio to keep the output voltage constant. Select the switching frequency. Determine a value for the inductor such that the variation in inductor current is no more than 40 percent of the average inductor current for all operating conditions. Determine a value of an ideal capacitor such that the output voltage ripple is no more than 2 percent. Determine the maximum capacitor equivalent series resistance for a 2 percent ripple.

■ **Solution**

Somewhat arbitrarily, choose 200 kHz for the switching frequency. The circuit must be analyzed for both input voltage extremes to determine the worst-case condition. For $V_s = 2.7$ V, the duty ratio is determined from Eq. (6-27).

$$D = 1 - \frac{V_s}{V_o} = 1 - \frac{2.7}{8} = 0.663$$

Average inductor current is determined from Eq. (6-28).

$$I_L = \frac{V_o I_o}{V_s} = \frac{8(1)}{2.7} = 2.96 \text{ A}$$

The variation in inductor current to meet the 40 percent specification is then $\Delta i_L = 0.4(2.96) = 1.19$ A. The inductance is then determined from Eq. (6-33).

$$L = \frac{V_s D}{\Delta i_L f} = \frac{2.7(0.663)}{1.19(200,000)} = 7.5 \text{ }\mu\text{H}$$

Repeating the calculations for $V_s = 4.2$ V,

$$D = 1 - \frac{V_s}{V_o} = 1 - \frac{4.2}{8} = 0.475$$

$$I_L = \frac{V_o I_o}{V_s} = \frac{8(1)}{4.2} = 1.90 \text{ A}$$

The variation in inductor current for this case is $\Delta i_L = 0.4(1.90) = 0.762$ A, and

$$L = \frac{V_s D}{\Delta i_L f} = \frac{4.2(0.475)}{0.762(200{,}000)} = 13.1 \ \mu\text{H}$$

The inductor must be 13.1 µH to satisfy the specifications for the total range of input voltages.

Equation (6-35), using the maximum value of D, gives the minimum capacitance as

$$C = \frac{D}{R(\Delta V_o/V_o)f} = \frac{D}{(V_o/I_o)(\Delta V_o/V_o)f} = \frac{0.663}{(8/1)(0.02)(200{,}000)} = 20.7 \ \mu\text{F}$$

The maximum ESR is determined from Eq. (6-36), using the maximum peak-to-peak variation in capacitor current. The peak-to-peak variation in capacitor current is the same as maximum inductor current. The average inductor current varies from 2.96 A at $V_s = 2.7$ V to 1.90 A at $V_s = 4.2$ V. The variation in inductor current is 0.762 A for $V_s = 4.2$ A, but it must be recalculated for $V_s = 2.7$ V using the 13.1-µH value selected, yielding

$$\Delta i_L = \frac{V_s D}{Lf} = \frac{2.7(0.663)}{13.1(10)^{-6}(200{,}000)} = 0.683 \ \text{A}$$

Maximum inductor current for each case is then computed as

$$I_{L,\text{max},2.7\text{V}} = I_L + \frac{\Delta i_L}{2} = 2.96 + \frac{0.683}{2} = 3.30 \ \text{A}$$

$$I_{L,\text{max},4.2\text{V}} = I_L + \frac{\Delta i_L}{2} = 1.90 + \frac{0.762}{2} = 2.28 \ \text{A}$$

This shows that the largest peak-to-peak current variation in the capacitor will be 3.30 A. The output voltage ripple due to the capacitor ESR must be no more than $(0.02)(8) = 0.16$ V. Using Eq. (6-36),

$$\Delta V_{o,\text{ESR}} = \Delta i_C r_C = I_{L,\text{max}} r_C = 3.3 r_C = 0.16 \ \text{V}$$

which gives

$$r_C = \frac{0.16 \ \text{V}}{3.3 \ \text{A}} = 48 \ \text{m}\Omega$$

In practice, a capacitor that has an ESR of 48 mΩ or less could have a capacitance value much larger than the 20.7 µF calculated.

Inductor Resistance

Inductors should be designed to have small resistance to minimize power loss and maximize efficiency. The existence of a small inductor resistance does not substantially change the analysis of the buck converter as presented previously in this chapter. However, inductor resistance affects performance of the boost converter, especially at high duty ratios.

For the boost converter, recall that the output voltage for the ideal case is

$$V_o = \frac{V_s}{1 - D} \qquad (6\text{-}37)$$

To investigate the effect of inductor resistance on the output voltage, assume that the inductor current is approximately constant. The source current is the same as the inductor current, and average diode current is the same as average load current. The power supplied by the source must be the same as the power absorbed by the load and the inductor resistance, neglecting other losses.

$$P_s = P_o + P_{r_L}$$
$$V_s I_L = V_o I_D + I_L^2 r_L \qquad (6\text{-}38)$$

where r_L is the series resistance of the inductor. The diode current is equal to the inductor current when the switch is off and is zero when the switch is on. Therefore, the average diode current is

$$I_D = I_L(1 - D) \qquad (6\text{-}39)$$

Substituting for I_D into Eq. (6-38),

$$V_s I_L = V_o I_L(1 - D) + I_L^2 r_L$$

which becomes

$$V_s = V_o(1 - D) + I_L r_L \qquad (6\text{-}40)$$

In terms of V_o from Eq. (6-39), I_L is

$$I_L = \frac{I_D}{1 - D} \approx \frac{V_o/R}{1 - D} \qquad (6\text{-}41)$$

Substituting for I_L into Eq. (6-40),

$$V_s = \frac{V_o r_L}{R(1 - D)} + V_o(1 - D)$$

Solving for V_o,

$$\boxed{V_o = \left(\frac{V_s}{1 - D}\right)\left(\frac{1}{1 + r_L/[R(1 - D)^2]}\right)} \qquad (6\text{-}42)$$

The preceding equation is similar to that for an ideal converter but includes a correction factor to account for the inductor resistance. Figure 6-10a shows the output voltage of the boost converter with and without inductor resistance.

The inductor resistance also has an effect on the power efficiency of converters. Efficiency is the ratio of output power to output power plus losses. For the boost converter

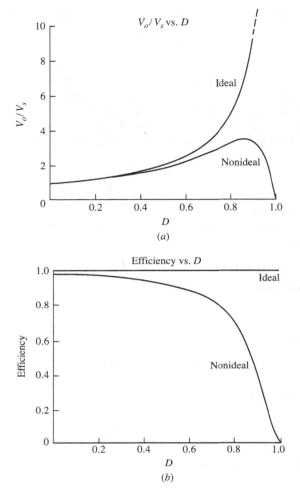

Figure 6-10 Boost converter for a nonideal inductor. (*a*) Output voltage; (*b*) Boost converter efficiency.

$$\eta = \frac{P_o}{P_o + P_{\text{loss}}} = \frac{V_o^2/R}{V_o^2/R + I_L^2 r_L} \tag{6-43}$$

Using Eq. (6-41) for I_L,

$$\eta = \frac{V_o^2/R}{V_o^2/R + (V_o/R)^2/(1-D)r_L} = \frac{1}{1 + r_L[R(1-D)^2]} \tag{6-44}$$

As the duty ratio increases, the efficiency of the boost converter decreases, as indicated in Fig. 6-10*b*.

6.6 THE BUCK-BOOST CONVERTER

Another basic switched-mode converter is the buck-boost converter shown in Fig. 6-11. The output voltage of the buck-boost converter can be either higher or lower than the input voltage.

Voltage and Current Relationships

Assumptions made about the operation of the converter are as follows:

1. The circuit is operating in the steady state.
2. The inductor current is continuous.
3. The capacitor is large enough to assume a constant output voltage.
4. The switch is closed for time DT and open for $(1-D)T$.
5. The components are ideal.

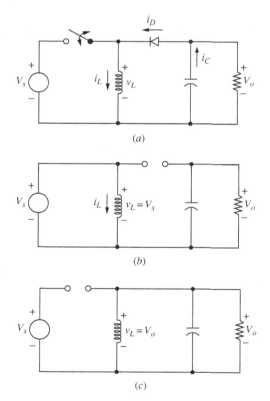

Figure 6-11 Buck-boost converter. (*a*) Circuit; (*b*) Equivalent circuit for the switch closed; (*c*) Equivalent circuit for the switch open.

Analysis for the Switch Closed When the switch is closed, the voltage across the inductor is

$$v_L = V_s = L\frac{di_L}{dt}$$

$$\frac{di_L}{dt} = \frac{V_s}{L}$$

The rate of change of inductor current is a constant, indicating a linearly increasing inductor current. The preceding equation can be expressed as

$$\frac{\Delta i_L}{\Delta t} = \frac{\Delta i_L}{DT} = \frac{V_s}{L}$$

Solving for Δi_L when the switch is closed gives

$$(\Delta i_L)_{\text{closed}} = \frac{V_s DT}{L} \qquad (6\text{-}45)$$

Analysis for the Switch Open When the switch is open, the current in the inductor cannot change instantaneously, resulting in a forward-biased diode and current into the resistor and capacitor. In this condition, the voltage across the inductor is

$$v_L = V_o = L\frac{di_L}{dt}$$

$$\frac{di_L}{dt} = \frac{V_o}{L}$$

Again, the rate of change of inductor current is constant, and the change in current is

$$\frac{\Delta i_L}{\Delta t} = \frac{\Delta i_L}{(1-D)T} = \frac{V_o}{L}$$

Solving for Δi_L,

$$(\Delta i_L)_{\text{open}} = \frac{V_o(1-D)T}{L} \qquad (6\text{-}46)$$

For steady-state operation, the net change in inductor current must be zero over one period. Using Eqs. (6-45) and (6-46),

$$(\Delta i_L)_{\text{closed}} + (\Delta i_L)_{\text{open}} = 0$$

$$\frac{V_s DT}{L} + \frac{V_o(1-D)T}{L} = 0$$

Solving for V_o,

$$\boxed{V_o = -V_s\left(\frac{D}{1-D}\right)} \qquad (6\text{-}47)$$

The required duty ratio for specified input and output voltages can be expressed as

$$D = \frac{|V_o|}{V_s + |V_o|} \qquad (6\text{-}48)$$

The average inductor voltage is zero for periodic operation, resulting in

$$V_L = V_s D + V_o(1 - D) = 0$$

Solving for V_o yields the same result as Eq. (6-47).

Equation (6-47) shows that the output voltage has opposite polarity from the source voltage. *Output voltage magnitude of the buck-boost converter can be less than that of the source or greater than the source, depending on the duty ratio of the switch.* If $D > 0.5$, the output voltage is larger than the input; and if $D < 0.5$, the output is smaller than the input. Therefore, this circuit combines the capabilities of the buck and boost converters. Polarity reversal on the output may be a disadvantage in some applications, however. Voltage and current waveforms are shown in Fig. 6-12.

Note that the source is never connected directly to the load in the buck-boost converter. Energy is stored in the inductor when the switch is closed and transferred to the load when the switch is open. Hence, the buck-boost converter is also referred to as an *indirect* converter.

Power absorbed by the load must be the same as that supplied by the source, where

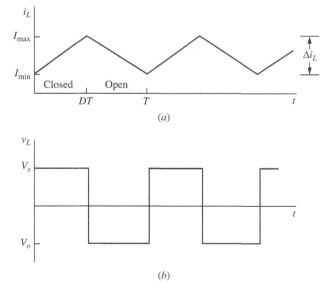

Figure 6-12 Buck-boost converter waveforms.
(*a*) Inductor current; (*b*) Inductor voltage; (*c*) Diode current; (*d*) Capacitor current.

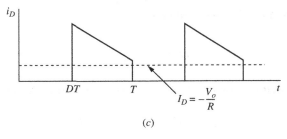

(c)

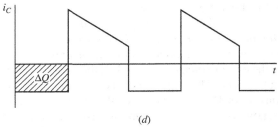

(d)

Figure 6-12 *(continued)*

$$P_o = \frac{V_o^2}{R}$$

$$P_s = V_s I_s$$

$$\frac{V_o^2}{R} = V_s I_s$$

Average source current is related to average inductor current by

$$I_s = I_L D$$

resulting in

$$\frac{V_o^2}{R} = V_s I_L D$$

Substituting for V_o using Eq. (6-47) and solving for I_L, we find

$$I_L = \frac{V_o^2}{V_s RD} = \frac{P_o}{V_s D} = \frac{V_s D}{R(1-D)^2} \qquad (6\text{-}49)$$

Maximum and minimum inductor currents are determined using Eqs. (6-45) and (6-49).

$$I_{\max} = I_L + \frac{\Delta i_L}{2} = \frac{V_s D}{R(1-D)^2} + \frac{V_s DT}{2L} \qquad (6\text{-}50)$$

$$I_{\min} = I_L - \frac{\Delta i_L}{2} = \frac{V_s D}{R(1-D)^2} - \frac{V_s DT}{2L} \qquad (6\text{-}51)$$

For continuous current, the inductor current must remain positive. To determine the boundary between continuous and discontinuous current, I_{min} is set to zero in Eq. (6-51), resulting in

$$(Lf)_{min} = \frac{(1-D)^2 R}{2} \quad (6-52)$$

or

$$\boxed{L_{min} = \frac{(1-D)^2 R}{2f}} \quad (6-53)$$

where f is the switching frequency.

Output Voltage Ripple

The output voltage ripple for the buck-boost converter is computed from the capacitor current waveform of Fig. 6-12d.

$$|\Delta Q| = \left(\frac{V_o}{R}\right) DT = C \Delta V_o$$

Solving for ΔV_o,

$$\Delta V_o = \frac{V_o DT}{RC} = \frac{V_o D}{RCf}$$

or

$$\boxed{\frac{\Delta V_o}{V_o} = \frac{D}{RCf}} \quad (6-54)$$

As is the case with other converters, the equivalent series resistance of the capacitor can contribute significantly to the output ripple voltage. The peak-to-peak variation in capacitor current is the same as the maximum inductor current. Using the capacitor model shown in Fig. 6-6, where $I_{L,max}$ is determined from Eq. (6-50),

$$\Delta V_{o,ESR} = \Delta i_C r_C = I_{L,max} r_C \quad (6-55)$$

EXAMPLE 6-6

Buck-Boost Converter

The buck-boost circuit of Fig. 6-11 has these parameters:

$V_s = 24$ V
$D = 0.4$
$R = 5\ \Omega$
$L = 20\ \mu H$
$C = 80\ \mu F$
$f = 100$ kHz

Determine the output voltage, inductor current average, maximum and minimum values, and the output voltage ripple.

■ **Solution**

Output voltage is determined from Eq. (6-47).

$$V_o = -V_s\left(\frac{D}{1-D}\right) = -24\left(\frac{0.4}{1-0.4}\right) = -16 \text{ V}$$

Inductor current is described by Eqs. (6-49) to (6-51).

$$I_L = \frac{V_s D}{R(1-D)^2} = \frac{24(0.4)}{5(1-0.4)^2} = 5.33 \text{ A}$$

$$\Delta i_L = \frac{V_s DT}{L} = \frac{24(0.4)}{20(10)^{-6}(100,000)} = 4.8 \text{ A}$$

$$I_{L,\max} = I_L + \frac{\Delta i_L}{2} = 5.33 + \frac{4.8}{2} = 7.33 \text{ A}$$

$$I_{L,\min} = I_L - \frac{\Delta i_L}{2} = 5.33 - \frac{4.8}{2} = 2.93 \text{ A}$$

Continuous current is verified by $I_{\min} > 0$.
Output voltage ripple is determined from Eq. (6-54).

$$\frac{\Delta V_o}{V_o} = \frac{D}{RCf} = \frac{0.4}{(5)(80)(10)^{-6}(100,000)} = 0.01 = 1\%$$

6.7 THE ĆUK CONVERTER

The Ćuk switching topology is shown in Fig. 6-13a. Output voltage magnitude can be either larger or smaller than that of the input, and there is a polarity reversal on the output.

The inductor on the input acts as a filter for the dc supply to prevent large harmonic content. Unlike the previous converter topologies where energy transfer is associated with the inductor, energy transfer for the Ćuk converter depends on the capacitor C_1.

The analysis begins with these assumptions:

1. Both inductors are very large and the currents in them are constant.
2. Both capacitors are very large and the voltages across them are constant.
3. The circuit is operating in steady state, meaning that voltage and current waveforms are periodic.
4. For a duty ratio of D, the switch is closed for time DT and open for $(1-D)T$.
5. The switch and the diode are ideal.

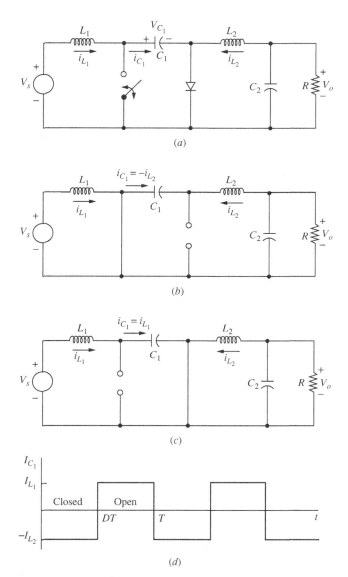

Figure 6-13 The Ćuk converter. (*a*) Circuit; (*b*) Equivalent circuit for the switch closed; (*c*) Equivalent circuit for the switch open; (*d*) Current in L_1 for a large inductance.

The average voltage across C_1 is computed from Kirchhoff's voltage law around the outermost loop. The average voltage across the inductors is zero for steady-state operation, resulting in

$$V_{C_1} = V_s - V_o$$

With the switch closed, the diode is off and the current in capacitor C_1 is

$$(i_{C_1})_{\text{closed}} = -I_{L_2} \tag{6-56}$$

With the switch open, the currents in L_1 and L_2 force the diode on. The current in capacitor C_1 is

$$(i_{C_1})_{\text{open}} = I_{L_1} \tag{6-57}$$

The power absorbed by the load is equal to the power supplied by the source:

$$-V_o I_{L_2} = V_s I_{L_1} \tag{6-58}$$

For periodic operation, the average capacitor current is zero. With the switch on for time DT and off for $(1-D)T$,

$$[(i_{C_1})_{\text{closed}}]DT + [(i_{C_1})_{\text{open}}](1-D)T = 0$$

Substituting using Eqs. (6-56) and (6-57),

$$-I_{L_2}DT + I_{L_1}(1-D)T = 0$$

or

$$\frac{I_{L_1}}{I_{L_2}} = \frac{D}{1-D} \tag{6-59}$$

Next, the average power supplied by the source must be the same as the average power absorbed by the load,

$$P_s = P_o$$

$$V_s I_{L_1} = -V_o I_{L_2} \tag{6-60}$$

$$\frac{I_{L_1}}{I_{L_2}} = -\frac{V_o}{V_s}$$

Combining Eqs. (6-59) and (6-60), the relationship between the output and input voltages is

$$\boxed{V_o = -V_s \left(\frac{D}{1-D}\right)} \tag{6-61}$$

The negative sign indicates a polarity reversal between output and input.

Note that the components on the output (L_2, C_2, and R) are in the same configuration as the buck converter and that the inductor current has the same form as for the buck converter. Therefore, the ripple, or variation in output voltage, is the same as for the buck converter:

$$\boxed{\frac{\Delta V_o}{V_o} = \frac{1-D}{8 L_2 C_2 f^2}} \tag{6-62}$$

The output ripple voltage will be affected by the equivalence series resistance of the capacitor as it was in the convertors discussed previously.

The ripple in C_1 can be estimated by computing the change in v_{C1} in the interval when the switch is open and the currents i_{L_1} and i_{C_1} are the same. Assuming the current in L_1 to be constant at a level I_{L_1} and using Eqs. (6-60) and (6-61), we have

$$\Delta v_{C1} \approx \frac{1}{C_1} \int_{DT}^{T} I_{L_1} d(t) = \frac{I_{L_1}}{C_1}(1-D)T = \frac{V_s}{RC_1 f}\left(\frac{D^2}{1-D}\right)$$

or

$$\Delta v_{C1} \approx \frac{V_o D}{RC_1 f} \qquad (6\text{-}63)$$

The fluctuations in inductor currents can be computed by examining the inductor voltages while the switch is closed. The voltage across L_1 with the switch closed is

$$v_{L1} = V_s = L_1 \frac{di_{L1}}{dt} \qquad (6\text{-}64)$$

In the time interval DT when the switch is closed, the change in inductor current is

$$\frac{\Delta i_{L1}}{DT} = \frac{V_s}{L_1}$$

or

$$\Delta i_{L1} = \frac{V_s DT}{L_1} = \frac{V_s D}{L_1 f} \qquad (6\text{-}65)$$

For inductor L_2, the voltage across it when the switch is closed is

$$v_{L2} = V_o + (V_s - V_o) = V_s = L_2 \frac{di_{L2}}{dt} \qquad (6\text{-}66)$$

The change in i_{L_2} is then

$$\Delta i_{L2} = \frac{V_s DT}{L_2} = \frac{V_s D}{L_2 f} \qquad (6\text{-}67)$$

For continuous current in the inductors, the average current must be greater than one-half the change in current. Minimum inductor sizes for continuous current are

$$L_{1,\text{min}} = \frac{(1-D)^2 R}{2Df}$$

$$L_{2,\text{min}} = \frac{(1-D)R}{2f} \qquad (6\text{-}68)$$

EXAMPLE 6-7

Ćuk Converter Design

A Ćuk converter has an input of 12 V and is to have an output of -18 V supplying a 40-W load. Select the duty ratio, the switching frequency, the inductor sizes such that the change in inductor currents is no more than 10 percent of the average inductor current, the output ripple voltage is no more than 1 percent, and the ripple voltage across C_1 is no more than 5 percent.

■ **Solution**

The duty ratio is obtained from Eq. (6-61),

$$\frac{V_o}{V_s} = -\frac{D}{1-D} = \frac{-18}{12} = -1.5$$

or

$$D = 0.6$$

Next, the switching frequency needs to be selected. Higher switching frequencies result in smaller current variations in the inductors. Let $f = 50$ kHz. The average inductor currents are determined from the power and voltage specifications.

$$I_{L2} = \frac{P_o}{-V_o} = \frac{40 \text{ W}}{18 \text{ V}} = 2.22 \text{ A}$$

$$I_{L1} = \frac{P_s}{V_s} = \frac{40 \text{ W}}{12 \text{ V}} = 3.33 \text{ A}$$

The change in inductor currents is computed from Eqs. (6-65) and (6-67).

$$\Delta i_L = \frac{V_s D}{L f}$$

The 10 percent limit in changes in inductor currents requires

$$L_2 \geq \frac{V_s D}{f \Delta i_{L2}} = \frac{(12)(0.6)}{(50{,}000)(0.222)} = 649 \text{ }\mu\text{H}$$

$$L_1 \geq \frac{V_s D}{f \Delta i_{L1}} = \frac{(12)(0.6)}{(50{,}000)(0.333)} = 432 \text{ }\mu\text{H}$$

From Eq. (6-62), the output ripple specification requires

$$C_2 \geq \frac{1-D}{(\Delta V_o/V_o) 8 L_2 f^2} = \frac{1-0.6}{(0.01)(8)(649)(10)^{-6}(50{,}000)^2} = 3.08 \text{ }\mu\text{F}$$

Average voltage across C_1 is $V_s - V_o = 12 - (-18) = 30$ V, so the maximum change in v_{C_1} is $(30)(0.05) = 1.5$ V.

The equivalent load resistance is

$$R = \frac{V_o^2}{P} = \frac{(18)^2}{40} = 8.1 \text{ }\Omega$$

Now C_1 is computed from the ripple specification and Eq. (6-63).

$$C_1 \geq \frac{V_o D}{R f \Delta v_{C_1}} = \frac{(18)(0.6)}{(8.1)(50{,}000)(1.5)} = 17.8 \text{ }\mu\text{F}$$

6.8 THE SINGLE-ENDED PRIMARY INDUCTANCE CONVERTER (SEPIC)

A converter similar to the Ćuk is the single-ended primary inductance converter (SEPIC), as shown in Fig. 6-14. The SEPIC can produce an output voltage that is either greater or less than the input but with no polarity reversal.

To derive the relationship between input and output voltages, these initial assumptions are made:

1. Both inductors are very large and the currents in them are constant.
2. Both capacitors are very large and the voltages across them are constant.
3. The circuit is operating in the steady state, meaning that voltage and current waveforms are periodic.
4. For a duty ratio of D, the switch is closed for time DT and open for $(1 - D)T$.
5. The switch and the diode are ideal.

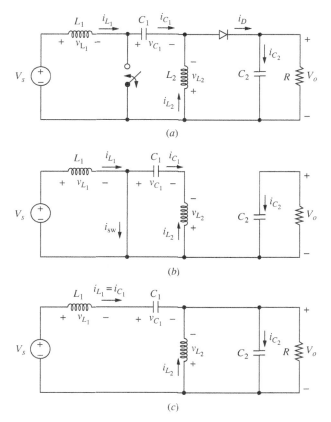

Figure 6-14 (*a*) SEPIC circuit; (*b*) Circuit with the switch closed and the diode off; (*c*) Circuit with the switch open and the diode on.

The inductor current and capacitor voltage restrictions will be removed later to investigate the fluctuations in currents and voltages. The inductor currents are assumed to be continuous in this analysis. Other observations are that the average inductor voltages are zero and that the average capacitor currents are zero for steady-state operation.

Kirchhoff's voltage law around the path containing V_s, L_1, C_1, and L_2 gives

$$-V_s + v_{L_1} + v_{C_1} - v_{L_2} = 0$$

Using the average of these voltages,

$$-V_s + 0 + V_{C_1} - 0 = 0$$

showing that the average voltage across the capacitor C_1 is

$$V_{C_1} = V_s \tag{6-69}$$

When the switch is closed, the diode is off, and the circuit is as shown in Fig. 6-14b. The voltage across L_1 for the interval DT is

$$v_{L_1} = V_s \tag{6-70}$$

When the switch is open, the diode is on, and the circuit is as shown in Fig. 6-14c. Kirchhoff's voltage law around the outermost path gives

$$-V_s + v_{L_1} + v_{C_1} + V_o = 0 \tag{6-71}$$

Assuming that the voltage across C_1 remains constant at its average value of V_s [Eq. (6-69)],

$$-V_s + v_{L_1} + V_s + V_o = 0 \tag{6-72}$$

or

$$v_{L_1} = -V_o \tag{6-73}$$

for the interval $(1 - D)T$. Since the average voltage across an inductor is zero for periodic operation, Eqs. (6-70) and (6-73) are combined to get

$$(v_{L_1,\,\text{sw closed}})(DT) + (v_{L_1,\,\text{sw open}})(1 - D)T = 0$$

$$V_s(DT) - V_o(1 - D)T = 0$$

where D is the duty ratio of the switch. The result is

$$\boxed{V_o = V_s\left(\frac{D}{1-D}\right)} \tag{6-74}$$

which can be expressed as

$$\boxed{D = \frac{V_o}{V_o + V_s}} \tag{6-75}$$

6.8 The Single-Ended Primary Inductance Converter (SEPIC)

This result is similar to that of the buck-boost and Ćuk converter equations, with the important distinction that there is no polarity reversal between input and output voltages. The ability to have an output voltage greater or less than the input with no polarity reversal makes this converter suitable for many applications.

Assuming no losses in the converter, the power supplied by the source is the same as the power absorbed by the load.

$$P_s = P_o$$

Power supplied by the dc source is voltage times the average current, and the source current is the same as the current in L_1.

$$P_s = V_s I_s = V_s I_{L_1}$$

Output power can be expressed as

$$P_o = V_o I_o$$

resulting in

$$V_s I_{L_1} = V_o I_o$$

Solving for average inductor current, which is also the average source current,

$$I_{L_1} = I_s = \frac{V_o I_o}{V_s} = \frac{V_o^2}{V_s R} \quad (6\text{-}76)$$

The variation in i_{L_1} when the switch is closed is found from

$$v_{L_1} = V_s = L_1 \left(\frac{di_{L_1}}{dt}\right) = L_1 \left(\frac{\Delta i_{L_1}}{\Delta t}\right) = L_1 \left(\frac{\Delta i_{L_1}}{DT}\right) \quad (6\text{-}77)$$

Solving for Δi_{L_1},

$$\Delta i_{L_1} = \frac{V_s DT}{L_1} = \frac{V_s D}{L_1 f} \quad (6\text{-}78)$$

For L_2, the average current is determined from Kirchhoff's current law at the node where C_1, L_2, and the diode are connected.

$$i_{L_2} = i_D - i_{C_1}$$

Diode current is

$$i_D = i_{C_2} + I_o$$

which makes

$$i_{L_2} = i_{C_2} + I_o - i_{C_1}$$

The average current in each capacitor is zero, so the average current in L_2 is

$$I_{L_2} = I_o \quad (6\text{-}79)$$

The variation in i_{L2} is determined from the circuit when the switch is closed. Using Kirchhoff's voltage law around the path of the closed switch, C_1, and L_2 with the voltage across C_1 assumed to be a constant V_s, gives

$$v_{L2} = v_{C_1} = V_s = L_2\left(\frac{di_{L2}}{dt}\right) = L_2\left(\frac{\Delta i_{L2}}{\Delta t}\right) = L_2\left(\frac{\Delta i_{L2}}{DT}\right)$$

Solving for Δi_{L2}

$$\Delta i_{L2} = \frac{V_s DT}{L_2} = \frac{V_s D}{L_2 f} \tag{6-80}$$

Applications of Kirchhoff's current law show that the diode and switch currents are

$$i_D = \begin{cases} 0 & \text{when switch is closed} \\ i_{L_1} + i_{L_2} & \text{when switch is open} \end{cases}$$

$$i_{sw} = \begin{cases} i_{L_1} + i_{L_2} & \text{when switch is closed} \\ 0 & \text{when switch is open} \end{cases} \tag{6-81}$$

Current waveforms are shown in Fig. 6-15.

Kirchhoff's voltage law applied to the circuit of Fig. 6-14c, assuming no voltage ripple across the capacitors, shows that the voltage across the switch when it is open is $V_s + V_o$. From Fig. 6-14b, the maximum reverse bias voltage across the diode when it is off is also $V_s + V_o$.

The output stage consisting of the diode, C_2, and the load resistor is the same as in the boost converter, so the output ripple voltage is

$$\boxed{\Delta V_o = \Delta V_{C_2} = \frac{V_o D}{R C_2 f}} \tag{6-82}$$

Solving for C_2,

$$\boxed{C_2 = \frac{D}{R(\Delta V_o / V_o) f}} \tag{6-83}$$

The voltage variation in C_1 is determined form the circuit with the switch closed (Fig. 6-14b). Capacitor current i_{C_1} is the opposite of i_{L2}, which has previously been determined to have an average value of I_o. From the definition of capacitance and considering the magnitude of charge,

$$\Delta V_{C_1} = \frac{\Delta Q_{C_1}}{C} = \frac{I_o \Delta t}{C} = \frac{I_o DT}{C}$$

6.8 The Single-Ended Primary Inductance Converter (SEPIC)

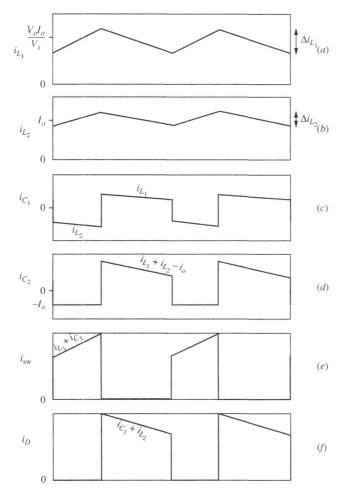

Figure 6-15 Currents in the SEPIC converter. (a) L_1; (b) L_2; (c) C_1; (d) C_2; (e) switch; (f) diode.

Replacing I_o with V_o/R,

$$\Delta V_{C_1} = \frac{V_o D}{R C_1 f} \qquad (6\text{-}84)$$

Solving for C_1,

$$C_1 = \frac{D}{R(\Delta V_{C_1}/V_o)f} \qquad (6\text{-}85)$$

The effect of equivalent series resistance of the capacitors on voltage variation is usually significant, and the treatment is the same as with the converters discussed previously.

EXAMPLE 6-8

SEPIC Circuit

The SEPIC circuit of Fig. 6-14a has the following parameters:

$$V_s = 9 \text{ V}$$
$$D = 0.4$$
$$f = 100 \text{ kHz}$$
$$L_1 = L_2 = 90 \text{ }\mu\text{H}$$
$$C_1 = C_2 = 80 \text{ }\mu\text{F}$$
$$I_o = 2 \text{ A}$$

Determine the output voltage; the average, maximum, and minimum inductor currents; and the variation in voltage across each capacitor.

■ **Solution**

The output voltage is determined from Eq. (6-74).

$$V_o = V_s\left(\frac{D}{1-D}\right) = 9\left(\frac{0.4}{1-0.4}\right) = 6 \text{ V}$$

The average current in L_1 is determined from Eq. (6-76).

$$I_{L_1} = \frac{V_o I_o}{V_s} = \frac{6(2)}{9} = 1.33 \text{ A}$$

From Eq. (6-78)

$$\Delta i_{L_1} = \frac{V_s D}{L_1 f} = \frac{9(0.4)}{90(10)^{-6}(100{,}000)} = 0.4 \text{ A}$$

Maximum and minimum currents in L_1 are then

$$I_{L_1,\text{max}} = I_{L_1} + \frac{\Delta i_{L_1}}{2} = 1.33 + \frac{0.4}{2} = 1.53 \text{ A}$$

$$I_{L_1,\text{min}} = I_{L_1} - \frac{\Delta i_{L_1}}{2} = 1.33 - \frac{0.4}{2} = 1.13 \text{ A}$$

For the current in L_2, the average is the same as the output current $I_o = 2$ A. The variation in I_{L_2} is determined from Eq. (6-80)

$$\Delta i_{L_2} = \frac{V_s D}{L_2 f} = \frac{9(0.4)}{90(10)^{-6}(100{,}000)} = 0.4 \text{ A}$$

resulting in maximum and minimum current magnitudes of

$$I_{L_2,\text{max}} = 2 + \frac{0.4}{2} = 2.2 \text{ A}$$

$$I_{L_2,\text{min}} = 2 - \frac{0.4}{2} = 1.8 \text{ A}$$

Using an equivalent load resistance of 6 V/2 A = 3 Ω, the ripple voltages in the capacitors are determined from Eqs. (6-82) and (6-84).

$$\Delta V_o = \Delta V_{C_2} = \frac{V_o D}{RC_2 f} = \frac{6(0.4)}{(3)80(10)^{-6}(100{,}000)} = 0.1 \text{ V}$$

$$\Delta V_{C_1} = \frac{V_o D}{RC_1 f} = \frac{6(0.4)}{(3)80(10)^{-6}(100{,}000)} = 0.1 \text{ V}$$

In Example 6-8, the values of L_1 and L_2 are equal, which is not a requirement. However, when they are equal, the rates of change in the inductor currents are identical [Eqs. (6-78) and (6-80)]. The two inductors may then be wound on the same core, making a 1:1 transformer. Figure 6-16 shows an alternative representation of the SEPIC converter.

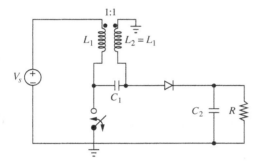

Figure 6-16 A SEPIC circuit using mutually coupled inductors.

6.9 INTERLEAVED CONVERTERS

Interleaving, also called *multiphasing*, is a technique that is useful for reducing the size of filter components. An interleaved buck converter is shown in Fig. 6-17a. This is equivalent to a parallel combination of two sets of switches, diodes, and inductors connected to a common filter capacitor and load. The switches are operated 180° out of phase, producing inductor currents that are also 180° out of phase. The current entering the capacitor and load resistance is the sum of the inductor currents, which has a smaller peak-to-peak variation and a frequency twice as large as individual inductor currents. This results in a smaller peak-to-peak variation in capacitor current than would be achieved with a single buck converter, requiring less capacitance for the same output ripple voltage. The variation in current coming from the source is also reduced. Figure 6-17b shows the current waveforms.

The output voltage is obtained by taking Kirchhoff's voltage law around either path containing the voltage source, a switch, an inductor, and the output voltage. The voltage across the inductor is $V_s - V_o$ with the switch closed and

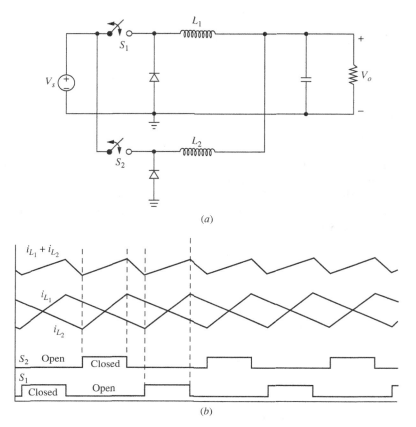

Figure 6-17 (*a*) An interleaved buck converter; (*b*) The switching scheme and current waveforms.

is $-V_o$ with the switch open. These are the same as for the buck converter of Fig. 6-3a discussed previously, resulting in

$$V_o = V_s D$$

where D is the duty ratio of each switch.

Each inductor supplies one-half of the load current and output power, so the average inductor current is one-half of what it would be for a single buck converter.

More than two converters can be interleaved. The phase shift between switch closing is 360°/n, where *n* is the number of converters in the parallel configuration. Interleaving can be done with the other converters in this chapter and with the converters that are described in Chap. 7. Figure 6-18 shows an interleaved boost converter.

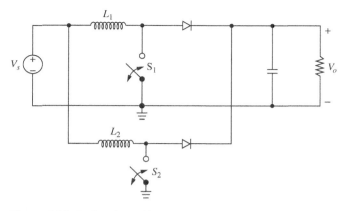

Figure 6-18 An interleaved boost converter.

6.10 NONIDEAL SWITCHES AND CONVERTER PERFORMANCE

Switch Voltage Drops

All the preceding calculations were made with the assumption that the switches were ideal. Voltage drops across conducting transistors and diodes may have a significant effect on converter performance, particularly when the input and output voltages are low. Design of dc-dc converters must account for nonideal components. The buck converter is used to illustrate the effects of switch voltage drops.

Referring again to the analysis of the buck converter of Fig. 6-3a, the input-output voltage relationship was determined using the inductor voltage and current. With nonzero voltage drops across conducting switches, the voltage across the inductor with the switch closed becomes

$$v_L = V_s - V_o - V_Q \tag{6-86}$$

where V_Q is the voltage across the conducting switch. With the switch open, the voltage across the diode is V_D and the voltage across the inductor is

$$v_L = -V_o - V_D \tag{6-87}$$

The average voltage across the inductor is zero for the switching period.

$$V_L = (V_s - V_o - V_Q)D + (-V_o - V_D)(1 - D) = 0$$

Solving for V_o,

$$\boxed{V_o = V_s D - V_Q D - V_D(1 - D)} \tag{6-88}$$

which is lower than $V_o = V_s D$ for the ideal case.

Switching Losses

In addition to the on-state voltage drops and associated power losses of the switches, other losses occur in the switches as they turn on and off. Figure 6-19a illustrates switch on-off transitions. For this case, it is assumed that the changes in voltage and current are linear and that the timing sequence is as shown. The instantaneous power dissipated in the switch is shown in Fig. 6-19a. Another possible switch on-off transition is shown in Fig. 6-12b. In this case, the voltage and current transitions do not occur simultaneously. This may be closer to actual switching situations, and switching power loss is larger for this case. (See Chap. 10 for additional information.)

The energy loss in one switching transition is the area under the power curve. Since the average power is energy divided by the period, higher switching frequencies result in higher switching losses. One way to reduce switching losses is to modify the circuit to make switching occur at zero voltage and/or zero current. This is the approach of the resonant converter, which is discussed in Chap. 9.

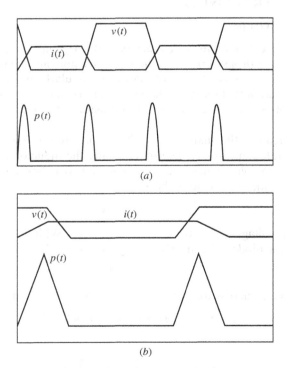

Figure 6-19 Switch voltage, current, and instantaneous power. (a) Simultaneous voltage and current transition; (b) Worst-case transition.

6.11 DISCONTINUOUS-CURRENT OPERATION

Continuous current in the inductor was an important assumption in the previous analyses for dc-dc converters. Recall that continuous current means that the current in the inductor remains positive for the entire switching period. Continuous current is not a necessary condition for a converter to operate, but a different analysis is required for the discontinuous-current case.

Buck Converter with Discontinuous Current

Figure 6-20 shows the inductor and source currents for discontinuous-current operation for the buck converter of Fig 6-3a. The relationship between output and input voltages is determined by first recognizing that the average inductor voltage is zero for periodic operation. From the inductor voltage shown in Fig. 6-20c,

$$(V_s - V_o)DT - V_o D_1 T = 0$$

which is rearranged to get

$$(V_s - V_o)D = V_o D_1 \qquad (6\text{-}89)$$

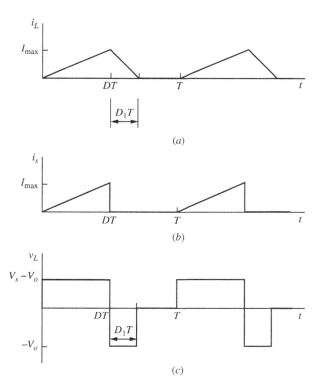

Figure 6-20 Buck converter discontinuous current.
(*a*) Inductor current; (*b*) Source current; (*c*) Inductor voltage.

$$\frac{V_o}{V_s} = \left(\frac{D}{D+D_1}\right) \qquad (6\text{-}90)$$

Next, the average inductor current equals the average resistor current because the average capacitor current is zero. With the output voltage assumed constant,

$$I_L = I_R = \frac{V_o}{R}$$

Computing the average inductor current from Fig. 6-20a,

$$I_L = \frac{1}{T}\left(\frac{1}{2}I_{max}\,DT + \frac{1}{2}I_{max}\,D_1 T\right) = \frac{1}{2}I_{max}(D+D_1)$$

which results in

$$\frac{1}{2}I_{max}(D+D_1) = \frac{V_o}{R} \qquad (6\text{-}91)$$

Since the current starts at zero, the maximum current is the same as the change in current over the time that the switch is closed. With the switch closed, the voltage across the inductor is

$$v_L = V_s - V_o$$

which results in

$$\frac{di_L}{dt} = \frac{V_s - V_o}{L} = \frac{\Delta i_L}{\Delta t} = \frac{\Delta i_L}{DT} = \frac{I_{max}}{DT} \qquad (6\text{-}92)$$

Solving for I_{max} and using Eq. (6-89) for $(V_s - V_o)D$,

$$I_{max} = \Delta i_L = \left(\frac{V_s - V_o}{L}\right)DT = \frac{V_o D_1 T}{L} \qquad (6\text{-}93)$$

Substituting for I_{max} in Eq. (6-91),

$$\frac{1}{2}I_{max}(D+D_1) = \frac{1}{2}\left(\frac{V_o D_1 T}{L}\right)(D+D_1) = \frac{V_o}{R} \qquad (6\text{-}94)$$

which gives

$$D_1^2 + DD_1 - \frac{2L}{RT} = 0$$

Solving for D_1,

$$D_1 = \frac{-D + \sqrt{D^2 + 8L/RT}}{2} \qquad (6\text{-}95)$$

Substituting for D_1 in Eq. (6-90),

$$V_o = V_s\left(\frac{D}{D+D_1}\right) = V_s\left[\frac{2D}{D+\sqrt{D^2 + 8L/RT}}\right] \quad (6\text{-}96)$$

The boundary between continuous and discontinuous current occurs when $D_1 = 1 - D$. Recall that another condition that occurs at the boundary between continuous and discontinuous current is $I_{min} = 0$ in Eq. (6-12).

EXAMPLE 6-9

Buck Converter with Discontinuous Current

For the buck converter of Fig. 6-3a,

$V_s = 24$ V
$L = 200$ μH
$R = 20$ Ω
$C = 1000$ μF
$f = 10$ kHz   switching frequency
$D = 0.4$

(a) Show that the inductor current is discontinuous, (b) Determine the output voltage V_o.

■ **Solution**

(a) For discontinuous current, $D_1 < 1 - D$, and D_1 is calculated from Eq. (6-95).

$$D_1 = \frac{-D + \sqrt{D^2 + 8L/RT}}{2}$$

$$= \frac{1}{2}\left(-0.4 + \sqrt{0.4^2 + \frac{8(200)(10)^{-6}(10,000)}{20}}\right) = 0.29$$

Comparing D_1 to $1 - D$, $0.29 < (1 - 0.4)$ shows that the inductor current is discontinuous. Alternatively, the minimum inductor current computed from Eq. (6-12) is $I_{min} = -0.96$ A. Since negative inductor current is not possible, inductor current must be discontinuous.

(b) Since D_1 is calculated and discontinuous current is verified, the output voltage can be computed from Eq. (6-96).

$$V_o = V_s\left(\frac{D}{D+D_1}\right) = 20\left(\frac{0.4}{0.4 + 0.29}\right) = 13.9 \text{ V}$$

Figure 6-21 shows the relationship between output voltage and duty ratio for the buck converter of Example 6-9. All parameters except D are those of Example 6-9. Note the linear relationship between input and output for continuous current and the nonlinear relationship for discontinuous current. For a given duty ratio, the output voltage is greater for discontinuous-current operation than it would be if current were continuous.

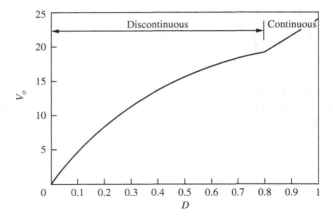

Figure 6-21 V_o versus duty ratio for the buck converter of Example 6-9.

Boost Converter with Discontinuous Current

The boost converter will also operate for discontinuous inductor current. In some cases, the discontinuous-current mode is desirable for control reasons in the case of a regulated output. The relationship between output and input voltages is determined from two relationships:

1. The average inductor voltage is zero.
2. The average current in the diode is the same as the load current.

The inductor and diode currents for discontinuous current have the basic waveforms as shown in Fig. 6-22a and c. When the switch is on, the voltage across the inductor is V_s. When the switch is off and the inductor current is positive, the inductor voltage is $V_s - V_o$. The inductor current decreases until it reaches zero and is prevented from going negative by the diode. With the switch open and the diode off, the inductor current is zero. The average voltage across the inductor is

$$V_s DT + (V_s - V_o)D_1 T = 0$$

which results in

$$V_o = V_s \left(\frac{D + D_1}{D_1} \right) \tag{6-97}$$

The average diode current (Fig. 6-22c) is

$$I_D = \frac{1}{T}\left(\frac{1}{2} I_{max} D_1 T\right) = \frac{1}{2} I_{max} D_1 \tag{6-98}$$

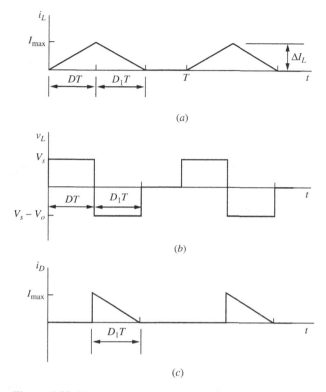

Figure 6-22 Discontinuous current in the boost converter. (a) Inductor current; (b) Inductor voltage; (c) Diode current.

Current I_{max} is the same as the change in inductor current when the switch is closed.

$$I_{max} = \Delta i_L = \frac{V_s DT}{L} \qquad (6\text{-}99)$$

Substituting for I_{max} in Eq. (6-98) and setting the result equal to the load current,

$$I_D = \frac{1}{2}\left(\frac{V_s DT}{L}\right)D_1 = \frac{V_o}{R} \qquad (6\text{-}100)$$

Solving for D_1,

$$D_1 = \left(\frac{V_o}{V_s}\right)\left(\frac{2L}{RDT}\right) \qquad (6\text{-}101)$$

Substituting the preceding expression for D_1 into Eq. (6-97) results in the quadratic equation

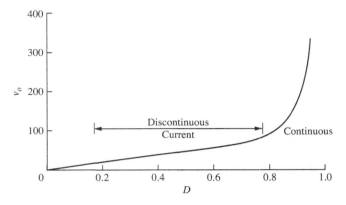

Figure 6-23 Output voltage of boost converter.

$$\left(\frac{V_o}{V_s}\right)^2 - \frac{V_o}{V_s} - \frac{D^2 RT}{2L} = 0$$

Solving for V_o/V_s,

$$\frac{V_o}{V_s} = \frac{1}{2}\left(1 + \sqrt{1 + \frac{2D^2 RT}{L}}\right) \quad (6\text{-}102)$$

The boundary between continuous and discontinuous current occurs when $D_1 = 1 - D$. Another condition at the boundary is when $I_{\min}$ in Eq. (6-30) is zero.

Whether the boost converter is operating in the continuous or discontinuous mode depends on the combination of circuit parameters, including the duty ratio. As the duty ratio for a given boost converter is varied, the converter may go into and out of the discontinuous mode. Figure 6-23 shows the output voltage for a boost converter as the duty ratio is varied.

EXAMPLE 6-10

Boost Converter with Discontinuous Current

The boost converter of Fig. 6-8a has parameters

$V_s = 20$ V
$D = 0.6$
$L = 100$ μH
$R = 50$ Ω
$C = 100$ μF
$f = 15$ kHz

(*a*) Verify that the inductor current is discontinuous, (*b*) determine the output voltage, and (*c*) determine the maximum inductor current.

■ Solution

(a) First assume that the inductor current is continuous and compute the minimum from Eq. (6-30), resulting in $I_{min} = -1.5$ A. Negative inductor current is not possible, indicating discontinuous current.

(b) Equation (6-102) gives the output voltage

$$V_o = \frac{V_s}{2}\left(1 + \sqrt{1 + \frac{2D^2R}{Lf}}\right) = \frac{20}{2}\left[1 + \sqrt{1 + \frac{2(0.6)^2(50)}{100(10)^{-6}(15,000)}}\right] = 60 \text{ V}$$

Note that a boost converter with the same duty ratio operating with continuous current would have an output of 50 V.

(c) The maximum inductor current is determined from Eq. (6-99).

$$I_{max} = \frac{V_s D}{Lf} = \frac{(20)(0.6)}{100(10)^{-6}(15,000)} = 8 \text{ A}$$

6.12 SWITCHED-CAPACITOR CONVERTERS

In switched-capacitor converters, capacitors are charged in one circuit configuration and then reconnected in a different configuration, producing an output voltage different from the input. Switched-capacitor converters do not require an inductor and are also known as *inductorless converters* or *charge pumps*. Switched-capacitor converters are useful for applications that require small currents, usually less than 100 MA. Applications include use in RS-232 data signals that require both positive and negative voltages for logic levels; in flash memory circuits, where large voltages are needed to erase stored information; and in drivers for LEDs and LCD displays.

The basic types of switched-capacitor converters are the step-up (boost), the inverting, and the step-down (buck) circuits. The following discussion introduces the concepts of switched-capacitor converters.

The Step-Up Switched-Capacitor Converter

A common application of a switched-capacitor converter is the step-up (boost) converter. The basic principle is shown in Fig. 6-24a. A capacitor is first connected across the source to charge it to V_s. The charged capacitor is then connected in series with the source, producing an output voltage of $2V_s$.

A switching scheme to accomplish this is shown in Fig. 6-24b. The switch pair labeled 1 is closed and opened in a phase sequence opposite to that of switch pair 2. Switch pair 1 closes to charge the capacitor and then opens. Switch pair 2 then closes to produce an output of $2V_s$.

The switches can be implemented with transistors, or they can be implemented with transistors and diodes, as shown in Fig. 6-24c. Transistor M_1 is turned on, and C_1 is charged to V_s through D_1. Next, M_1 is turned off and M_2 is turned on. Kirchhoff's voltage law around the path of the source, the charged

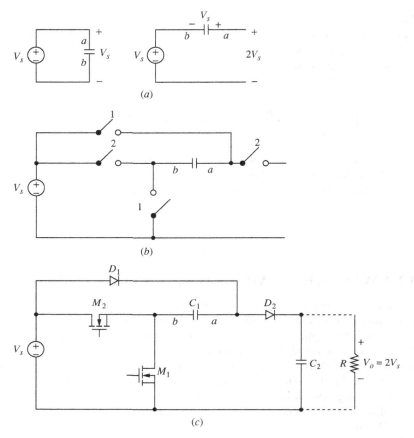

Figure 6-24 A switched-capacitor step-up converter. (*a*) A capacitor is charged and then reconnected to produce a voltage of twice that of the source; (*b*) A switch arrangement; (*c*) An implementation using transistors and diodes and showing a second capacitor C_2 to sustain the output voltage during switching.

capacitor C_1, and V_o shows that $V_o = 2V_s$. The capacitor C_2 on the output is required to sustain the output voltage and to supply load current when C_1 is disconnected from the load. With C_2 included, it will take several switching cycles to charge it and achieve the final output voltage. With the resistor connected, current will flow from the capacitors, but the output voltage will be largely unaffected if the switching frequency is sufficiently high and capacitor charges are replenished in short time intervals. The output will be less than $2V_s$ for real devices because of voltage drops in the circuit.

Converters can be made to step up the input voltage to values greater than $2V_s$. In Fig. 6-25*a*, two capacitors are charged and then reconnected to create a

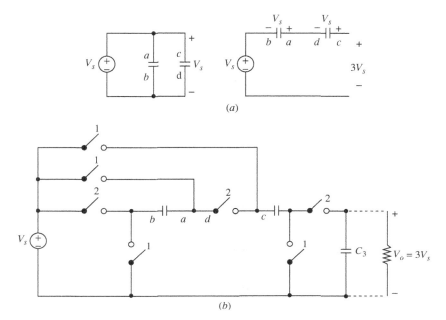

Figure 6-25 A step-up switched-capacitor converter to produce 3 times the source voltage. (*a*) Each capacitor is charged to V_s and reconnected to produce an output of $3V_s$; (*b*) A switch arrangement also shows an output capacitor to sustain the output voltage during switching.

voltage of $3V_s$. A switching arrangement to implement this circuit is shown in Fig. 6-25*b*. Switch sets 1 and 2 open and close alternately. The circuit includes an output capacitor C_3 to sustain the voltage across the load during the switching cycle.

The Inverting Switched-Capacitor Converter

The inverting switched-capacitor converter is useful for producing a negative voltage from a single voltage source. For example, -5 V can be made from a 5-V source, thereby creating a +/− 5-V supply. The basic concept is shown in Fig. 6-26*a*. A capacitor is charged to the source voltage and then connected to the output with opposite polarity.

A switching scheme to accomplish this is shown in Fig. 6-26*b*. Switch pairs 1 and 2 open and close in opposite phase sequence. Switch pair 1 closes to charge the capacitor and then opens. Switch pair 2 then closes to produce an output of $-V_s$.

A switch configuration to implement the inverting circuit is shown in Fig. 6-26*c*. An output capacitor C_2 is included to sustain the output and supply current to the load during the switching cycle. Transistor M_1 is turned on, charging C_1 to V_s through D_1. Transistor M_1 is turned off and M_2 is turned on, charging C_2 with a

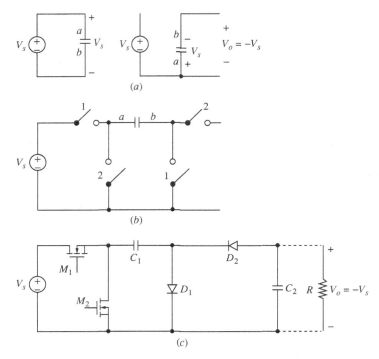

Figure 6-26 The inverting switched-capacitor converter. (*a*) The capacitor is charged to V_s and then reconnected to produce an output of $-V_s$; (*b*) A switch arrangement; (*c*) An implementation using transistors and diodes and showing a second capacitor to sustain the output voltage during switching.

polarity that is positive on the bottom. After several switching cycles, the output voltage is $-V_s$.

The Step-Down Switched-Capacitor Converter

A step-down (buck) switched-capacitor converter is shown in Fig. 6-27. In Fig. 6-27*a*, two capacitors of equal value are connected in series, resulting in a voltage of $V_s/2$ across each. The capacitors are then reconnected in parallel, making the output voltage $V_s/2$. A switching scheme to accomplish this is shown in Fig. 6-27*b*. Switch pairs 1 and 2 open and close in opposite phase sequence. With the resistor connected, current will flow from the capacitors, but the output voltage will be unaffected if the switching frequency is sufficiently high and capacitor charges are replenished in short time intervals.

A switch configuration to implement the inverting circuit is shown in Fig. 6-27*c*. Transistor M_1 is turned on, and both capacitors charge through D_1.

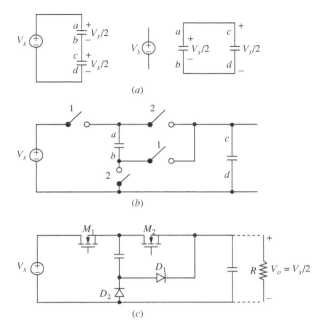

Figure 6-27 The step-down switched-capacitor converter. (a) The capacitors are in series and each is charged to $V_s/2$, followed by the capacitors in parallel, with the output voltage at $V_s/2$; (b) A switch arrangement; (c) An implementation using transistors and diodes.

Transistor M_1 is turned off, and M_2 is turned on, connecting the capacitors in parallel through D_1. And D_2 is forward-biased as the capacitors discharge into the load resistor.

6.13 PSPICE SIMULATION OF DC-DC CONVERTERS

The circuit model to be used for PSpice simulation of the dc-dc converters discussed in this chapter depends on the ultimate goal of the simulation. To predict the behavior of a circuit with the goal of producing the periodic voltage and current waveforms requires a circuit model that includes a switch. A voltage-controlled switch is convenient for this application. If the circuit includes an ideal diode and lossless inductors and capacitors, the simulation results will be first-order approximations of circuit behavior, much the same as the analytical work done previously in this chapter. By including parasitic elements and using nonideal switching devices in the circuit model, the simulation will be useful to investigate how a real circuit is expected to depart from the ideal.

Another simulation goal may be to predict the dynamic behavior of a dc-dc converter for changes in the source voltage or load current. A disadvantage of using the cycle-to-cycle switched model is that the time for overall circuit transients may be orders of magnitude larger than the switching period, thereby making the program execution time quite long. A circuit model that does not include the cycle-by-cycle details but does simulate the large-scale dynamic behavior by using averaging techniques may be preferred. PSpice simulations for both cycle-to-cycle and large-scale dynamic behavior are discussed in this section.

A Switched PSpice Model

A voltage-controlled switch is a simple way to model a transistor switch that would actually be used in a physical converter. The voltage-controlled switch has an on resistance that could be selected to match the transistor's, or the on resistance could be chosen negligibly small to simulate an ideal switch. A pulse voltage source acts as the control for the switch.

When periodic closing and opening of the switch in a dc-dc converter begins, a transient response precedes the steady-state voltages and currents described earlier in this chapter. The following example illustrates a PSpice simulation for a buck converter using idealized models for circuit components.

EXAMPLE 6-11

Buck Converter Simulation Using Idealized Components

Use PSpice to verify the buck converter design in Example 6-3.
The buck converter has the following parameters:

$V_s = 3.3$ V
$L = 1$ μH
$C = 667$ μF   with an ESR of 15 mΩ
$R = 0.3$ Ω   for a load current of 4 A
$D = 0.364$ for an output of 1.2 V
Switching frequency = 500 kHz

■ Solution

A PSpice model for the buck converter is shown in Fig. 6-28. A voltage-controlled switch (Sbreak) is used for the switching transistor, with the on resistance R_{on} set to 1 mΩ to approximate an ideal device. An ideal diode is simulated by letting the diode parameter n (the emission coefficient in the diode equation) be 0.001. The switch is controlled by a pulse voltage source. The parameter statements file facilitates modification of the circuit file for other buck converters. Initial conditions for the inductor current and capacitor voltage are assumed to be zero to demonstrate the transient behavior of the circuit.

Figure 6-29a shows the Probe output for inductor current and capacitor voltage. Note that there is a transient response of the circuit before the steady-state periodic condition is reached. From the steady-state portion of the Probe output shown in Fig. 6-29b, the maximum and minimum values of the output voltage are 1.213 and 1.1911 V, respectively, for a peak-to-peak variation of about 22 mV, agreeing well with the 24-mV design objective. The maximum and minimum inductor currents are about 4.77 and 3.24 A, agreeing well with the 4.8- and 3.2-A design objectives.

6.13 PSpice Simulation of DC-DC Converters

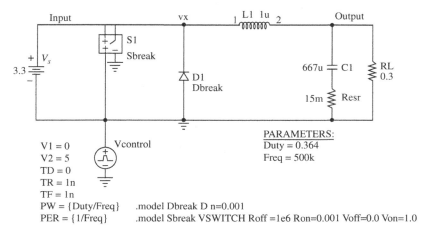

Figure 6-28 PSpice circuit for the buck converter.

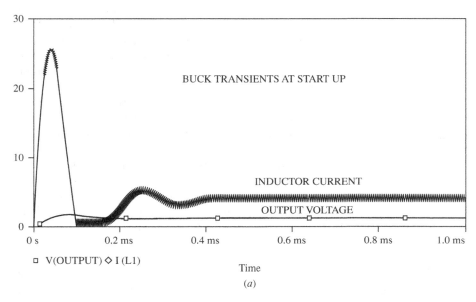

Figure 6-29 Probe output for Example 6-11 (*a*) showing the transient at start-up and (*b*) in steady state.

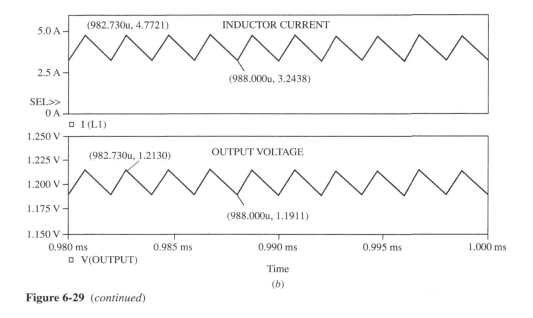

Figure 6-29 (*continued*)

An Averaged Circuit Model

PSpice simulation of the dc-dc buck converter in Example 6-11 includes both the large scale transient behavior and the cycle-to-cycle waveforms of voltage and current. If the goal of a simulation is to determine the large-scale transient behavior, the cycle-to-cycle response merely adds to the execution time of the program. A more time-efficient way to simulate the transient behavior of dc-dc converters is to use a circuit model that produces the *average* values of voltages and currents only, rather than including the detailed variations around the averages. In general, transient behavior for dc-dc converters can be predicted by analyzing linear networks, with the response equal to the average value of the switching waveforms. The discussion that follows is focused on the buck converter operating in the continuous-current mode.

The transient behavior of the average output voltage can be described using linear circuit analysis. The input v_x to the *RLC* circuit of the buck converter of Fig. 6-3a has an average value of $V_x = V_s D$. The response of the *RLC* circuit to a step input voltage of $v_x(t) = (V_s D)u(t)$ represents the average of the output voltage and current waveforms when the converter is turned on. This represents the same large-scale transient that was present in the PSpice simulation shown in Fig. 6-29a.

For complete simulation of the large-scale behavior of a dc-dc converter, it is desirable to include the proper voltage and current relationships between the source and the load. Taking the buck converter as an example, the relationship between average voltage and current at the input and output for continuous inductor current is given by

6.13 PSpice Simulation of DC-DC Converters   255

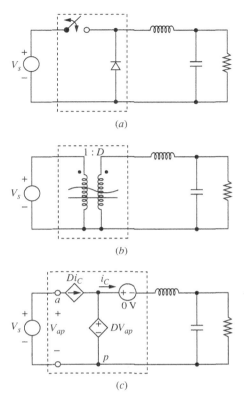

Figure 6-30 (*a*) Buck converter with switch; (*b*) Circuit model for averaged buck converter; (*c*) PSpice circuit.

$$\frac{V_o}{V_s} = \frac{I_s}{I_o} = D \tag{6-103}$$

Since $V_o = V_s D$ and $I_o = I_s/D$, the switch in a model for computing average voltage and current is the same as a "transformer" which has a turns ratio of 1:D. Circuit models for a buck converter using a 1:D transformer and a PSpice circuit for implementing the averaged model are shown in Fig. 6-30. The circuit symbol for the transformer indicates that the model is valid for both ac and dc signals.

The following example illustrates the use of the PSpice model to simulate the response of average voltage and current for a buck converter.

EXAMPLE 6-12

Averaged Buck Converter

Use the averaged circuit of Fig. 6-30*c* to simulate the buck converter having parameters

$V_s = 10$ V
$D = 0.2$
$L = 400$ μH
$C = 400$ μF
$R = 2$ Ω
$f = 5$ kHz

Use initial conditions of zero for inductor current and capacitor voltage.

■ **Solution**

The PSpice implementation of the averaged model is shown in Fig. 6-31a. The simulation results from both a switched model and for the averaged model are shown in Fig. 6-31b. Note that the switched model shows the cycle-to-cycle variation, while the average model shows only the averaged values.

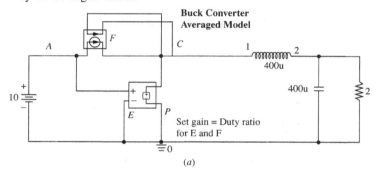

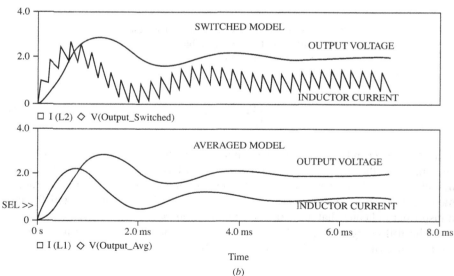

Figure 6-31 (a) PSpice implementation of the averaged buck converter model; (b) Probe output for both the switched model and the averaged model.

The averaged model can be quite useful in investigating the dynamic behavior of the converter when it is subjected to changes in operating parameters. Such an analysis is essential when the output is regulated through a feedback loop which is designed to keep the output at a set level by adjusting the duty ratio of the switch to accommodate variations in the source or the load. Closed-loop response is discussed in Chap. 7 on dc power supplies.

The following example illustrates the use of the averaged circuit model to simulate a step change in load resistance.

EXAMPLE 6-13

Step Change in Load

Use the averaged buck converter model to determine the dynamic response when the load resistance is changed. The circuit parameters are

$V_s = 50$ V
$L = 1$ mH   with a series resistance of 0.4 Ω
$C = 100$ μF   with an equivalent series resistance of 0.5 Ω
$R = 4\ \Omega$,   stepped to 2 Ω and back to 4 Ω
$D = 0.4$
Switching frequency = 5 kHz

■ Solution

Step changes in load are achieved by switching a second 4-Ω resistor across the output at 6 ms and disconnecting it at 16 ms. The averaged model shows the transients associated with output voltage and inductor current (Fig. 6-32b). Also shown for comparison are the results of a different simulation using a switch, showing the cycle-to-cycle variations in voltage and current.

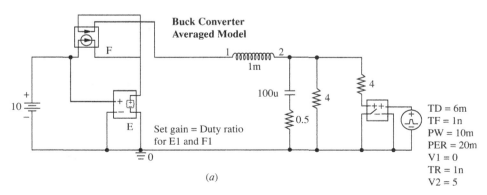

Figure 6-32 (a) PSpice implementation of the averaged model with a switched load; (b) Probe results for both the switched model and the averaged model.

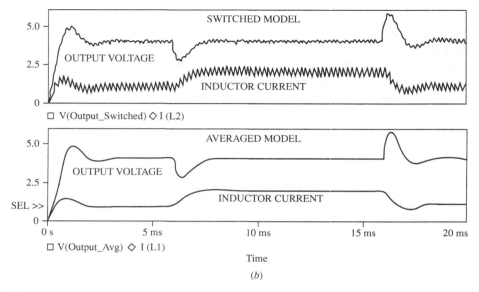

(b)

Figure 6-32 (*continued*)

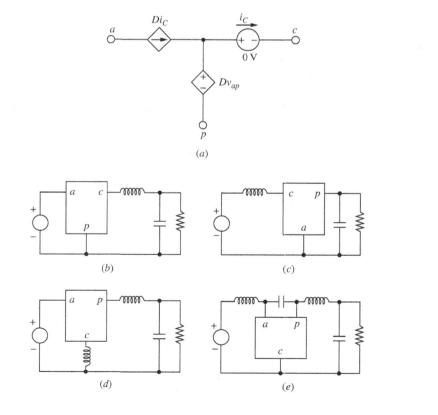

Figure 6-33
Averaged switch model in dc-dc converters. (*a*) PSpice averaged model for switch and diode; (*b*) Buck equivalent; (*c*) Boost equivalent; (*d*) Buck-boost equivalent; (*e*) Ćuk equivalent.

The averaged switch model can be used to simulate the other dc-dc converters discussed in this chapter. Figure 6-33 shows how the average switch model is used in the boost, buck-boost, and Ćuk converters for continuous-current operation. The designation of the switch terminals a, p, and c represents active, passive, and common terminals.

6.14 Summary

- A switched-mode dc-dc converter is much more efficient than a linear converter because of reduced losses in the electronic switch.
- A buck converter has an output voltage less than the input.
- A boost converter has an output voltage greater than the input.
- Buck-boost and Ćuk converters can have output voltages greater than or less than the input, but there is a polarity reversal.
- A SEPIC (single-ended primary-inductor converter) can have an output voltage greater than or less than the input with no polarity reversal.
- Output voltage is generally reduced from the theoretical value when switch drops and inductor resistances are included in the analysis.
- Capacitor equivalent series resistance (ESR) may produce an output voltage ripple much greater than that of the capacitance alone.
- Interleaved converters have parallel switch/inductor paths to reduce the current variation in the output capacitor.
- Discontinuous-current modes for dc-dc converters are possible and sometimes desirable, but input-output relationships are different from those for the continuous-current modes.
- Switched-capacitor converters charge capacitors in one configuration and then use switches to reconnect the capacitors to produce an output voltage different from the input.
- PSpice can be used to simulate dc-dc converters by using a voltage-controlled switch or by using an averaged circuit model.

6.15 Bibliography

S. Ang and A. Oliva, *Power-Switching Converters*, 2d ed., Taylor & Francis, Boca Raton, Fla., 2005.

C. Basso, *Switch-Mode Power Supplies,* McGraw-Hill, New York, 2008.

B. K. Bose, *Power Electronics and Motor Drives: Advances and Trends*, Elsevier/Academic Press, Boston, 2006.

R. W. Erickson and D. Maksimović, *Fundamentals of Power Electronics*, 2d ed., Kluwer Academic, Boston, 2001.

W. Gu, "Designing a SEPIC Converter," National Semiconductor Application Note 1484, 2007, http://www.national.com/an/AN/AN-1484.pdf.

P. T. Krein, *Elements of Power Electronics*, Oxford University Press, New York, 1998.

D. Maksimović, and S. Dhar, "Switched-Capacitor DC-DC Converters for Low-Power On-Chip Applications," *IEEE Annual Power Electronics Specialists Conference*, vol. 1, pp. 54–59, 1999.

R. D. Middlebrook and, S. Ćuk, *Advances in Switched-Mode Power Conversion*, vols. I and II, TESLAco, Pasadena, Calif., 1981.

N. Mohan, T. M. Undeland, and W. P. Robbins, *Power Electronics: Converters, Applications, and Design,* 3d ed., Wiley, New York, 2003.

A. I. Pressman, K. Billings, and T. Morey, *Switching Power Supply Design*, McGraw-Hill, New York, 2009.

M. H. Rashid, *Power Electronics: Circuits, Devices, and Systems*, 3d ed., Prentice-Hall, Upper Saddle River, N.J., 2004.

"SEPIC Equations and Component Ratings," MAXIM Application Note 1051, 2002, http://www.maxim-ic.com/an1051.

V. Vorperian, "Simplified Analysis of PWM Converters Using Model of PWM Switch", *IEEE Transactions on Aerospace and Electronic Systems*, May 1990.

Problems

Linear Converters

6-1. What is the relationship between V_o/V_s and efficiency for the linear converter described in Sec. 6.1?

6-2. A dc power supply must step down a 100-V. source to 30 V. The output power is 100 W. (*a*) Determine the efficiency of the linear converter of Fig. 6-1 when it is used for this application. (*b*) How much energy is lost in the transistor in 1 yr? (*c*) Using the electric rate in your area, what is the cost of the energy loss for 1 yr?

Basic Switched Converter

6-3. The basic dc-dc converter of Fig. 6-2*a* has a source of 100 V and a load resistance of 10 Ω. The duty ratio of the switch is $D = 0.6$, and the switching frequency is 1 kHz. Determine (*a*) the average voltage across the load, (*b*) the rms voltage across the load, and (*c*) the average power absorbed by the load. (*d*) What would happen if the switching frequency were increased to 2 kHz?

Buck Converter

6-4. The buck converter of Fig. 6-3*a* has the following parameters: $V_s = 24$ V, $D = 0.65$, $L = 25$ μH, $C = 15$ μF, and $R = 10$ Ω. The switching frequency is 100 kHz. Determine (*a*) the output voltage, (*b*) the maximum and minimum inductor currents, and (*c*) the output voltage ripple.

6-5. The buck converter of Fig. 6-3*a* has the following parameters: $V_s = 15$ V, $D = 0.6$, $L = 10$ μH, $C = 50$ μF, and $R = 5$ Ω. The switching frequency is 150 kHz. Determine (*a*) the output voltage, (*b*) the maximum and minimum inductor currents, and (*c*) the output voltage ripple.

6-6. The buck converter of Fig. 6-3*a* has an input of 50 V and an output of 25 V. The switching frequency is 100 kHz, and the output power to a load resistor is 125 W. (*a*) Determine the duty ratio. (*b*) Determine the value of inductance to limit the peak inductor current to 6.25 A. (*c*) Determine the value of capacitance to limit the output voltage ripple to 0.5 percent.

6-7. A buck converter has an input of 6 V and an output of 1.5 V. The load resistor is 3 Ω, the switching frequency is 400 kHz, $L = 5$ μH, and $C = 10$ μF. (*a*) Determine the duty ratio. (*b*) Determine the average, peak, and rms inductor currents. (*c*) Determine the average source current. (*d*) Determine the peak and average diode current.

6-8. The buck converter of Fig. 6-3*a* has $V_s = 30$ V, $V_o = 20$ V, and a switching frequency of 40 kHz. The output power is 25 W. Determine the size of the inductor such that the minimum inductor current is 25 percent of the average inductor current.

6-9. A buck converter has an input voltage that varies between 50 and 60 V and a load that varies between 75 and 125 W. The output voltage is 20 V. For a switching frequency of 100 kHz, determine the minimum inductance to provide for continuous current for every operating possibility.

6-10. A buck converter has an input voltage that varies between 10 and 15 V and a load current that varies between 0.5 A and 1.0 A. The output voltage is 5 V. For a switching frequency of 200 kHz, determine the minimum inductance to provide for continuous current for every operating possibility.

6-11. Design a buck converter such that the output voltage is 15 V when the input is 48 V. The load is 8 Ω. Design for continuous inductor current. The output voltage ripple must be no greater than 0.5 percent. Specify the switching frequency and the value of each of the components. Assume ideal components.

6-12. Specify the voltage and current ratings for each of the components in the design of Prob. 6-11.

6-13. Design a buck converter to produce an output of 15 V from a 24-V source. The load is 2 A. Design for continuous inductor current. Specify the switching frequency and the values of each of the components. Assume ideal components.

6-14. Design a buck converter that has an output of 12 V from an input of 18 V. The output power is 10 W. The output voltage ripple must be no more than 100 mV p-p. Specify the duty ratio, switching frequency, and inductor and capacitor values. Design for continuous inductor current. Assume ideal components.

6-15. The voltage V_x in Fig. 6-3a for the buck converter with continuous inductor current is the pulsed waveform of Fig. 6-2c. The Fourier series for this waveform has a dc term of $V_s D$. The ac terms have a fundamental frequency equal to the switching frequency and amplitudes given by

$$V_n = \frac{\sqrt{2}V_s}{n\pi}\sqrt{1 - \cos(2\pi n D)} \quad n = 1, 2, 3, \ldots$$

Using ac circuit analysis, determine the amplitude of the first ac term of the Fourier series for voltage across the load for the buck converter in Example 6-1. Compare your result with the peak-to-peak voltage ripple determined in the example. Comment on your results.

6-16. (a) If the equivalent series resistance of the capacitor in the buck converter in Example 6-2 is 0.5 Ω, recompute the output voltage ripple. (b) Recompute the required capacitance to limit the output voltage ripple to 0.5 percent if the ESR of the capacitor is given by $r_C = 50(10)^{-6}/C$, where C is in farads.

Boost Converter

6-17. The boost converter of Fig. 6-8 has parameter $V_s = 20$ V, $D = 0.6$, $R = 12.5\ \Omega$, $L = 10\ \mu H$, $C = 40\ \mu F$, and the switching frequency is 200 kHz. (a) Determine the output voltage. (b) Determine the average, maximum, and minimum inductor currents. (c) Determine the output voltage ripple. (d) Determine the average current in the diode. Assume ideal components.

6-18. For the boost converter in Prob. 6-17, sketch the inductor and capacitor currents. Determine the rms values of these currents.

6-19. A boost converter has an input of 5 V and an output of 25 W at 15 V. The minimum inductor current must be no less than 50 percent of the average. The output voltage ripple must be less than 1 percent. The switching frequency is 300 kHz. Determine the duty ratio, minimum inductor value, and minimum capacitor value.

6-20. Design a boost converter to provide an output of 18 V from a 12-V source. The load is 20 W. The output voltage ripple must be less than 0.5 percent. Specify the duty ratio, the switching frequency, the inductor size and rms current rating, and the capacitor size and rms current rating. Design for continuous current. Assume ideal components.

6-21. The ripple of the output voltage of the boost converter was determined assuming that the capacitor current was constant when the diode was off. In reality, the current is a decaying exponential with a time constant RC. Using the capacitance and resistance values in Example 6-4, determine the change in output voltage while the switch is closed by evaluating the voltage decay in the RC circuit. Compare it to that determined from Eq. (6-34).

6-22. For the boost converter with a nonideal inductor, produce a family of curves of V_o/V_s similar to Fig. 6-10a for $r_L/R = 0.1, 0.3, 0.5$, and 0.7.

Buck-boost Converter

6-23. The buck-boost converter of Fig. 6-11 has parameters $V_s = 12$ V, $D = 0.6$, $R = 10\,\Omega$, $L = 10\,\mu\text{H}$, $C = 20\,\mu\text{F}$, and a switching frequency of 200 kHz. Determine (a) the output voltage, (b) the average, maximum, and minimum inductor currents, and (c) the output voltage ripple.

6-24. Sketch the inductor and capacitor currents for the buck-boost converter in Prob. 6-23. Determine the rms values of these currents.

6-25. The buck-boost converter of Fig. 6-11 has $V_s = 24$ V, $V_o = -36$ V, and a load resistance of $10\,\Omega$. If the switching frequency is 100 kHz, (a) determine the inductance such that the minimum current is 40 percent of the average and (b) determine the capacitance required to limit the output voltage ripple to 0.5 percent.

6-26. Design a buck-boost converter to supply a load of 75 W at 50 V from a 40-V source. The output ripple must be no more than 1 percent. Specify the duty ratio, switching frequency, inductor size, and capacitor size.

6-27. Design a dc-dc converter to produce a -15-V output from a source that varies from 12 to 18 V. The load is a 15-Ω resistor.

6-28. Design a buck-boost converter that has a source that varies from 10 to 14 V. The output is regulated at -12 V. The load varies from 10 to 15 W. The output voltage ripple must be less than 1 percent for any operating condition. Determine the range of the duty ratio of the switch. Specify values of the inductor and capacitor, and explain how you made your design decisions.

Ćuk Converter

6-29. The Ćuk converter of Fig. 6-13a has parameters $V_s = 12$ V, $D = 0.6$, $L_1 = 200\,\mu\text{H}$, $L_2 = 100\,\mu\text{H}$, $C_1 = C_2 = 2\,\mu\text{F}$, and $R = 12\,\Omega$, and the switching frequency is 250 kHz. Determine (a) the output voltage, (b) the average and the peak-to-peak variation of the currents in L_1 and L_2, and (c) the peak-to-peak variation in the capacitor voltages.

6-30. The Ćuk converter of Fig. 6-13a has an input of 20 V and supplies an output of 1.0 A at 10 V. The switching frequency is 100 kHz. Determine the values of L_1

and L_2 such that the peak-to-peak variation in inductor currents is less than 10 percent of the average.

6-31. Design a Ćuk converter that has a in input of 25 V and an output of −30 V. The load is 60 W. Specify the duty ratio, switching frequency, inductor values, and capacitor values. The maximum change in inductor currents must be 20 percent of the average currents. The ripple voltage across C_1 must be less than 5 percent, and the output ripple voltage must be less than 1 percent.

SEPIC Circuit

6-32. The SEPIC circuit of Fig. 6-14a has $V_s = 5$ V, $V_o = 12$ V, $C_1 = C_2 = 50$ μF, $L_1 = 10$ μH, and $L_2 = 20$ μH. The load resistor is 4 Ω. Sketch the currents in L_1 and L_2, indicating average, maximum, and minimum values. The switching frequency is 100 kHz.

6-33. The SEPIC circuit of Fig. 6-14a has $V_s = 3.3$ V, $D = 0.7$, $L_1 = 4$ μH, and $L_2 = 10$ μH. The load resistor is 5 Ω. The switching frequency is 300 kHz. (a) Determine the maximum and minimum values of the currents in L_1 and L_2. (b) Determine the variation in voltage across each capacitor.

6-34. The relationship between input and output voltages for the SEPIC circuit of Fig. 6-14a expressed in Eq. (6-74) was developed using the average voltage across L_1. Derive the relationship using the average voltage across L_2.

6-35. A SEPIC circuit has an input voltage of 15 V and is to have an output of 6 V. The load resistance is 2 Ω, and the switching frequency is 250 kHz. Determine values of L_1 and L_2 such that the variation in inductor current is 40 percent of the average value. Determine values of C_1 and C_2 such that the variation in capacitor voltage is 2 percent.

6-36. A SEPIC circuit has an input voltage of 9 V and is to have an output of 2.7 V. The output current is 1 A, and the switching frequency is 300 kHz. Determine values of L_1 and L_2 such that the variation in inductor current is 40 percent of the average value. Determine values of C_1 and C_2 such that the variation in capacitor voltage is 2 percent.

Nonideal Effects

6-37. The boost converter of Example 6-4 has a capacitor with an equivalent series resistance of 0.6 Ω. All other parameters are unchanged. Determine the output voltage ripple.

6-38. Equation (6-88) expresses the output voltage of a buck converter in terms of input, duty ratio, and voltage drops across the nonideal switch and diode. Derive an expression for the output voltage of a buck-boost converter for a nonideal switch and diode.

Discontinuous Current

6-39. The buck converter of Example 6-2 was designed for a 10-Ω load. (a) What is the limitation on the load resistance for continuous-current operation? (b) What would be the range of output voltage for a load resistance range of 5 to 20 Ω? (c) Redesign the converter so inductor current remains continuous for a load resistance range of 5 to 20 Ω.

6-40. The boost converter of Example 6-4 was designed for a 50-Ω load. (*a*) What is the limitation on the load resistance for continuous-current operation? (*b*) What would be the range of output voltage for a load resistance range of 25 to 100 Ω? (*c*) Redesign the converter so inductor current remains continuous for a load resistance range of 25 to 100 Ω.

6-41. Section 6.11 describes the buck and boost converters for discontinuous-current operation. Derive an expression for the output voltage of a buck-boost converter when operating in the discontinuous-current mode.

Switched-capacitor Converters

6-42. Capacitors C_1 and C_2 in Fig. P6-42 are equal in value. In the first part of the switching cycle, the switches labeled 1 are closed while the switches labeled 2 are open. In the second part of the cycle, switches 1 are opened and then switches 2 are closed. Determine the output voltage V_o at the end of the switching cycle. *Note*: A third capacitor would be placed from V_o to ground to sustain the output voltage during subsequent switching cycles.

PSpice

6-43. Simulate the buck converter of Example 6-11, but use the IRF150 MOSFET from the PSpice device library for the switch. Use an idealized gate drive circuit of a pulsed voltage source and small resistance. Use the default model for the diode. Use Probe to graph $p(t)$ versus t for the switch for steady-state conditions. Determine the average power loss in the switch.

6-44. Simulate the buck converter of Example 6-1 using PSpice. (*a*) Use an ideal switch and ideal diode. Determine the output ripple voltage. Compare your PSpice results with the analytic results in Example 6-1. (*b*) Determine the steady-state output voltage and voltage ripple using a switch with an on resistance of 2 Ω and the default diode model.

6-45. Show that the equivalent circuits for the PSpice averaged models in Fig. 6-33 satisfy the average voltage and current input-output relationships for each of the converters.

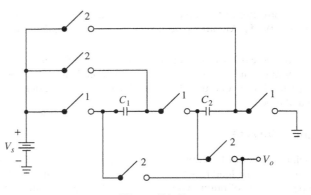

Figure P6-42

CHAPTER 7

DC Power Supplies

7.1 INTRODUCTION

A basic disadvantage of the dc-dc converters discussed in Chap. 6 is the electrical connection between the input and the output. If the input supply is grounded, that same ground will be present on the output. A way to isolate the output from the input electrically is with a transformer. If the dc-dc converter has a first stage that rectifies an ac power source to dc, a transformer could be used on the ac side. However, not all applications require ac to dc conversion as a first stage. Moreover, a transformer operating at a low frequency (50 or 60 Hz) requires a large magnetic core and is therefore relatively large, heavy, and expensive.

A more efficient method of providing electrical isolation between input and output of a dc-dc converter is to use a transformer in the switching scheme. The switching frequency is much greater than the ac power-source frequency, enabling the transformer to be small. Additionally, the transformer turns ratio provides increased design flexibility in the overall relationship between the input and the output of the converter. With the use of multiple transformer windings, switching converters can be designed to provide multiple output voltages.

7.2 TRANSFORMER MODELS

Transformers have two basic functions: to provide electrical isolation and to step up or step down time-varying voltages and currents. A two-winding transformer is depicted in Fig. 7-1a. An idealized model for the transformer, as shown in Fig. 7-1b, has input-output relationships

$$\frac{v_1}{v_2} = \frac{N_1}{N_2}$$
$$\frac{i_1}{i_2} = \frac{N_2}{N_1}$$
(7-1)

The dot convention is used to indicate relative polarity between the two windings. When the voltage at the dotted terminal on one winding is positive, the voltage at the dotted terminal on the other winding is also positive. When current enters the dotted terminal on one winding, current leaves the dotted terminal on the other winding.

A more complete transformer model is shown in Fig. 7-1c. Resistors r_1 and r_2 represent resistances of the conductors, L_1 and L_2 represent leakage inductances of the windings, L_m represents magnetizing inductance, and r_m represents core loss. The ideal transformer is incorporated into this model to represent the voltage and current transformation between primary and secondary.

In some applications in this chapter, the ideal transformer representation is sufficient for preliminary investigation of a circuit. The ideal model assumes that the series resistances and inductances are zero and that the shunt elements are infinite. A somewhat better approximation for power supply applications includes the magnetizing inductance L_m, as shown in Fig. 7-1d. The value of L_m is an important design parameter for the flyback converter.

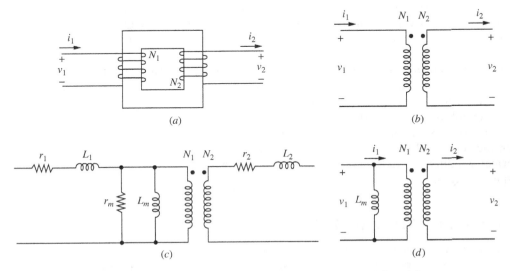

Figure 7-1 (a) Transformer; (b) Ideal model; (c) Complete model; (d) Model used for most power electronics circuits.

The leakage inductances L_1 and L_2 are usually not crucial to the general operation of the power electronics circuits described in this chapter, but they are important when considering switching transients. Note that in ac power system applications, the leakage inductance is normally the important analysis and design parameter.

For periodic voltage and current operation for a transformer circuit, the magnetic flux in the core must return to its starting value at the end of each switching period. Otherwise, flux will increase in the core and eventually cause saturation. A saturated core cannot support a voltage across a transformer winding, and this will lead to device currents that are beyond the design limits of the circuit.

7.3 THE FLYBACK CONVERTER

Continuous-Current Mode

A dc-dc converter that provides isolation between input and output is the flyback circuit of Fig. 7-2a. In a first analysis, Fig. 7-2b uses the transformer model which includes the magnetizing inductance L_m, as in Fig. 7-1d. The effects of

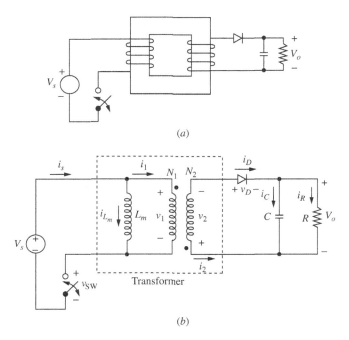

Figure 7-2 (a) Flyback converter; (b) Equivalent circuit using a transformer model that includes the magnetizing inductance; (c) Circuit for the switch on; (d) Circuit for the switch off.

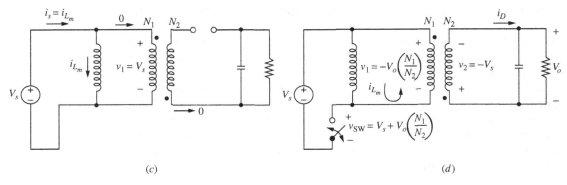

(c)　　　　　　　　　　　　　(d)

Figure 7-2 (*continued*)

losses and leakage inductances are important when considering switch performance and protection, but the overall operation of the circuit is best understood with this simplified transformer model. Note the polarity of the transformer windings in Fig. 7-2.

Additional assumptions for the analysis are made:

1. The output capacitor is very large, resulting in a constant output voltage V_o.
2. The circuit is operating in the steady state, implying that all voltages and currents are periodic, beginning and ending at the same points over one switching period.
3. The duty ratio of the switch is D, being closed for time DT and open for $(1 - D)T$.
4. The switch and diode are ideal.

The basic operation of the flyback converter is similar to that of the buck-boost converter described in Chap. 6. Energy is stored in L_m when the switch is closed and is then transferred to the load when the switch is open. The circuit is analyzed for both switch positions to determine the relationship between input and output.

Analysis for the Switch Closed On the source side of the transformer (Fig. 7-2c),

$$v_1 = V_s = L_m \frac{di_{L_m}}{dt}$$

$$\frac{di_{L_m}}{dt} = \frac{\Delta i_{L_m}}{\Delta t} = \frac{\Delta i_{L_m}}{DT} = \frac{V_s}{L_m}$$

Solving for the change in current in the transformer magnetizing inductance,

$$(\Delta i_{L_m})_{\text{closed}} = \frac{V_s DT}{L_m} \qquad (7\text{-}2)$$

On the load side of the transformer,

$$v_2 = v_1\left(\frac{N_2}{N_1}\right) = V_s\left(\frac{N_2}{N_1}\right)$$

$$v_D = -V_o - V_s\left(\frac{N_2}{N_1}\right) < 0$$

$$i_2 = 0$$

$$i_1 = 0$$

Since the diode is off, $i_2 = 0$, which means that $i_1 = 0$. So while the switch is closed, current is increasing linearly in the magnetizing inductance L_m, and there is no current in the windings of the ideal transformer in the model. Remember that in the actual transformer, this means that the current is increasing linearly in the physical primary winding, and no current exists in the secondary winding.

Analysis for the Switch Open When the switch opens (Fig. 7-2d), the current cannot change instantaneously in the inductance L_m, so the conduction path must be through the primary turns of the ideal transformer. The current i_{L_m} enters the undotted terminal of the primary and must exit the undotted terminal of the secondary. This is allowable since the diode current is positive. Assuming that the output voltage remains constant at V_o, the transformer secondary voltage v_2 becomes $-V_o$. The secondary voltage transforms back to the primary, establishing the voltage across L_m at

$$v_1 = -V_o\left(\frac{N_1}{N_2}\right)$$

Voltages and currents for an open switch are

$$v_2 = -V_o$$

$$v_1 = v_2\left(\frac{N_1}{N_2}\right) = -V_o\left(\frac{N_1}{N_2}\right)$$

$$L_m \frac{di_{L_m}}{dt} = v_1 = -V_o\left(\frac{N_1}{N_2}\right)$$

$$\frac{di_{L_m}}{dt} = \frac{\Delta i_{L_m}}{\Delta t} = \frac{\Delta i_{L_m}}{(1-D)T} = \frac{-V_o}{L_m}\left(\frac{N_1}{N_2}\right)$$

Solving for the change in transformer magnetizing inductance with the switch open,

$$(\Delta i_{L_m})_{\text{open}} = \frac{-V_o(1-D)T}{L_m}\left(\frac{N_1}{N_2}\right) \quad (7\text{-}3)$$

Since the net change in inductor current must be zero over one period for steady-state operation, Eqs. (7-2) and (7-3) show

$$(\Delta i_{L_m})_{\text{closed}} + (\Delta i_{L_m})_{\text{open}} = 0$$

$$\frac{V_s DT}{L_m} - \frac{V_o(1-D)T}{L_m}\left(\frac{N_1}{N_2}\right) = 0$$

Solving for V_o,

$$\boxed{V_o = V_s\left(\frac{D}{1-D}\right)\left(\frac{N_2}{N_1}\right)} \tag{7-4}$$

Note that the relation between input and output for the flyback converter is similar to that of the buck-boost converter but includes the additional term for the transformer ratio.

Other currents and voltages of interest while the switch is open are

$$i_D = -i_1\left(\frac{N_1}{N_2}\right) = i_{L_m}\left(\frac{N_1}{N_2}\right)$$

$$v_{\text{sw}} = V_s - v_1 = V_s + V_o\left(\frac{N_1}{N_2}\right)$$

$$i_R = \frac{V_o}{R} \tag{7-5}$$

$$i_C = i_D - i_R = i_{L_m}\left(\frac{N_1}{N_2}\right) - \frac{V_o}{R}$$

Note that v_{sw}, the voltage across the open switch, is greater than the source voltage. If the output voltage is the same as the input and the turns ratio is 1, for example, the voltage across the switch will be twice the source voltage. Circuit currents are shown in Fig. 7-3.

The power absorbed by the load resistor must be the same as that supplied by the source for the ideal case, resulting in

$$P_s = P_o$$

or

$$V_s I_s = \frac{V_o^2}{R} \tag{7-6}$$

The average source current I_s is related to the average of the magnetizing inductance current I_{L_m} by

$$I_s = \frac{(I_{L_m})DT}{T} = I_{L_m}D \tag{7-7}$$

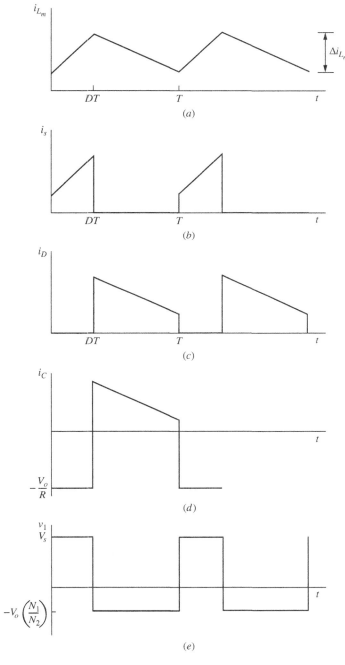

Figure 7-3 Flyback converter current and voltage waveforms. (*a*) Magnetizing inductance current; (*b*) Source current; (*c*) Diode current; (*d*) Capacitor current; (*e*) Transformer primary voltage.

Substituting for I_s in Eq. (7-6) and solving for I_{L_m},

$$V_s I_{L_m} D = \frac{V_o^2}{R}$$

$$I_{L_m} = \frac{V_o^2}{V_s D R} \qquad (7\text{-}8)$$

Using Eq. (7-4) for V_s, the average inductor current is also expressed as

$$I_{L_m} = \frac{V_s D}{(1-D)^2 R}\left(\frac{N_2}{N_1}\right)^2 = \frac{V_o}{(1-D)R}\left(\frac{N_2}{N_1}\right) \qquad (7\text{-}9)$$

The maximum and minimum values of inductor current are obtained from Eqs. (7-9) and (7-2).

$$\begin{aligned} I_{L_m,\max} &= I_{L_m} + \frac{\Delta i_{L_m}}{2} \\ &= \frac{V_s D}{(1-D)^2 R}\left(\frac{N_2}{N_1}\right)^2 + \frac{V_s DT}{2L_m} \end{aligned} \qquad (7\text{-}10)$$

$$\begin{aligned} I_{L_m,\min} &= I_{L_m} - \frac{\Delta i_{L_m}}{2} \\ &= \frac{V_s D}{(1-D)^2 R}\left(\frac{N_2}{N_1}\right)^2 - \frac{V_s DT}{2L_m} \end{aligned} \qquad (7\text{-}11)$$

Continuous-current operation requires that $I_{L_m,\min} > 0$ in Eq. (7-11). At the boundary between continuous and discontinuous current,

$$I_{L_m,\min} = 0$$

$$\frac{V_s D}{(1-D)^2 R}\left(\frac{N_2}{N_1}\right)^2 = \frac{V_s DT}{2L_m} = \frac{V_s D}{2L_m f}$$

where f is the switching frequency. Solving for the minimum value of L_m that will allow continuous current,

$$\boxed{(L_m)_{\min} = \frac{(1-D)^2 R}{2f}\left(\frac{N_1}{N_2}\right)^2} \qquad (7\text{-}12)$$

In a flyback converter design, L_m is selected to be larger than $L_{m,\min}$ to ensure continuous current operation. A convenient expression relating inductance and current variation is found from Eq. (7-2).

$$L_m = \frac{V_s DT}{\Delta i_{L_m}} = \frac{V_s D}{\Delta i_{L_m} f} \qquad (7\text{-}13)$$

The output configuration for the flyback converter is the same as for the buck-boost converter, so the output ripple voltages for the two converters are also the same.

$$\boxed{\frac{\Delta V_o}{V_o} = \frac{D}{RCf}} \qquad (7\text{-}14)$$

As with the converters described in Chap. 6, the equivalent series resistance of the capacitor can contribute significantly to the output voltage ripple. The peak-to-peak variation in capacitor current is the same as the maximum current in the diode and the transformer secondary. Using Eq. (7-5), the voltage ripple due to the ESR is

$$\Delta V_{o,\text{ESR}} = \Delta i_C r_C = I_{L_m,\max}\left(\frac{N_1}{N_2}\right) r_C \qquad (7\text{-}15)$$

EXAMPLE 7-1

Flyback Converter

A flyback converter of Fig. 7-2 has the following circuit parameters:

$V_s = 24$ V
$N_1/N_2 = 3.0$
$L_m = 500$ μH
$R = 5\ \Omega$
$C = 200$ μF
$f = 40$ kHz
$V_o = 5$ V

Determine (*a*) the required duty ratio D; (*b*) the average, maximum, and minimum values for the current in L_m; and (*c*) the output voltage ripple. Assume that all components are ideal.

■ **Solution**

(*a*) Rearranging Eq. (7-4) yields

$$V_o = V_s\left(\frac{D}{1-D}\right)\left(\frac{N_2}{N_1}\right)$$

$$D = \frac{1}{(V_s/V_o)(N_2/N_1)+1} = \frac{1}{(24/5)(1/3)+1} = 0.385$$

(*b*) Average current in L_m is determined from Eq. (7-8).

$$I_{L_m} = \frac{V_o^2}{V_s D R} = \frac{5^2}{(24)(0.385)(5)} = 540 \text{ mA}$$

The change in i_{L_m} can be calculated from Eq. (7-2).

$$\Delta i_{L_m} = \frac{V_s D}{L_m f} = \frac{(24)(0.385)}{500(10)^{-6}(40{,}000)} = 460 \text{ mA}$$

274 CHAPTER 7 DC Power Supplies

Maximum and minimum inductor currents can be computed from

$$I_{L_m,\max} = I_{L_m} + \frac{\Delta i_{L_m}}{2} = 540 + \frac{460}{2} = 770 \text{ mA}$$

$$I_{L_m,\min} = I_{L_m} - \frac{\Delta i_{L_m}}{2} = 540 - \frac{460}{2} = 310 \text{ mA}$$

Equations (7-10) and (7-11), which are derived from the above computation, could also be used directly to obtain the maximum and minimum currents. Note that a positive $I_{L_m,\min}$ verifies continuous current in L_m.

(c) Output voltage ripple is computed from Eq. (7-14).

$$\frac{\Delta V_o}{V_o} = \frac{D}{RCf} = \frac{0.385}{(5)[200(10)^{-6}](40{,}000)} = 0.0096 = 0.96\%$$

EXAMPLE 7-2

Flyback Converter Design, Continuous-Current Mode

Design a converter to produce an output voltage of 36 V from a 3.3-V source. The output current is 0.1 A. Design for an output ripple voltage of 2 percent. Include ESR when choosing a capacitor. Assume for this problem that the ESR is related to the capacitor value by $r_C = 10^{-5}/C$.

■ **Solution**

Considering a boost converter for this application and calculating the required duty ratio from Eq. (6-27),

$$D = 1 - \frac{V_s}{V_o} = 1 - \frac{3.3}{36} = 0.908$$

The result of a high duty ratio will likely be that the converter will not function as desired because of losses in the circuit (Fig. 6-10). Therefore, a boost converter would not be a good choice. A flyback converter is much better suited for this application.

As a somewhat arbitrary design decision, start by letting the duty ratio be 0.4. From Eq. (7-4), the transformer turns ratio is calculated to be

$$\left(\frac{N_2}{N_1}\right) = \frac{V_o}{V_s}\left(\frac{1-D}{D}\right) = \frac{36}{3.3}\left(\frac{1-0.4}{0.4}\right) = 16.36$$

Rounding, let $N_2/N_1 = 16$. Recalculating the duty ratio using a turns ratio of 16 gives $D = 0.405$.

To determine L_m, first compute the average current in L_m from Eq. (7-9), using $I_o = V_o/R$.

$$I_{L_m} = \frac{V_o}{(1-D)R}\left(\frac{N_2}{N_1}\right) = \frac{I_o}{1-D}\left(\frac{N_2}{N_1}\right) = \left(\frac{0.1}{1-0.405}\right)16 = 2.69 \text{ A}$$

Let the current variation in L_m be 40 percent of the average current: $\Delta i_{L_m} = 0.4(2.69) = 1.08$ A. As another somewhat arbitrary choice, let the switching frequency be 100 kHz. Using Eq. (7-13),

$$L_m = \frac{V_s D}{\Delta i_{L_m} f} = \frac{3.3(0.405)}{1.08(100,000)} = 12.4 \ \mu H$$

Maximum and minimum currents in L_m are found from Eqs. (7-10) and (7-11) as 3.23 and 2.15 A, respectively.

The output voltage ripple is to be limited to 2 percent, which is $0.02(36) = 0.72$ V. Assume that the primary cause of the voltage ripple will be the voltage drop across the equivalent series resistance $\Delta i_C r_C$. The peak-to-peak variation in capacitor current is the same as in the diode and the transformer secondary and is related to current in L_m by

$$\Delta i_C = I_{L_m, \max}\left(\frac{N_1}{N_2}\right) = (3.23 \text{ A})\left(\frac{1}{16}\right) = 0.202 \text{ A}$$

Using Eq. (7-15),

$$r_C = \frac{\Delta V_{o,\text{ESR}}}{\Delta i_C} = \frac{0.72 \text{ V}}{0.202 \text{ A}} = 3.56 \ \Omega$$

Using the relationship between ESR and capacitance given in this problem,

$$C = \frac{10^{-5}}{r_C} = \frac{10^{-5}}{3.56} = 2.8 \ \mu F$$

The ripple voltage due to the capacitance only is obtained from Eq. (7-14) as

$$\frac{\Delta V_o}{V_o} = \frac{D}{RCf} = \frac{0.405}{(36 \text{ V}/0.1 \text{ A})[2.8(10)^{-6}](100,000)} = 0.004 = 0.04\%$$

showing that the assumption that the voltage ripple is primarily due to the ESR was correct. A standard value of 3.3 μF would be a good choice. Note that the designer should consult manufacturers' specifications for ESR when selecting a capacitor.

The turns ratio of the transformer, current variation, and switching frequency were selected somewhat arbitrarily, and many other combinations are suitable.

Discontinuous-Current Mode in the Flyback Converter

For the discontinuous-current mode for the flyback converter, the current in the transformer increases linearly when the switch is closed, just as it did for the continuous-current mode. However, when the switch is open, the current in the transformer magnetizing inductance decreases to zero before the start of the next switching cycle, as shown in Fig. 7-4. While the switch is closed, the increase in inductor current is described by Eq. (7-2). Since the current starts at zero, the maximum value is also determined from Eq. (7-2).

$$I_{L_m, \max} = \frac{V_s DT}{L_m} \qquad (7\text{-}16)$$

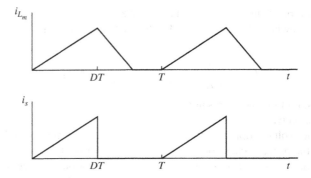

Figure 7-4 Discontinuous current for the flyback converter.

The output voltage for discontinuous-current operation can be determined by analyzing the power relationships in the circuit. If the components are ideal, the power supplied by the dc source is the same as the power absorbed by the load resistor. Power supplied by the source is the dc voltage times average source current, and load power is V_o^2/R:

$$P_s = P_o$$
$$V_s I_s = \frac{V_o^2}{R} \tag{7-17}$$

Average source current is the area under the triangular waveform of Fig. 7-4b divided by the period, resulting in

$$I_s = \left(\frac{1}{2}\right)\left(\frac{V_s DT}{L_m}\right)(DT)\left(\frac{1}{T}\right) = \frac{V_s D^2 T}{2 L_m} \tag{7-18}$$

Equating source power and load power [Eq. (7-17)],

$$\frac{V_s^2 D^2 T}{2 L_m} = \frac{V_o^2}{R} \tag{7-19}$$

Solving for V_o for discontinuous-current operation in the flyback converter,

$$\boxed{V_o = V_s D \sqrt{\frac{TR}{2 L_m}} = V_s D \sqrt{\frac{R}{2 L_m f}}} \tag{7-20}$$

EXAMPLE 7-3

Flyback Converter, Discontinuous Current

For the flyback converter in Example 7-1, the load resistance is increased from 5 to 20 Ω with all other parameters remaining unchanged. Show that the magnetizing inductance current is discontinuous, and determine the output voltage.

Solution

Using $L_m = 500$ μH, $f = 40$ kHz, $N_1/N_2 = 3$, $D = 0.385$, and $R = 20$ Ω, the minimum inductor current from Eq. (7-11) is calculated as

$$I_{L_m,\text{min}} = \frac{V_s D}{(1-D)^2 R}\left(\frac{N_2}{N_1}\right)^2 - \frac{V_s DT}{2L_m}$$

$$= \frac{(24)(0.385)}{(1-0.385)^2(20)}\left(\frac{1}{3}\right)^2 - \frac{(24)(0.385)}{2(500)(10)^{-6}(40{,}000)} = -95 \text{ mA}$$

Since negative current in L_m is not possible, i_{L_m} must be discontinuous. Equivalently, the minimum inductance for continuous current can be calculated from Eq. (7-12).

$$(L_m)_{\text{min}} = \frac{(1-D)^2 R}{2f}\left(\frac{N_1}{N_2}\right)^2 = \frac{(1-0.385)^2 20}{2(40{,}000)}(3)^2 = 850 \text{ μH}$$

which is more than the 500 μH specified, also indicating discontinuous current. Using Eq. (7-20),

$$V_o = V_s D\sqrt{\frac{R}{2L_m f}} = (24)(0.385)\sqrt{\frac{20}{2(500)(10)^{-6}(40{,}000)}} = 6.53 \text{ V}$$

For the current in L_m in the discontinuous-current mode, the output voltage is no longer 5 V but increases to 6.53 V. Note that for any load that causes the current to be continuous, the output would remain at 5 V.

Summary of Flyback Converter Operation

When the switch is closed in the flyback converter of Fig. 7-2a, the source voltage is across the transformer magnetizing inductance L_m and causes i_{L_m} to increase linearly. Also while the switch is closed, the diode on the output is reverse-biased, and load current is supplied by the output capacitor. When the switch is open, energy stored in the magnetizing inductance is transferred through the transformer to the output, forward-biasing the diode and supplying current to the load and to the output capacitor. The input-output voltage relationship in the continuous-current mode of operation is like that of the buck-boost dc-dc converter but includes a factor for the turns ratio.

7.4 THE FORWARD CONVERTER

The forward converter, shown in Fig. 7-5a, is another magnetically coupled dc-dc converter. The switching period is T, the switch is closed for time DT and open for $(1 - D)T$. Steady-state operation is assumed for the analysis of the circuit, and the current in inductance L_x is assumed to be continuous.

The transformer has three windings: windings 1 and 2 transfer energy from the source to the load when the switch is closed; winding 3 is used to provide a path for the magnetizing current when the switch is open and to reduce the magnetizing current to zero before the start of each switching period. The transformer

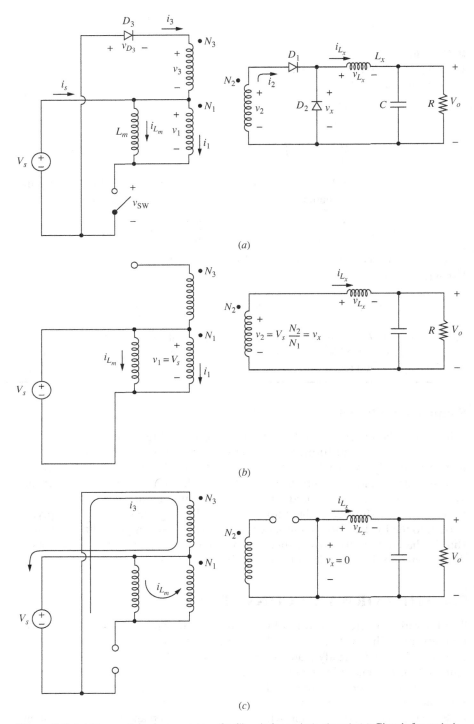

Figure 7-5 (*a*) Forward dc-dc converter; (*b*) Circuit for switch closed; (*c*) Circuit for switch open.

is modeled as three ideal windings with a magnetizing inductance L_m, which is placed across winding 1. Leakage inductance and losses are not included in this simplified transformer model.

For the forward converter, energy is transferred from the source to the load while the switch is closed. Recall that for the flyback converter, energy was stored in L_m when the switch was closed and transferred to the load when the switch was open. In the forward converter, L_m is not a parameter that is included in the input-output relationship and is generally made large.

Analysis for the Switch Closed The equivalent circuit for the forward converter with the switch closed is shown in Fig. 7-5b. Closing the switch establishes the voltage across transformer winding 1, resulting in

$$v_1 = V_s$$
$$v_2 = v_1 \left(\frac{N_2}{N_1}\right) = V_s \left(\frac{N_2}{N_1}\right) \qquad (7\text{-}21)$$
$$v_3 = v_1 \left(\frac{N_3}{N_1}\right) = V_s \left(\frac{N_3}{N_1}\right)$$

The voltage across D_3 is

$$V_{D_3} = -V_s - v_3 < 0$$

showing that D_3 is off. A positive v_2 forward-biases D_1 and reverse-biases D_2.

The relationship between input and output voltages can be determined by examining the current in inductor L_x. Assuming the output is held at a constant V_o,

$$v_{L_x} = v_2 - V_o = V_s\left(\frac{N_2}{N_1}\right) - V_o = L_x \frac{di_{L_x}}{dt}$$

$$\frac{di_{L_x}}{dt} = \frac{V_s(N_2/N_1) - V_o}{L_x} = \frac{\Delta i_{L_x}}{\Delta t} = \frac{\Delta i_{L_x}}{DT}$$

$$(\Delta i_{L_x})_{\text{closed}} = \left[V_s\left(\frac{N_2}{N_1}\right) - V_o\right]\frac{DT}{L_x} \qquad (7\text{-}22)$$

The voltage across the magnetizing inductance L_m is also V_s, resulting in

$$\Delta i_{L_m} = \frac{V_s DT}{L_m} \qquad (7\text{-}23)$$

Equations (7-22) and (7-23) show that the current is increasing linearly in both L_x and L_m while the switch is closed. The current in the switch and in the physical transformer primary is

$$i_{\text{sw}} = i_1 + i_{L_m} \qquad (7\text{-}24)$$

Analysis for the Switch Open Figure 7-5c shows the circuit with the switch open. The currents in L_x and L_m do not change instantaneously when the switch is opened. Continuity of i_{Lm} establishes $i_1 = -i_{Lm}$. Looking at the transformation from winding 1 to 2, current out of the dotted terminal on 1 would establish current into the dotted terminal on 2, but diode D_1 prevents current in that direction.

For the transformation from winding 1 to 3, current out of the dotted terminal of winding 1 forces current into the dotted terminal of winding 3. Diode D_3 is then forward-biased to provide a path for winding 3 current, which must go back to the source.

When D_3 is on, the voltage across winding 3 is established at

$$v_3 = -V_s$$

With v_3 established, v_1 and v_2 become

$$v_1 = v_3\left(\frac{N_1}{N_3}\right) = -V_s\left(\frac{N_1}{N_3}\right)$$
$$v_2 = v_3\left(\frac{N_2}{N_3}\right) = -V_s\left(\frac{N_2}{N_3}\right) \tag{7-25}$$

With D_1 off and positive current in L_x, D_2 must be on. With D_2 on, the voltage across L_x is

$$v_{L_x} = -V_o = L_x \frac{di_{L_x}}{dt}$$

resulting in

$$\frac{di_{L_x}}{dt} = \frac{-V_o}{L} = \frac{\Delta i_{L_x}}{\Delta t} = \frac{\Delta i_{L_x}}{(1-D)T}$$
$$(\Delta i_{L_x})_{\text{open}} = \frac{-V_o(1-D)T}{L_x} \tag{7-26}$$

Therefore, the inductor current decreases linearly when the switch is open.

For steady-state operation, the net change in inductor current over one period must be zero. From Eq. (7-22) and (7-26),

$$(\Delta i_{L_x})_{\text{closed}} + (\Delta i_{L_x})_{\text{open}} = 0$$

$$\left[V_s\left(\frac{N_2}{N_1}\right) - V_o\right]\frac{DT}{L_x} - \frac{V_o(1-D)T}{L_x} = 0$$

Solving for V_o,

$$\boxed{V_o = V_s D\left(\frac{N_2}{N_1}\right)} \tag{7-27}$$

Note that the relationship between input and output voltage is similar to that for the buck dc-dc converter except for the added term for the turns ratio. Current in L_x must be continuous for Eq. (7-27) to be valid.

7.4 The Forward Converter

Meanwhile, the voltage across L_m is v_1, which is negative, resulting in

$$v_{L_m} = v_1 = -V_s\left(\frac{N_1}{N_3}\right) = L_m \frac{di_{L_m}}{dt} \quad (7\text{-}28)$$

$$\frac{di_{L_m}}{dt} = -\frac{V_s}{L_m}\left(\frac{N_1}{N_3}\right)$$

The current in L_m should return to zero before the start of the next period to reset the transformer core (return the magnetic flux to zero). When the switch opens, Eq. (7-28) shows that i_{L_m} decreases linearly. Since D_3 will prevent i_{L_m} from going negative, Eq. (7-28) is valid as long as i_{L_m} is positive. From Eq. (7-28),

$$\frac{\Delta i_{Lm}}{\Delta t} = -\frac{V_s}{L_m}\left(\frac{N_1}{N_3}\right) \quad (7\text{-}29)$$

For i_{L_m} to return to zero after the switch is opened, the decrease in current must equal the increase in current given by Eq. (7-22). Letting ΔT_x be the time for i_{L_m} to decrease from the peak back to zero,

$$\frac{\Delta i_{L_m}}{\Delta T_x} = -\frac{V_s DT}{L_m} = -\frac{V_s}{L_m}\left(\frac{N_1}{N_3}\right) \quad (7\text{-}30)$$

Solving for ΔT_x,

$$\Delta T_x = DT\left(\frac{N_3}{N_1}\right) \quad (7\text{-}31)$$

The time at which the current i_{L_m} reaches zero t_0, is

$$t_0 = DT + \Delta T_x = DT + DT\left(\frac{N_3}{N_1}\right) = DT\left(1 + \frac{N_3}{N_1}\right) \quad (7\text{-}32)$$

Because the current must reach zero before the start of the next period,

$$t_0 < T$$

$$sDT\left(1 + \frac{N_3}{N_1}\right) < T \quad (7\text{-}33)$$

$$D\left(1 + \frac{N_3}{N_1}\right) < 1$$

For example, if the ratio $N_3/N_1 = 1$ (a common practice), then the duty ratio D must be less than 0.5. The voltage across the open switch is $V_s - v_1$, resulting in

$$v_{sw} = \begin{cases} V_s - v_1 = V_s - \left(-V_s\dfrac{N_1}{N_3}\right) = V_s\left(1 + \dfrac{N_1}{N_3}\right) & \text{for } DT < t < t_0 \\ V_s & \text{for } t_0 < t < T \end{cases} \quad (7\text{-}34)$$

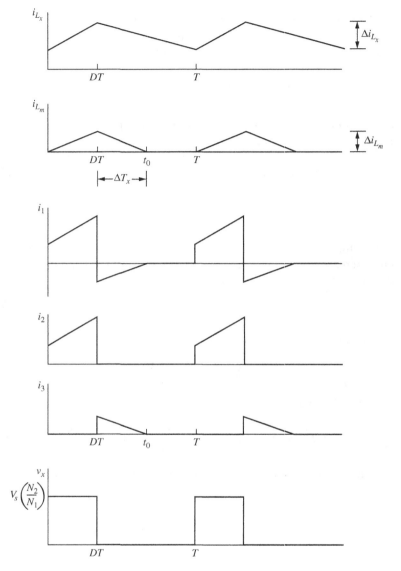

Figure 7-6 Current and voltage waveforms for the forward converter.

Forward converter current and voltage waveforms are shown in Fig. (7-6).

The circuit configuration on the output of the forward converter is the same as that for the buck converter, so the output voltage ripple based on an ideal capacitance is also the same.

7.4 The Forward Converter

$$\boxed{\frac{\Delta V_o}{V_o} = \frac{1-D}{8L_xCf^2}} \qquad (7\text{-}35)$$

The equivalent series resistance of the capacitor often dominates the output voltage ripple. The peak-to-peak voltage variation due to the ESR is

$$\Delta V_{o,\text{ESR}} = \Delta i_C r_C = \Delta i_{L_x} r_C = \left[\frac{V_o(1-D)}{L_x f}\right] r_C \qquad (7\text{-}36)$$

where Eq. (7-26) is used for Δi_{L_x}.

Summary of Forward Converter Operation

When the switch is closed, energy is transferred from the source to the load through the transformer. The voltage on the transformer secondary is a pulsed waveform, and the output is analyzed like that of the buck dc-dc converter. Energy stored in the magnetizing inductance while the switch is closed can be returned to the input source via a third transformer winding while the switch is open.

EXAMPLE 7-4

Forward Converter

The forward converter of Fig. 7-5a has the following parameters:

$V_s = 48$ V
$R = 10$ Ω
$L_x = 0.4$ mH, $L_m = 5$ mH
$C = 100$ μF
$f = 35$ kHz
$N_1/N_2 = 1.5$, $N_1/N_3 = 1$
$D = 0.4$

(a) Determine the output voltage, the maximum and minimum currents in L_x, and the output voltage ripple. (b) Determine the peak current in the transformer primary winding. Verify that the magnetizing current is reset to zero during each switching period. Assume all components are ideal.

■ **Solution**
(a) The output voltage is determined from Eq. (7-27).

$$V_o = V_s D\left(\frac{N_2}{N_1}\right) = 48(0.4)\left(\frac{1}{1.5}\right) = 12.8 \text{ V}$$

Average current in L_x is the same as the current in the load.

$$I_{L_x} = \frac{V_o}{R} = \frac{12.8}{10} = 1.28 \text{ A}$$

The change in i_{Lx} is determined from Eq. (7-22) or (7-26). Using Eq. (7-26),

$$\Delta i_{L_x} = \frac{V_o(1-D)}{L_x f} = \frac{12.8(1-0.4)}{0.4(10)^{-3}(35{,}000)} = 0.55 \text{ A}$$

Maximum and minimum currents in L_x are then

$$I_{L_x,\max} = I_{L_x} + \frac{\Delta i_{L_x}}{2} = 1.28 + \frac{0.55}{2} = 1.56 \text{ A}$$

$$I_{L_x,\min} = I_{L_x} - \frac{\Delta i_{L_x}}{2} = 1.28 - \frac{0.55}{2} = 1.01 \text{ A}$$

(b) Current in the primary winding of the transformer is the sum of the reflected current from the secondary and the magnetizing currents. The peak secondary current is the same as $I_{L_x,\max}$. The peak magnetizing current is obtained from Eq. (7-23).

$$I_{L_m,\max} = \Delta i_{L_m} = \frac{V_s DT}{L_m} = \frac{48(0.4)}{5(10)^{-3}(35{,}000)} = 0.11 \text{ A}$$

The peak current in the transformer primary is therefore

$$I_{\max} = I_{L_x,\max}\left(\frac{N_2}{N_1}\right) + I_{L_m,\max} = 1.56\left(\frac{1}{1.5}\right) + 0.11 = 1.15 \text{ A}$$

The time for the magnetizing current to return to zero after the switch is opened is determined from Eq. (7-31).

$$\Delta T_x = DT\left(\frac{N_3}{N_1}\right) = \frac{0.4(1)}{35{,}000} = 11.4 \text{ }\mu\text{s}$$

Since the switch is closed for $DT = 11.4$ μs, the time at which the magnetizing current reaches zero is 22.8 μs [Eq. (7-32)], which is less than the switching period of 28.6 μs.

EXAMPLE 7-5

Forward Converter Design

Design a forward converter such that the output is 5 V when the input is 170 V. The output current is 5 A. The output voltage ripple must not exceed 1 percent. Choose the transformer turns ratio, duty ratio, and switching frequency. Choose L_x such that the current in it is continuous. Include the ESR when choosing a capacitor. For this problem, use $r_C = 10^{-5}/C$.

■ **Solution**

Let the turns ratio $N_1/N_3 = 1$. This results in a maximum duty ratio of 0.5 for the switch. For margin, let $D = 0.35$. From Eq. (7-27),

$$\frac{N_1}{N_2} = \frac{V_s D}{V_o} = \frac{170(0.35)}{5} = 11.9$$

Rounding, let $N_1/N_2 = 12$. Recalculating D for $N_1/N_2 = 12$ yields

$$D = \frac{V_o}{V_s}\left(\frac{N_1}{N_2}\right) = \left(\frac{5}{170}\right)(12) = 0.353$$

The inductor L_x and the capacitor are selected using the same design criteria as discussed for the buck converter in Chap. 6. For this design, let $f = 300$ kHz. The average current in L_x is 5 A, the same as average current in the load since the average current in the capacitor is zero. Let the variation in inductor current be 2 A, which is 40 percent of the average value. From Eq. (7-26),

$$L_x = \frac{V_o(1-D)T}{\Delta i_{L_x}} = \frac{V_o(1-D)}{0.4 I_{L_x} f} = \frac{5(1-0.353)}{0.4(5)(300{,}000)} = 5.39 \ \mu\text{H}$$

A standard value of 5.6 μH is suitable for this design and would result in a slightly smaller Δi_{L_x}.

For a 1 percent output voltage ripple,

$$\Delta v_o \leq (0.01)(5) = 0.05 \ \text{V}$$

The capacitor size is determined by assuming that the voltage ripple is produced primarily by the equivalent series resistance, or

$$\Delta V_o \approx \Delta V_{o,\text{ESR}} = \Delta i_C r_C = (2 \ \text{A})(r_C) = 0.05 \ \text{V}$$

$$r_C = \frac{0.05 \ \text{V}}{2 \ \text{A}} = 0.025 \ \Omega = 25 \ \text{m}\Omega$$

The designer would now search for a capacitor having a 25-mΩ or lower ESR. Using $r_C = 10^{-5}/C$ given in this problem,

$$C = \frac{10^{-5}}{0.025} = 400 \ \mu\text{F}$$

A standard value of 470 μF is suitable.

7.5 THE DOUBLE-ENDED (TWO-SWITCH) FORWARD CONVERTER

The forward converter discussed in Sec. 7.4 has a single transistor switch and is referred to as a single-ended converter. The double-ended (two-switch) forward converter shown in Fig. 7-7 is a variation of the forward converter. In this circuit, the switching transistors are turned on and off simultaneously. When the switches are on, the voltage across the primary transformer winding is V_s. The voltage across the secondary winding is positive, and energy is transferred to the load, as it was for the forward converter discussed in Sec. 7.4. Also when the switches are on, the current in the magnetizing inductance is increasing. When the switches turn off, diode D_1 prevents i_{Lm} from flowing in the secondary (and hence primary) winding of the transformer and forces the magnetizing current to flow in diodes D_3 and D_4 and back to the source. This establishes the primary voltage at $-V_s$, causing a linear decrease in magnetizing current. If the duty ratio of the switches is less than 0.5, the transformer core resets (the magnetic flux returns to zero) during every cycle.

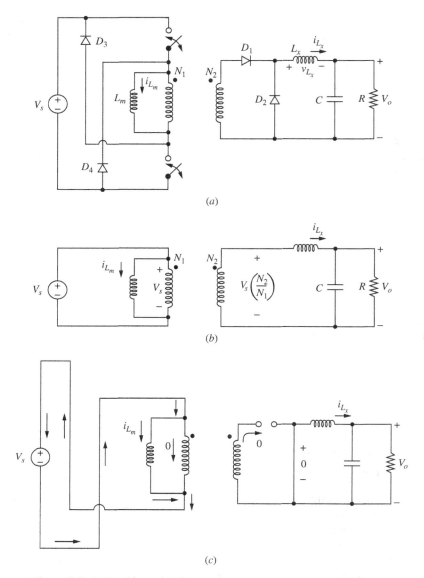

Figure 7-7 (*a*) Double-ended forward converter; (*b*) Circuit for the switches closed; (*c*) Circuit for the switches open.

The output voltage is the same as for the single-ended forward converter [Eq. (7-27)]. An advantage of the double-ended forward converter is that the voltage across an off transistor is V_s rather than $V_s(1 + N_1/N_3)$ as it was for the single-ended forward converter. This is an important feature for high-voltage applications.

7.6 THE PUSH-PULL CONVERTER

Another dc-dc converter that has transformer isolation is the push-pull converter shown in Fig. 7-8a. As with the forward converter, the transformer magnetizing inductance is not a design parameter. The transformer is assumed to be ideal for

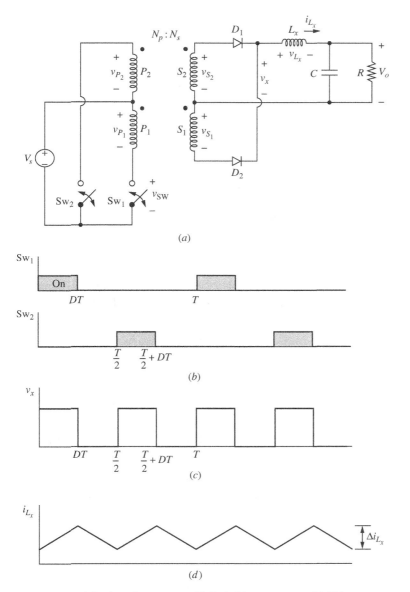

Figure 7-8 (a) Push-pull converter; (b) Switching sequence; (c) Voltage v_x; (d) Current in L_x.

this analysis. Switches Sw_1 and Sw_2 turn on and off with the switching sequence shown in Fig. 7-8b. Analysis proceeds by analyzing the circuit with either switch closed and then with both switches open.

Switch Sw_1 Closed Closing Sw_1 establishes the voltage across primary winding P_1 at

$$v_{P_1} = V_s \tag{7-37}$$

The voltage across P_1 is transformed to the three other windings, resulting in

$$\begin{aligned} v_{S_1} &= V_s\left(\frac{N_S}{N_P}\right) \\ v_{S_2} &= V_s\left(\frac{N_S}{N_P}\right) \\ v_{P_2} &= V_s \\ v_{Sw_2} &= 2V_s \end{aligned} \tag{7-38}$$

Diode D_1 is forward-biased, D_2 is reverse-biased, and

$$v_x = v_{S_2} = V_s\left(\frac{N_S}{N_P}\right) \tag{7-39}$$

$$v_{L_x} = v_x - V_o = V_s\left(\frac{N_S}{N_P}\right) - V_o$$

Assuming a constant output voltage V_o, the voltage across L_x is a constant, resulting in a linearly increasing current in L_x. In the interval when Sw_1 is closed, the change in current in L_x is

$$\frac{\Delta i_{L_x}}{\Delta t} = \frac{\Delta i_{L_x}}{DT} = \frac{V_s(N_S/N_P) - V_o}{L_x}$$

$$(\Delta i_{L_x})_{\text{closed}} = \left[\frac{V_s(N_S/N_P) - V_o}{L_x}\right]DT \tag{7-40}$$

Switch Sw_2 Closed Closing Sw_2 establishes the voltage across primary winding P_2 at

$$v_{P_2} = -V_s \tag{7-41}$$

The voltage across P_2 is transformed to the three other windings, resulting in

$$\begin{aligned} v_{P_1} &= -V_s \\ v_{S_1} &= -V_s\left(\frac{N_S}{N_P}\right) \\ v_{S_2} &= -V_s\left(\frac{N_S}{N_P}\right) \\ v_{S_1} &= 2V_s \end{aligned} \tag{7-42}$$

Diode D_2 is forward-biased, D_1 is reverse-biased, and

$$v_x = -v_{S_2} = V_s\left(\frac{N_S}{N_P}\right)$$

$$v_{L_x} = v_x - V_o = V_s\left(\frac{N_S}{N_P}\right) - V_o$$

(7-43)

which is a positive pulse. The current in L_x increases linearly while Sw_2 is closed, and Eq. (7-40) applies.

Both Switches Open With both switches open, the current in each of the primary windings is zero. The current in the filter inductor L_x must maintain continuity, resulting in both D_1 and D_2 becoming forward-biased. Inductor current divides evenly between the transformer secondary windings. The voltage across each secondary winding is zero, and

$$v_x = 0$$

$$v_{L_x} = v_x - V_o = -V_o$$

(7-44)

The voltage across L_x is $-V_o$, resulting in a linearly decreasing current in L_x. The change in current while both switches are open is

$$\frac{\Delta i_{L_x}}{\Delta t} = \frac{\Delta i_{L_x}}{T/2 - DT} = -\frac{V_o}{L_x}$$

Solving for Δi_{L_x},

$$(\Delta i_{L_x})_{\text{open}} = -\left(\frac{V_o}{L_x}\right)\left(\frac{1}{2} - D\right)T \qquad (7\text{-}45)$$

Since the net change in inductor current over one period must be zero for steady-state operation,

$$(\Delta i_{L_x})_{\text{closed}} + (\Delta i_{L_x})_{\text{open}} = 0$$

(7-46)

$$\left[\frac{V_s(N_S/N_P) - V_o}{L_x}\right]DT + \left(\frac{V_o}{L_x}\right)\left(\frac{1}{2} - D\right)T = 0$$

Solving for V_o,

$$\boxed{V_o = 2V_s\left(\frac{N_S}{N_P}\right)D} \qquad (7\text{-}47)$$

where D is the duty ratio of *each* switch. The above analysis assumes continuous current in the inductor. Note that the result is similar to that for the buck converter, discussed in Chap. 6. Ripple voltage on the output is derived in a manner similar to the buck converter. The output ripple for the push-pull converter is

$$\boxed{\frac{\Delta V_o}{V_o} = \frac{1 - 2D}{32L_xCf^2}} \qquad (7\text{-}48)$$

CHAPTER 7 DC Power Supplies

As with the other converters analyzed previously, the equivalent series resistance of the capacitor is usually responsible for most of the voltage output ripple. Recognizing that $\Delta i_C = \Delta i_{L_x}$ and using Eq. (7-45),

$$\Delta V_{o,ESR} = \Delta i_C r_C = \Delta i_{L_x} r_C = \left[\frac{V_o(\frac{1}{2} - D)}{L_x f} \right] r_C \qquad (7\text{-}49)$$

The preceding analysis neglected the magnetizing inductance of the transformer. If L_m were included in the equivalent circuit, i_{L_m} would increase linearly when Sw_1 was closed, circulate while both Sw_1 and Sw_2 were open, and decrease linearly when Sw_2 was closed. Because Sw_1 and Sw_2 are closed for equal intervals, the net change in i_{L_m} is zero, and the transformer core is reset during each period in the ideal case. In actual applications of the push-pull converter, control techniques are used to ensure that the core is reset.

Summary of Push-Pull Operation

Pulses of opposite polarity are produced on the primary and secondary windings of the transformer by switching Sw_1 and Sw_2 (Fig. 7-8). The diodes on the secondary rectify the pulse waveform and produce a waveform v_x at the input of the low-pass filter, as shown in Fig. 7-8c. The output is analyzed like that of the buck converter discussed in Chap. 6.

EXAMPLE 7-6

Push-Pull Converter

A push-pull converter has the following parameters:

$V_s = 30$ V
$N_P/N_S = 2$
$D = 0.3$
$L_x = 0.5$ mH
$R = 6 \, \Omega$
$C = 50 \, \mu\text{F}$
$f = 10$ kHz

Determine V_o, the maximum and minimum values of i_{L_x}, and the output ripple voltage. Assume all components are ideal.

■ **Solution**

Using Eq. (7-47), the output voltage is

$$V_o = 2V_s \left(\frac{N_S}{N_P}\right) D = (2)(30)\left(\frac{1}{2}\right)(0.3) = 9.0 \text{ V}$$

Average inductor current is the same as average load current,

$$I_{L_x} = \frac{V_o}{R} = \frac{9}{6} = 1.5 \text{ A}$$

The change in i_{L_x} is determined from Eq. (7-45).

$$\Delta i_{L_x} = \frac{V_o(\tfrac{1}{2} - D)T}{L_x} = \frac{9(0.5 - 0.3)}{0.5(10)^{-3}(10{,}000)} = 0.36 \text{ A}$$

resulting in maximum and minimum currents of

$$I_{L_x,\max} = I_{L_x} + \frac{\Delta i_{L_x}}{2} = 1.68 \text{ A}$$

$$I_{L_x,\min} = I_{L_x} - \frac{\Delta i_{L_x}}{2} = 1.32 \text{ A}$$

Output voltage ripple is determined from Eq. (7-48).

$$\frac{\Delta V_o}{V_o} = \frac{1 - 2D}{32f^2 L_x C} = \frac{1 - 2(0.3)}{32(10{,}000)^2(0.5)(10)^{-3}(50)(10)^{-6}}$$
$$= 0.005 = 0.5\%$$

7.7 FULL-BRIDGE AND HALF-BRIDGE DC-DC CONVERTERS

The full-bridge and half-bridge converters shown in Figs. 7-9 and 7-10 are similar in operation to the push-pull converter. Assuming that the transformer is ideal, the full-bridge converter of Fig. 7-9a has switch pairs (Sw_1, Sw_2) and (Sw_3, Sw_4) alternate closing. When Sw_1 and Sw_2 are closed, the voltage across the transformer primary is V_s. When Sw_3 and Sw_4 are closed, the transformer primary voltage is $-V_s$. For an ideal transformer, having all switches open will make $v_p = 0$. With a proper switching sequence, the voltage v_p across the transformer primary is the alternating pulse waveform shown in Fig. 7-9c. Diodes D_1 and D_2 on the transformer secondary rectify this waveform to produce the voltage v_x as shown in Fig. 7-9d. This v_x is identical to the v_x shown in Fig. 7-8c for the push-pull converter. Hence the output of the full-bridge converter is analyzed as for the push-pull converter, resulting in

$$\boxed{V_o = 2V_s\left(\frac{N_S}{V_P}\right)D} \qquad (7\text{-}50)$$

where D is the duty ratio of *each* switch pair.

Note that the maximum voltage across an open switch for the full-bridge converter is V_s, rather than $2V_s$ as for the push-pull and single-ended forward converters. Reduced voltage stress across an open switch is important when the input voltage is high, giving the full-bridge converter an advantage.

The half-bridge converter of Fig. 7-10a has capacitors C_1 and C_2 which are large and equal in value. The input voltage is equally divided between the

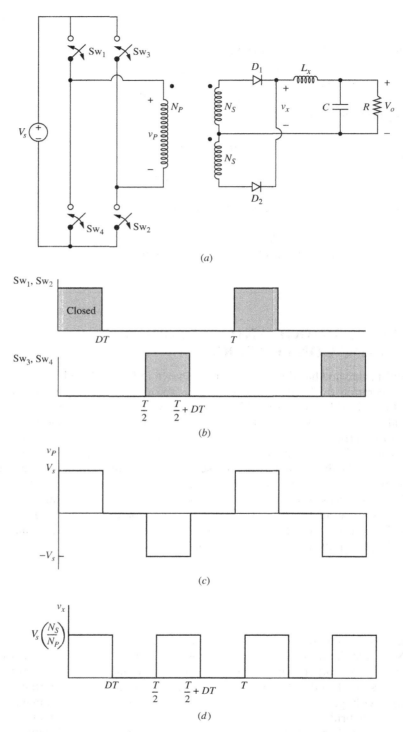

Figure 7-9 (*a*) Full-bridge converter; (*b*) Switching sequence; (*c*) Voltage on the transformer primary; (*d*) Voltage v_x.

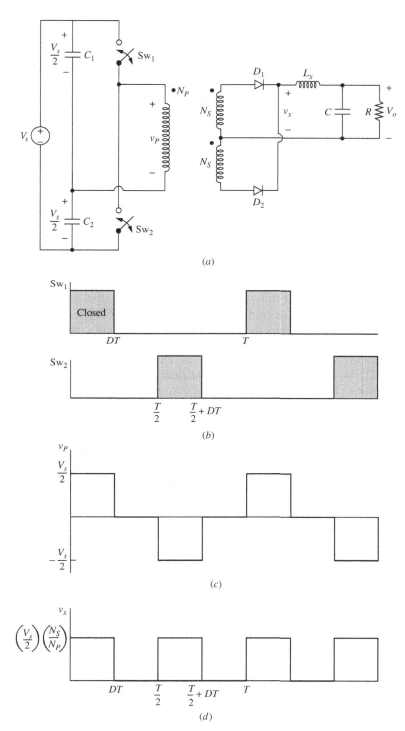

Figure 7-10 (*a*) Half-bridge converter; (*b*) Switching sequence; (*c*) Voltage on the transformer primary; (*d*) Voltage v_x.

capacitors. Switches Sw_1 and Sw_2 close with the sequence shown, producing an alternating voltage pulse v_P on the transformer primary. The rectified secondary voltage v_x has the waveform shown in Fig. 7-10d. Voltage v_x is the same form as for the push-pull and the full-bridge converters, but the amplitude is one-half the value. The relationship between the input and output voltages for the half-bridge converter is

$$V_o = V_s \left(\frac{N_S}{N_P}\right) D \qquad (7\text{-}51)$$

where D is the duty ratio of *each* switch. The voltage across an open switch for the half-bridge converter is V_s.

7.8 CURRENT-FED CONVERTERS

The converters described thus far in this chapter are called *voltage-fed converters*. Another method of controlling output is to establish a constant source current and use the switches to direct the current. Current control has advantages over voltage control for some converters. A circuit that operates by switching current rather than voltage is called a *current-fed converter*. Figure 7-11 shows a circuit that is a modification of the push-pull converter. The inductor L_x has been moved from the output side of the transformer to the input side. A large inductor in this position establishes a nearly constant source current. Switch Sw_1 directs the current through winding P_1, and switch Sw_2 directs the current through winding P_2. With both switches closed, the current divides evenly between the windings. At least one switch must be closed to provide a current path.

The switching sequence and waveforms are shown in Fig. 7-11. The following analysis assumes that L_x is large and the current in it is a constant I_{L_x}. The transformer is assumed to be ideal.

Sw_1 Closed and Sw_2 Open The inductor current I_{L_x} flows through primary winding P_1 and through D_1 on the secondary when switch 1 is closed and switch 2 is open. D_1 is on, D_2 is off, and the following equations apply:

$$\begin{aligned}
i_{D_1} &= I_{L_x}\left(\frac{N_P}{N_S}\right) \\
v_{P_1} &= V_o\left(\frac{N_P}{N_S}\right) \\
v_{L_x} &= V_s - v_{P_1} = V_s - V_o\left(\frac{N_P}{N_S}\right) \\
v_{Sw_2} &= v_{P_1} + v_{P_2} = 2V_o\left(\frac{N_P}{N_S}\right)
\end{aligned} \qquad (7\text{-}52)$$

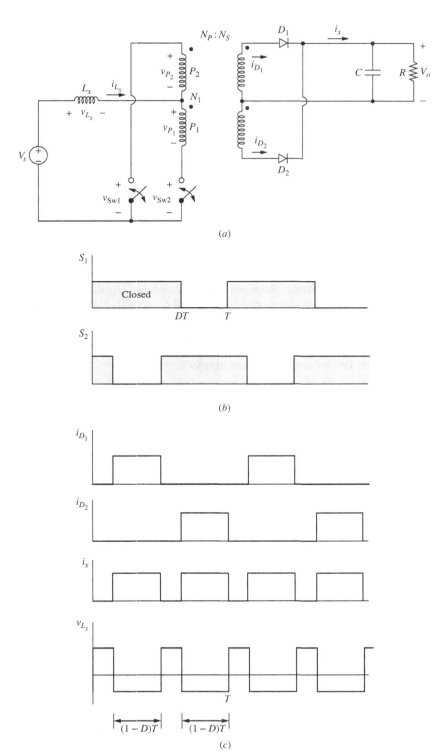

Figure 7-11 (a) A current-fed converter; (b) Switching sequence; (c) Current and voltage waveforms.

Sw₁ Open and Sw₂ Closed With switch 1 open and switch 2 closed, I_{L_x} flows through primary winding P_2 and through D_2 on the secondary. D_1 is off and D_2 is on, and the following equations apply:

$$i_{D_2} = I_{L_x}\left(\frac{N_P}{N_S}\right)$$

$$v_{P_2} = V_o\left(\frac{N_P}{N_S}\right)$$

$$v_{L_x} = V_s - V_o\left(\frac{N_P}{N_S}\right) \tag{7-53}$$

$$v_{Sw_1} = v_{P_1} + v_{P_2} = 2V_o\left(\frac{N_P}{N_S}\right)$$

Both Sw₁ and Sw₂ Closed With both switches closed, I_{L_x} divides evenly between the two primary windings, and both D_1 and D_2 are off. The voltage on each primary winding is zero:

$$v_{P_1} = v_{P_2} = 0$$

Inductor L_x then has the source voltage across it:

$$v_{L_x} = V_s \tag{7-54}$$

The average voltage across L_x must be zero for steady-state operation. During one switching period, $v_{L_x} = V_s - V_o(N_P/N_S)$ for two intervals of $(1 - D)T$ when only one switch is closed, and $v_{L_x} = V_s$ for the remaining time, which is $T - 2(1 - D)T = (2D - 1)T$. The average inductor voltage is thus expressed as

$$V_{L_x} = V_s(2D - 1)T + \left[V_s - V_o\left(\frac{N_P}{N_S}\right)\right]2(1 - D)T = 0 \tag{7-55}$$

Solving for V_o,

$$\boxed{V_o = \frac{V_s}{2(1 - D)}\left(\frac{N_S}{N_P}\right)} \tag{7-56}$$

where D is the duty ratio of *each* switch. This result is similar to that of the boost converter. Note that the duty ratio of each switch must be greater than 0.5 to prevent an open circuit in the path of the inductor current.

EXAMPLE 7-7

Current-Fed Converter

The current-fed converter of Fig. 7-11 has an input inductor L_x that is large enough to assume that the source current is essentially constant. The source voltage is 30 V, and the load resistor is 6 Ω. The duty ratio of each switch is 0.7, and the transformer has a turns ratio of $N_P/N_S = 2$. Determine (a) the output voltage, (b) the current in L_x, and (c) the maximum voltage across each switch.

■ **Solution**
(a) The output voltage is determined by using Eq. (7-56).

$$V_o = \frac{V_s}{2(1-D)}\left(\frac{N_S}{N_P}\right) = \frac{30}{2(1-0.7)}\left(\frac{1}{2}\right) = 25 \text{ V}$$

(b) To determine I_{L_x}, recognize that the power delivered to the load must be the same as that supplied by the source in the ideal case:

$$P_s = P_o$$

which can be expressed as

$$I_{L_x} V_s = \frac{V_o^2}{R}$$

Solving for I_{L_x},

$$I_{L_x} = \frac{V_o^2}{V_s R} = \frac{25^2}{30(6)} = 3.47 \text{ A}$$

(c) The maximum voltage across each switch is determined from Eqs. (7-52) and (7-53).

$$V_{\text{sw,max}} = 2V_o\left(\frac{N_P}{N_S}\right) = 2(25)(2) = 100 \text{ V}$$

7.9 MULTIPLE OUTPUTS

The dc power supply circuits discussed thus far in this chapter have only one output voltage. With additional transformer windings, multiple outputs are possible. Flyback and forward converters with two outputs are shown in Fig. 7-12.

Multiple outputs are useful when different output voltages are necessary. The duty ratio of the switch and the turns ratio of the primary to the specific secondary winding determine the output/input voltage ratio. An example is a single converter with three windings on the output producing voltages of 12, 5, and −5 V with respect to a common ground on the output side. Multiple outputs are possible with all the dc power supply topologies discussed in this chapter. Note, however, that only one of the outputs can be regulated with a feedback control loop. Other outputs will follow according to the duty ratio and the load.

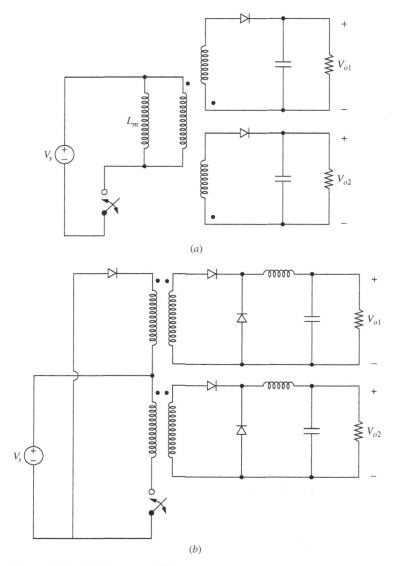

Figure 7-12 (a) Flyback and (b) forward converters with two outputs.

7.10 CONVERTER SELECTION

In theory, any power supply circuit can be designed for any application, depending on how much the designer is willing to spend for components and control circuitry. In practice, some circuits are much more suited to particular applications than others.

The flyback converter, having a low parts count, is a simple circuit to implement and is very popular for low-power applications. The main disadvantages are that the transformer core must be made large as power requirements increase, and the voltage stress across the switch is high ($2V_s$). Typical applications can go up to about 150 W, but the flyback converter is used most often for an output power of 10 W or less.

The forward converter is a popular circuit for low and medium power levels, up to about 500 W. It has one transistor as does the flyback, but it requires a smaller transformer core. Disadvantages are high voltage stress for the transistor and the extra cost of the filter inductor. The double-ended forward converter can be used to reduce the switch voltage stress, but the drive circuit for one of the transistors must be floating with respect to ground.

The push-pull converter is used for medium to high power requirements, typically up to 1000 W. Advantages include transistor drive circuits that have a common point and a relatively small transformer core because it is excited in both directions. Disadvantages include a high voltage stress for the transistors and potential core saturation problems caused by a dc imbalance in nonideal circuits.

The half-bridge converter is also used for medium power requirements, up to about 500 W, and has some of the same advantages as the push-pull. The voltage stress on the switches is limited to V_s.

The full-bridge converter is often the circuit of choice for high-power applications, up to about 2000 W. The voltage stress on the transistors is limited to V_s. Extra transistors and floating drive circuits are disadvantages.

A method of reducing switching losses is to use a resonant converter topology. Resonant converters switch at voltage or current zeros, thus reducing the switch power loss, enabling high switching frequencies and reduced component sizes. Resonant converters are discussed in Chap. 9.

7.11 POWER FACTOR CORRECTION

Power supplies often have an ac source as the input, and the first stage is a fullwave rectifier that converts the ac input to a dc voltage. Figure 7-13, as discussed in Chap. 4, is one such arrangement. The diodes conduct for only a small amount of time during each cycle, resulting in currents that are highly nonsinusoidal. The result is a large total harmonic distortion (THD) of current coming from the ac

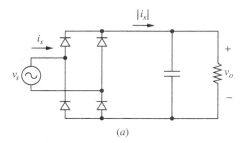

Figure 7-13 (a) Fullwave rectifier and (b) voltage and current waveforms. The source current is highly nonsinusoidal because the diodes conduct for a short time interval.

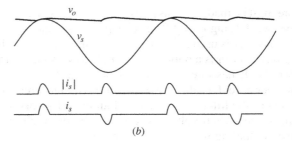

(b)

Figure 7-13 (*continued*)

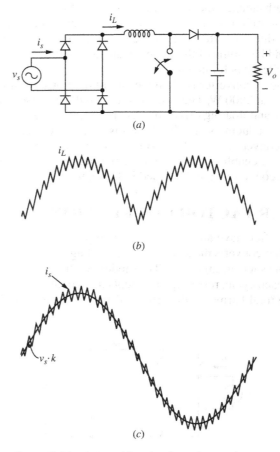

Figure 7-14 (*a*) A rectifier circuit used to produce a high power factor and low THD; (*b*) Current in the inductor for continuous-current mode (CCM) operation; (*c*) Current from the ac source.

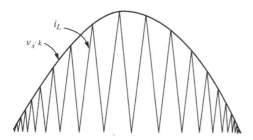

Figure 7-15 Discontinuous-current mode (DCM) power factor correction.

source. A large THD corresponds to a low power factor (see Chap. 2). The resistor represents any load on the output, which may be a dc-dc converter.

A way to improve the power factor (and reduce the THD) is with a power factor correction circuit, as shown in Fig. 7-14a. A boost converter is used to make the current in the inductor approximate a sinusoid. When the switch is closed, the inductor current increases. When the switch is open, the inductor current decreases. By using appropriate switching intervals, the inductor current can be made to follow the sinusoidal shape of the full-wave rectified input voltage.

The voltage on the output of the diode bridge is a full-wave rectified sinusoid. The current in the inductor is of the general form as shown in Fig. 7-14b, and the resulting current from the ac source is shown in Fig. 7-14c. This current is predominantly at the same frequency and phase angle as the voltage, making the power factor quite high and the THD quite low. This type of switching scheme is called continuous-current mode (CCM) power factor correction (PFC). In an actual implementation, the switching frequency would be much greater than is shown in the figure.

Another type of switching scheme produces a current like that shown in Fig. 7-15. In this scheme, the inductor current varies between zero and a peak that follows a sinusoidal shape. This type of switching scheme is called discontinuous-current mode (DCM) power factor correction. DCM is used with low-power circuits, while CCM is more suitable for high-power applications.

In both the CCM and DCM schemes, the output of the power factor correction (PFC) stage is a large dc voltage, usually on the order of 400 V. The output of the PFC stage will go to a dc-dc converter. For example, a forward converter can be used to step down the 400-V output of the PFC stage to 5 V.

Other converter topologies in addition to the boost converter can be used for power factor correction. The SEPIC and Ćuk converters are well suited for this purpose.

7.12 PSPICE SIMULATION OF DC POWER SUPPLIES

PSpice simulations of the magnetically coupled dc-dc converters discussed in this chapter are similar to those of the dc-dc converters of Chap. 6. For initial investigation, the switches can be implemented with voltage-controlled switches

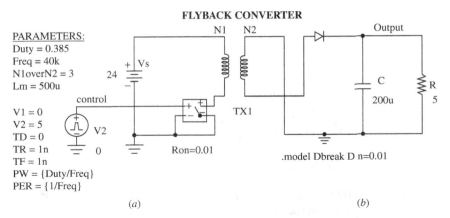

Figure 7-16 (*a*) The flyback converter circuit for simulation; (*b*) Probe output showing the transient and steady-state output voltage.

rather than with transistors, simplifying the switching and allowing examination of the overall circuit behavior.

Transformers can be modeled in PSpice as two or more inductances with ideal coupling. Since inductance is proportional to the square of the turns in a winding, the transformer turns ratio is

$$\frac{N_1}{N_2} = \sqrt{\frac{L_1}{L_2}} \qquad (7\text{-}57)$$

For the flyback converter, let $L_1 = L_m$ and determine L_2 from Eq. (7-57). For other converters where L_m is not a design parameter, let L_1 be any large value and determine L_2 accordingly. For two-winding transformers, the part XFRM_LINEAR can serve as a template.

Figures 7-16 and 7-17 show circuits for the flyback and forward converter topologies. The flyback simulation uses the XFRM_LINEAR part, and the forward simulation uses mutually coupled inductors. The switches and diodes are idealized by setting $R_{on} = 0.01\ \Omega$ for the switches and $n = 0.01$ for the diodes. Just as with the dc-dc converters in Chap. 6, transient voltages and currents precede the steady-state waveforms that were presented in the earlier discussion of the converters in this chapter.

7.13 POWER SUPPLY CONTROL

In ideal switching dc-dc converters, the output voltage is a function of the input voltage and duty ratio. In real circuits with nonideal components, the output is also a function of the load current because of resistances in the components. A power supply output is regulated by modulating the duty ratio to compensate for variations in the input or load. A feedback control system for power supply control compares output voltage to a reference and converts the error to a duty ratio.

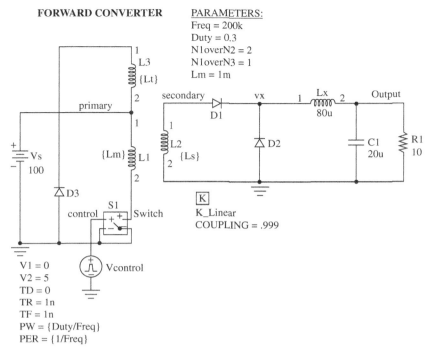

Figure 7-17 The forward converter circuit for simulation.

The buck converter operating in the continuous-current mode is used to illustrate the basics of power supply control. Figure 7-18*a* shows the converter and feedback loop consisting of

1. The switch, including the diode and drive circuit
2. The output filter
3. A compensated error amplifier
4. A pulse-width modulating circuit that converts the output of the compensated error amplifier to a duty ratio to drive the switch

The regulated converter is represented by the closed-loop system of Fig. 7-18*b*.

Control Loop Stability

Performance and stability of the control loop for regulating the output voltage for a converter can be determined from the open-loop characteristics:

1. The gain at low frequencies should be large so the steady-state error between the output and the reference signal is small.
2. The gain at the converter's switching frequency should be small.

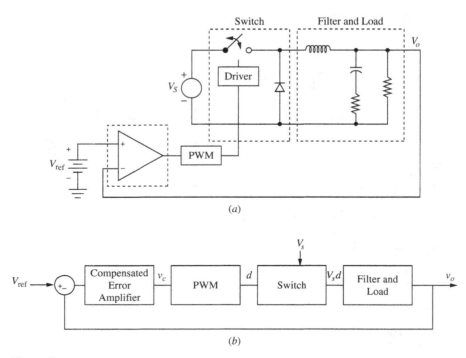

Figure 7-18 (a) Buck converter with feedback; (b) Control representation.

3. The open-loop phase shift at the crossover frequency (the frequency where the open-loop gain is unity) must lag by less than 180°. If the phase lag were 180°(or −180°), negative feedback provides a shift of another 180°, resulting in a total of 360° (or zero). A gain of magnitude 1 and phase of 360° around the loop make the loop unstable. The open-loop phase shift less than −180° at crossover is called the *phase margin*. A phase margin of at least 45° is a commonly used criterion for stability. Figure 7-19 illustrates the concept of phase margin. Note that phase margin is the angle between the phase shift and zero when the 180° phase angle of the inverting operational amplifier is included, which is convenient for use with PSpice analysis.

The transfer function of each block of the system in Fig. 7-18b must be developed to describe the control properties.

Small-Signal Analysis

Control loop analysis is based on the dynamic behavior of voltages, currents, and switching, unlike the steady-state analysis where the averaged circuit quantities are constants. Dynamic behavior can be described in terms of small-signal

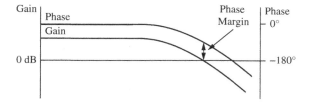

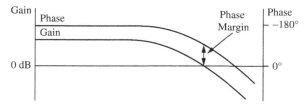

Figure 7-19 Phase margin. (*a*) In classical control theory, the phase margin is the angle difference between $-180°$ and the open-loop phase angle at the crossover frequency, where the open-loop gain magnitude is 0 dB; (*b*) The phase margin is between zero and the phase angle when the 180° phase angle of the inverting operational amplifier is included, which is convenient for PSpice simulation.

variations around a steady-state operating point. Output voltage, duty ratio, inductor current, source voltage, and other quantities are represented as

$$v_o = V_o + \tilde{v}_o$$
$$d = D + \tilde{d}$$
$$i_L = I_L + \tilde{i}_L$$
$$v_s = V_s + \tilde{v}_s$$
(7-58)

In these equations, the steady-state or dc term is represented by the uppercase letters, the ~ (tilde) quantity represents the ac term or small-signal perturbation, and the sum is the total quantity, represented by the lowercase letters.

Switch Transfer Function

For control purposes, the average values of voltages and currents are of greater interest than the instantaneous values that occur during the switching period. Equivalent representations of the switch in a buck converter are shown in Fig. 7-20. The relationship between input and output for the switch for a time-varying duty ratio is represented by the ideal transformation of $1:d$ shown in Fig. 7-20*b*. Here, d represents a time-varying duty ratio consisting of a dc (constant) component D plus a small-signal component $\tilde{d}$.

$$d = D + \tilde{d} \quad (7\text{-}59)$$

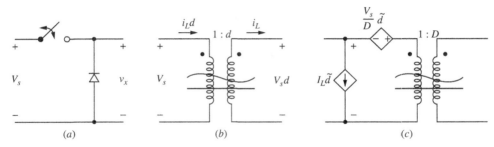

Figure 7-20 Switch models. (*a*) Switch and diode; (*b*) Model representing the transformation of average voltage and average current; (*c*) Model that separates steady-state and small-signal components.

An alternative representation of the switch shown in Fig. 7-20*c* separates the steady-state and small-signal components. The transformer secondary voltage v_x is related to the source voltage by

$$v_x = v_s d = (V_s + \tilde{v}_s)(D + \tilde{d}) = V_s D + \tilde{v}_s D + V_s \tilde{d} + \tilde{v}_s \tilde{d} \tag{7-60}$$

Neglecting the product of the small-signal terms,

$$v_x \doteq V_s D + \tilde{v}_s D + V_s \tilde{d} = v_s D + V_s \tilde{d} \tag{7-61}$$

Similarly, the current on the source side of the transformer is related to the secondary current by

$$i_s = i_L d = (I_L + \tilde{i}_L)(D + \tilde{d}) \doteq i_L D + I_L \tilde{d} \tag{7-62}$$

The circuit of Fig. 7-20*c*, with the transformer ratio fixed at D and the small-signal terms included with the dependent sources, satisfies the voltage and current requirements of the switch expressed in Eqs. (7-61) and (7-62).

Filter Transfer Function

The input to the buck converter filter is the switch output, which is $v_x = v_s d$ on an averaged circuit basis in the continuous current mode. The *RLC* filter of the buck converter has a transfer function developed from a straightforward application of circuit analysis in the *s* domain. From Fig. 7-21*a*, the transfer function of the filter with the load resistor is

$$\frac{v_o(s)}{v_x(s)} = \frac{v_o(s)}{V_s d(s)} = \frac{1}{LC[s^2 + s(1/RC) + 1/LC]} \tag{7-63}$$

or

$$\frac{v_o(s)}{d(s)} = \frac{V_s}{LC[s^2 + s(1/RC) + 1/LC]} \tag{7-64}$$

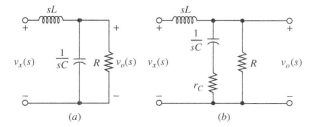

Figure 7-21 Circuits for deriving the filter transfer function (*a*) with an ideal capacitor and (*b*) with the ESR of the capacitor.

The above transfer function is based on ideal filter components. An equivalent series resistance (ESR) of r_C for a nonideal capacitor in Fig. 7-21*b* results in a filter transfer function of

$$\frac{v_o(s)}{d(s)} = \frac{V_s}{LC}\left[\frac{1 + sr_C R}{s^2(1 + r_C/R) + s(1/RC + r_C/L) + 1/LC}\right] \quad (7\text{-}65)$$

Since $r_C \ll R$ in practical circuits, the transfer function becomes approximately

$$\boxed{\frac{v_o(s)}{d(s)} \approx \frac{V_s}{LC}\left[\frac{1 + sr_C R}{s^2 + s(1/RC + r_C/L) + 1/LC}\right]} \quad (7\text{-}66)$$

The numerator of Eq. (7-66) shows that the ESR of the capacitor produces a zero in the transfer function, which may be important in determining system stability.

A general technique for establishing the switch and filter transfer function is state-space averaging. A development of this method is shown in App. B.

Pulse-Width Modulation Transfer Function

The pulse-width modulation (PWM) circuit converts the output from the compensated error amplifier to a duty ratio. The error amplifier output voltage v_c is compared to a sawtooth waveform with amplitude V_p, as shown in Fig. 7-22. The output of the PWM circuit is high while v_c is larger than the sawtooth and is zero when v_c is less than the sawtooth. If the output voltage falls below the reference, the error between the converter output and the reference signal increases, causing v_c to increase and the duty ratio to increase. Conversely, a rise in output voltage reduces the duty ratio. A transfer function for the PWM process is derived from the linear relation

$$d = \frac{v_c}{V_p} \quad (7\text{-}67)$$

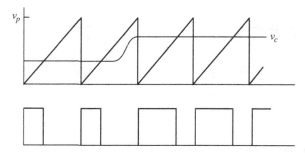

Figure 7-22 The PWM process. The output is high when v_c from the compensated error amplifier is higher than the sawtooth waveform.

The transfer function of the PWM circuit is therefore

$$\boxed{\frac{d(s)}{v_c(s)} = \frac{1}{V_p}} \tag{7-68}$$

Type 2 Error Amplifier with Compensation

The error amplifier compares the converter output voltage with a reference voltage to produce an error signal that is used to adjust the duty ratio of the switch. Compensation associated with the amplifier determines control loop performance and provides for a stable control system.

The transfer function of the compensated error amplifier should give a total loop characteristic consistent with the stability criteria described previously. Namely, the amplifier should have a high gain at low frequencies, a low gain at high frequencies, and an appropriate phase shift at the crossover frequency.

An amplifier that suits this purpose for many applications is shown in Fig. 7-23a. This is commonly referer to as a type 2 compensated error amplifier. (A type 1 amplifier is a simple integrator with one resistor on the input and one capacitor as feedback.). The amplifier is analyzed for the small-signal transfer function, so the dc reference voltage V_{ref} has no effect on the small-signal portion of the analysis. Furthermore, a resistor can be placed between the inverting input terminal and ground to act as a voltage divider to adjust the converter output voltage, and that resistor will have no effect on the small-signal analysis because the small-signal voltage at the noninverting terminal, and therefore at the inverting terminal, is zero.

The small-signal transfer function (with dc terms set to zero) of the amplifier is expressed in terms of input and feedback impedances Z_i and Z_f, where

$$Z_f = \left(R_2 + \frac{1}{sC_1}\right) \| \frac{1}{sC_2} = \frac{(R_2 + 1/sC_1)(1/sC_2)}{R_2 + 1/sC_1 + 1/sC_2} \tag{7-69}$$

$$Z_i = R_1$$

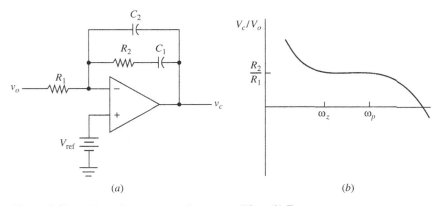

Figure 7-23 (a) Type 2 compensated error amplifier; (b) Frequency response.

The gain function $G(s)$ is expressed as the ratio of the compensated error amplifier small-signal output $\tilde{v}_c$ to the input, which is the converter output $\tilde{v}_o$.

$$G(s) = \frac{\tilde{v}_c(s)}{\tilde{v}_o(s)} = -\frac{Z_f}{Z_i} = -\frac{(R_2 + 1/sC_1)(1/sC_2)}{R_1(R_2 + 1/sC_1 + 1/sC_2)} \qquad (7\text{-}70)$$

Rearranging terms and assuming $C_2 \ll C_1$,

$$G(s) = \frac{\tilde{v}_c(s)}{\tilde{v}_o(s)} = -\frac{s + 1/R_2C_2}{R_1C_2 s[s + (C_1 + C_2)/R_2C_1C_2]} \approx -\frac{s + 1/R_2C_1}{R_1C_2 s(s + 1/R_2C_2)} \qquad (7\text{-}71)$$

The above transfer function has a pole at the origin and a zero and pole at

$$\omega_z = \frac{1}{R_2 C_1} \qquad (7\text{-}72)$$

$$\omega_p = \frac{C_1 + C_2}{R_2 C_1 C_2} \approx \frac{1}{R_2 C_2} \qquad (7\text{-}73)$$

The frequency response of this amplifier has the form shown in Fig. 7-23b. The values of R_1, R_2, C_1, and C_2 are chosen to make the overall control system have the desired attributes.

The combined frequency response of the transfer functions of the PWM circuit, the switch, and the output filter of a converter is shown in Fig. 7-24. The ESR of the filter capacitor puts a zero at $\omega = 1/r_c C$. A simulation program such as PSpice is useful to determine the frequency response. Otherwise, the transfer function may be evaluated with $s = j\omega$.

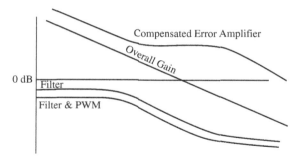

Figure 7-24 The control loop transfer function frequency response.

EXAMPLE 7-8

A Type 2 Amplifier Control Loop for a Buck Converter

The source voltage for a buck converter is $V_s = 6$ V, and the output voltage is to be regulated at 3.3 V. The load resistance is 2 Ω, $L = 100$ μH with negligible internal resistance, and $C = 75$ μF with an ESR of 0.4 Ω. The PWM circuit has a sawtooth voltage with peak value $V_p = 1.5$ V. A type 2 compensated error amplifier has $R_1 = 1$ kΩ, $R_2 = 2.54$ kΩ, $C_1 = 48.2$ nF, and $C_2 = 1.66$ nF. The switching frequency is 50 kHz. Use PSpice to determine the crossover frequency and the phase margin.

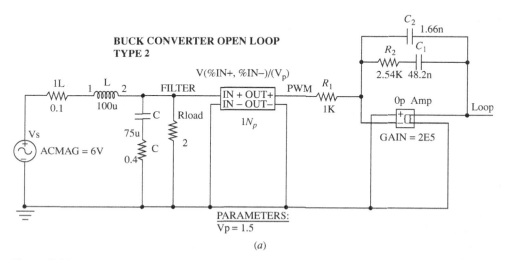

(a)

Figure 7-25 (a) PSpice circuit for simulating the open-loop response of a buck converter; (b) Probe output for Example 7-8 showing a crossover frequency of 6.83 kHz and a phase margin of approximately 45°.

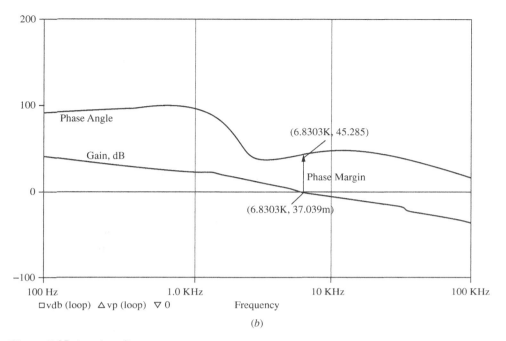

Figure 7-25 (*continued*).

■ **Solution**

A PSpice circuit for the filter, compensated error amplifier, and PWM converter is shown in Fig. 7-25a. The input voltage source is the ac source V_{ac}, the PWM function of $1/V_p$ is implemented with the dependent source EVALUE, and the ideal op-amp is implemented with a high-gain voltage-controlled voltage source.

The Probe output shown in Fig. 7-25b reveals the crossover frequency to be 6.83 kHz. The phase margin is the angle greater than zero (or 360°) because the operational amplifier contains the inversion (180°) for negative feedback (Fig. 7-19a). The Probe output shows the phase margin to be slightly larger than 45°. The gain is low, −23.8 dB, at the 50-kHz switching frequency. Therefore, this circuit meets the criteria for a stable control system.

Design of a Type 2 Compensated Error Amplifier

The midfrequency gain and the location of the pole and zero of the transfer function of the compensated error amplifier must be selected to provide the desired total open-loop crossover frequency and phase margin required for stability.

The transfer function of the compensated error amplifier in Eq. (7-71) can be expressed for $s = j\omega$ as

$$G(j\omega) = \frac{\tilde{v}_c(j\omega)}{\tilde{v}_o(j\omega)} = -\frac{j\omega + \omega_z}{R_1 C_2 j\omega (j\omega + \omega_p)} \qquad (7\text{-}74)$$

For the middle frequencies, $\omega_z \ll \omega \ll \omega_p$, resulting in

$$G(j\omega) = \frac{\tilde{v}_c(j\omega)}{\tilde{v}_o(j\omega)} \approx -\frac{j\omega}{R_1 C_2 j\omega\omega_p} = -\frac{1}{R_1 C_2 (1/R_2 C_2)} = -\frac{R_2}{R_1} \qquad (7\text{-}75)$$

The phase angle θ_{comp} of the compensated error amplifier transfer function of Eq. (7-74) is

$$\theta_{\text{comp}} = -180° + \tan^{-1}\left(\frac{\omega}{\omega_z}\right) - 90° - \tan^{-1}\left(\frac{\omega}{\omega_p}\right)$$

$$= -270° + \tan^{-1}\left(\frac{\omega}{\omega_z}\right) - \tan^{-1}\left(\frac{\omega}{\omega_p}\right) \qquad (7\text{-}76)$$

The $-180°$ is from the negative sign, and the $-90°$ is from the pole at the origin. Note that in this development the inverting amplifier phase shift of $-180°$ is included in Eq. (7-76). In some developments of this method, the inverting amplifier phase shift is omitted at this point and then included later.

The following is a design procedure for the type 2 compensated error amplifier.

1. Choose the desired crossover frequency of the total open-loop transfer function. This is usually around an order of magnitude less than the converter switching frequency. Some designers go as high as 25 percent of the switching frequency.
2. Determine the transfer function and frequency response of all elements in the control circuit except for the compensated error amplifier.
3. Determine the midfrequency gain of the compensated error amplifier required to achieve the overall desired crossover frequency. This establishes the R_2/R_1 ratio as in Eq. (7-75).
4. Choose the desired phase margin needed to ensure stability, typically greater than 45°. Having established R_1 and R_2 for the midfrequency gain, the pole and zero, ω_p and ω_z, are determined by C_1 and C_2. The phase angle θ_{comp} of the compensated error amplifier at the crossover frequency ω_{co} is

$$\theta_{\text{comp}} = -270° + \tan^{-1}\left(\frac{\omega_{co}}{\omega_z}\right) - \tan^{-1}\left(\frac{\omega_{co}}{\omega_p}\right) \qquad (7\text{-}77)$$

A procedure for selecting the pole and zero frequencies is the K factor method [see Venable (1983) and Basso (2008) in the Bibliography]. Using the K factor method, the value of K is determined as follows:
Let the zero and pole of the transfer function be at

$$\omega_z = \frac{\omega_{co}}{K} \qquad (7\text{-}78)$$

and

$$\omega_p = K\omega_{co} \qquad (7\text{-}79)$$

Then

$$K = \frac{\omega_{co}}{\omega_z} = \frac{\omega_p}{\omega_{co}} \quad (7\text{-}80)$$

The phase angle of the compensated error amplifier at crossover in Eq. (7-77) is then

$$\theta_{comp} = -270° + \tan^{-1} K - \tan^{-1}\left(\frac{1}{K}\right) \quad (7\text{-}81)$$

Using the trigonometric identity

$$\tan^{-1}(x) + \tan^{-1}\left(\frac{1}{x}\right) = 90° \quad (7\text{-}82)$$

gives

$$\tan^{-1}\left(\frac{1}{K}\right) = 90° - \tan^{-1}(K) \quad (7\text{-}83)$$

Equation (7-81) becomes

$$\theta_{comp} = -270° + \tan^{-1}(K) - \tan^{-1}\left(\frac{1}{K}\right) = 2\tan^{-1}(K) - 360° = 2\tan^{-1}(K) \quad (7\text{-}84)$$

Solving for K,

$$\boxed{K = \tan\left(\frac{\theta_{comp}}{2}\right)} \quad (7\text{-}85)$$

The angle θ_{comp} is the desired phase angle of the compensated error amplifier at the crossover frequency. From Eq. (7-84), the phase angle of the compensated error amplifier can range from 0 to 180° for $0 < K < \infty$.

The required phase angle of the compensated error amplifier to obtain the desired phase margin is determined, establishing the value of K. If the desired crossover frequency ω_{co} is known, then ω_z and ω_p are obtained from Eqs. (7-78) and (7-79). Then C_1 and C_2 are determined from Eqs. (7-71) and (7-72).

$$\omega_z = \frac{1}{R_2 C_1} = \frac{\omega_{co}}{K}$$
$$C_1 = \frac{K}{\omega_{co} R_2} = \frac{K}{2\pi f_{co} R_2} \quad (7\text{-}86)$$

$$\omega_p = \frac{1}{R_2 C_2} = K\omega_{co}$$
$$C_2 = \frac{1}{K\omega_{co} R_2} = \frac{1}{K 2\pi f_{co} R_2} \quad (7\text{-}87)$$

EXAMPLE 7-9

Design of a Type 2 Compensated Error Amplifier

For a buck converter shown in Fig. 7-26a,

$V_s = 10$ V   with an output of 5 V
$f = 100$ kHz
$L = 100$ μH   with a series resistance of 0.1 Ω
$C = 100$ μF   with an equivalent series resistance of 0.5 Ω
$R = 5$ Ω
$V_p = 3$ V   in PWM circuit

Design a type 2 compensated error amplifier that results in a stable control system.

■ Solution

1. The crossover frequency of the total open-loop transfer function (the frequency where the gain is 1, or 0 dB) should be well below the switching frequency. Let $f_{co} = 10$ kHz.

2. A PSpice simulation of the frequency response of the filter with load resistor (Fig. 7-26b) shows that the converter (V_s and the filter) gain at 10 kHz is -2.24 dB and the phase angle is $-101°$. The PWM converter has a gain of $1/V_p = 1/3 = -9.5$ dB. The combined gain of the filter and PWM converter is then -2.24 dB -9.54 dB $= -11.78$ dB.

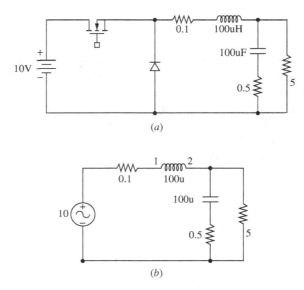

Figure 7-26 (a) Buck converter circuit; (b) The ac circuit for determining the frequency response of the converter.

3. The compensated error amplifier should therefore have a gain of +11.78 dB at 10 kHz to make the loop gain 0 dB. Converting the gain in decibels to a ratio of v_o/v_i,

$$11.78 \text{ dB} = 20\log\left(\frac{\tilde{v}_c}{\tilde{v}_o}\right)$$

$$\frac{\tilde{v}_c}{\tilde{v}_o} = 10^{11.78/20} = 3.88$$

Using Eq. (7-75), the magnitude of the midfrequency gain is

$$\frac{R_2}{R_1} = 3.88$$

Letting $R_1 = 1$ kΩ, R_2 is then 3.88 kΩ.

4. The phase angle of the compensated error amplifier at crossover must be adequate to give a phase margin of at least 45°. The required phase angle of the amplifier is

$$\theta_{\text{comp}} = \theta_{\text{phase margin}} - \theta_{\text{converter}} = 45° - (-101°) = 146°$$

A K factor of 3.27 is obtained from Eq. (7-85).

$$K = \tan\left(\frac{\theta_{\text{comp}}}{2}\right) = \tan\left(\frac{146°}{2}\right) = \tan(73°) = 3.27$$

Using Eq. (7-86) to get C_1,

$$C_1 = \frac{K}{2\pi f_{co} R_2} = \frac{3.27}{2\pi(10{,}000)(3880)} = 13.4 \text{ nF}$$

Using Eq. (7-87) to get C_2,

$$C_2 = \frac{1}{K 2\pi f_{co} R_2} = \frac{1}{3.27(2\pi)(10{,}000)(3880)} = 1.25 \text{ nF}$$

A PSpice simulation of the control loop gives a crossover frequency of 9.41 kHz and a phase margin of 46°, verifying the design.

PSpice Simulation of Feedback Control

Simulation is a valuable tool in the design and verification of a closed-loop control system for dc power supplies. Figure 7.27a shows a PSpice implementation using idealized switches and ETABLE sources for the op-amp and for the comparator in the PWM function. The input is 6 V, and the output is to be regulated at 3.3 V. The phase margin of this circuit is 45° when the load is 2 Ω, and slightly greater than 45° when the load changes to 2||4 Ω. The switching frequency is 100 kHz. A step change in load occurs at $t = 1.5$ ms. If the circuit were unregulated, the output voltage would change as the load current changed because of the inductor resistance. The control circuit adjusts the duty ratio to compensate for changes in operating conditions.

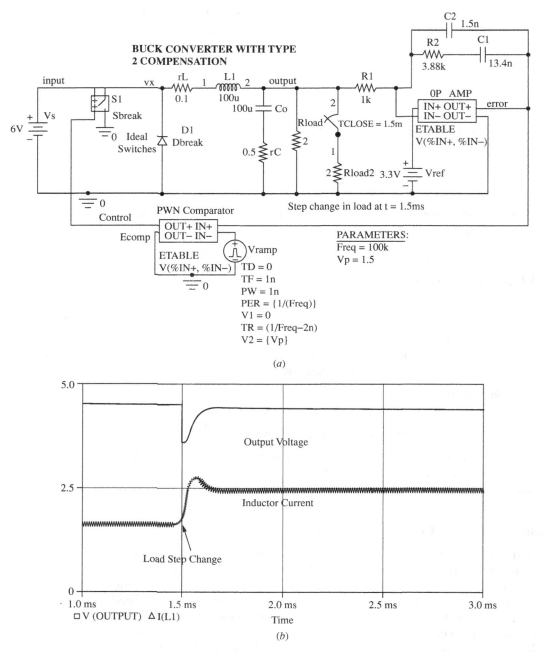

Figure 7-27 (*a*) PSpice circuit for a regulated buck converter; (*b*) Output voltage and inductor current for a step change in load.

Figure 7-27b shows the output voltage and inductor current, verifying that the control circuit is stable.

Type 3 Error Amplifier with Compensation

The type 2 compensation circuit described previously is sometimes not capable of providing sufficient phase angle difference to meet the stability criterion of a 45° phase margin. Another compensation circuit, known as the type 3 amplifier, is shown in Fig. 7-28a. The type 3 amplifier provides an additional phase angle boost compared to the type 2 circuit and is used when an adequate phase margin is not achievable using the type 2 amplifier.

The small-signal transfer function is expressed in terms of input and feedback impedances Z_i and Z_f,

$$G(s) = \frac{\tilde{v}_c(s)}{\tilde{v}_o(s)} = -\frac{Z_f}{Z_i} = -\frac{(R_2 + 1/sC_1) \| 1/sC_2}{R_1 \| (R_3 + 1/sC_3)} \quad (7\text{-}88)$$

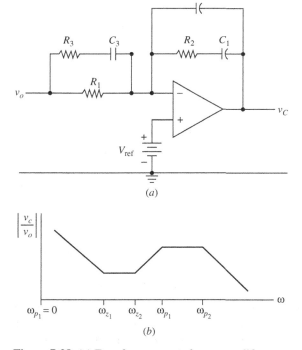

Figure 7-28 (a) Type 3 compensated error amplifier; (b) Bode magnitude plot.

resulting in

$$G(s) = -\frac{R_1 + R_3}{R_1 R_3 C_3} \frac{\left(s + \frac{1}{R_2 C_1}\right)\left(s + \frac{1}{(R_1 + R_3)C_3}\right)}{s\left(s + \frac{C_1 + C_2}{R_2 C_1 C_2}\right)\left(s + \frac{1}{R_3 C_3}\right)} \tag{7-89}$$

The reference voltage V_{ref} is purely dc and has no effect on the small-signal transfer function. Assuming $C_2 \ll C_1$ and $R_3 \ll R_1$,

$$G(s) \approx -\frac{1}{R_3 C_2} \frac{(s + 1/R_2 C_1)(s + 1/R_1 C_3)}{s(s + 1/R_2 C_2)(s + 1/R_3 C_3)} \tag{7-90}$$

An inspection of the transfer function of Eq. (7-90) shows that there are two zeros and three poles, including the pole at the origin. A particular placement of the poles and zeros produces the Bode plot of the transfer function shown in Fig. 7-28b.

$$G(j\omega) = -\frac{1}{R_3 C_2} \frac{(j\omega + \omega_{z_1})(j\omega + \omega_{z_2})}{j\omega(j\omega + \omega_{p_2})(j\omega + \omega_{p_3})} \tag{7-91}$$

The zeros and poles of the transfer function are

$$\begin{aligned}
\omega_{z_1} &= \frac{1}{R_2 C_2} \\
\omega_{z_2} &= \frac{1}{(R_1 + R_3)C_3} \approx \frac{1}{R_1 C_3} \\
\omega_{p_1} &= 0 \\
\omega_{p_2} &= \frac{C_1 + C_2}{R_2 C_1 C_2} \approx \frac{1}{R_2 C_2} \\
\omega_{p_3} &= \frac{1}{R_3 C_3}
\end{aligned} \tag{7-92}$$

The phase angle of the compensated error amplifier is

$$\begin{aligned}
\theta_{\text{comp}} &= -180° + \tan^{-1}\left(\frac{\omega}{\omega_{z_1}}\right) + \tan^{-1}\left(\frac{\omega}{\omega_{z_2}}\right) - 90° - \tan^{-1}\left(\frac{\omega}{\omega_{p_2}}\right) - \tan^{-1}\left(\frac{\omega}{\omega_{p_3}}\right) \\
&= -270° + \tan^{-1}\left(\frac{\omega}{\omega_{z_1}}\right) + \tan^{-1}\left(\frac{\omega}{\omega_{z_2}}\right) - \tan^{-1}\left(\frac{\omega}{\omega_{p_2}}\right) - \tan^{-1}\left(\frac{\omega}{\omega_{p_3}}\right)
\end{aligned} \tag{7-93}$$

The $-180°$ is from the negative sign, and the $-90°$ is from the pole at the origin.

Design of a Type 3 Compensated Error Amplifier

The K factor method can be used for the type 3 amplifier in a similar way as it was used in the type 2 circuit. Using the K factor method, the zeros are placed at the same frequency to form a double zero, and the second and third poles are placed at the same frequency to form a double pole:

$$\begin{aligned}
\omega_z &= \omega_{z_1} = \omega_{z_2} \\
\omega_p &= \omega_{p_2} = \omega_{p_3}
\end{aligned} \tag{7-94}$$

The first pole remains at the origin.
The double zeros and poles are placed at frequencies

$$\omega_z = \frac{\omega_{co}}{\sqrt{K}}$$
$$\omega_p = \omega_{co}\sqrt{K} \qquad (7\text{-}95)$$

The amplifier transfer function from Eq. (7-91) can then be written as

$$G(j\omega) = -\frac{1}{R_3 C_2} \frac{(j\omega + \omega_z)^2}{j\omega(j\omega + \omega_p)^2} \qquad (7\text{-}96)$$

At the crossover frequency ω_{co}, the gain is

$$G(j\omega_{co}) = -\frac{1}{R_3 C_2} \frac{(j\omega_{co} + \omega_z)^2}{j\omega_{co}(j\omega_{co} + \omega_p)^2} \qquad (7\text{-}97)$$

The phase angle of the amplifier at the crossover frequency is then

$$\theta_{comp} = -270° + 2\tan^{-1}\left(\frac{\omega_{co}}{\omega_z}\right) - 2\tan^{-1}\left(\frac{\omega_{co}}{\omega_p}\right) \qquad (7\text{-}98)$$

Using Eq. (7-95) for ω_z and ω_p,

$$\theta_{comp} = -270° + 2\tan^{-1}\left(\frac{\omega_{co}}{\omega_{co}/\sqrt{K}}\right) - 2\tan^{-1}\left(\frac{\omega_{co}}{\omega_{co}\sqrt{K}}\right) \qquad (7\text{-}99)$$

resulting in

$$\theta_{comp} = -270° + 2\tan^{-1}\sqrt{K} - 2\tan^{-1}\left(\frac{1}{\sqrt{K}}\right)$$
$$= -270° + 2\left[\tan^{-1}\sqrt{K} - \tan^{-1}\left(\frac{1}{\sqrt{K}}\right)\right] \qquad (7\text{-}100)$$

By using the identity

$$\tan^{-1}(x) + \tan^{-1}\left(\frac{1}{x}\right) = 90° \qquad (7\text{-}101)$$

making

$$\tan^{-1}\left(\frac{1}{\sqrt{K}}\right) = 90° - \tan^{-1}(\sqrt{K}) \qquad (7\text{-}102)$$

Eq. (7-100) becomes

$$\theta_{comp} = -270° + 2[\tan^{-1}\sqrt{K} + (-90° - \tan^{-1}\sqrt{K})]$$
$$\theta_{comp} = -450° + 4\tan^{-1}\sqrt{K} = -90° + 4\tan^{-1}\sqrt{K} \qquad (7\text{-}103)$$

Solving for K,

$$K = \tan\left(\frac{\theta_{comp} + 90°}{4}\right)^2 \quad (7\text{-}104)$$

From Eq. (7-103), the maximum angle of the compensated error amplifier is 270°. Recall that the maximum phase angle of the type 2 amplifier is 180°.

The phase angle of the compensated error amplifier is

$$\theta_{comp} = \theta_{phase\ margin} - \theta_{converter} \quad (7\text{-}105)$$

The minimum phase margin is usually 45°, and the phase angle of the converter at the desired crossover frequency can be determined from a PSpice simulation.

At the crossover frequency ω_{co},

$$G(j\omega_{co}) = -\frac{1}{R_3 C_2} \frac{(j\omega_{co} + \omega_z)^2}{j\omega_{co}(j\omega_{co} + \omega_p)^2}$$
$$\approx -\frac{1}{R_3 C_2} \frac{(j\omega_{co})^2}{j\omega_{co}(\omega_p)^2} = -\frac{1}{R_3 C_2} \frac{j\omega_{co}}{(\omega_p)^2} \quad (7\text{-}106)$$

Using Eqs. (7-95) and (7-92),

$$\omega_{co} = \sqrt{K}\omega_z = \frac{\sqrt{K}}{R_1 C_3} \quad (7\text{-}107)$$

$$\omega_p = \frac{1}{R_2 C_2} = \frac{1}{R_3 C_3} \quad \Rightarrow \quad \omega_p^2 = \frac{1}{R_2 C_2 R_3 C_3} \quad (7\text{-}108)$$

Equation (7-106) becomes

$$G(j\omega_{co}) = -\frac{1}{R_3 C_2} \frac{j\omega_{co}}{(\omega_p)^2} = -\frac{1}{R_3 C_2} \frac{j\sqrt{K}/R_1 C_3}{1/R_2 C_2 R_3 C_3} = -\frac{j\sqrt{K} R_2}{R_1} \quad (7\text{-}109)$$

In the design of a type 3 compensated error amplifier, first choose R_1 and then compute R_2 from Eq. (7-109). Other component values can then be determined from

$$\omega_z = \frac{\omega_{co}}{\sqrt{K}} = \frac{1}{R_2 C_1} = \frac{1}{R_1 C_3} \quad (7\text{-}110)$$

and

$$\omega_p = \omega_{co}\sqrt{K} = \frac{1}{R_2 C_2} = \frac{1}{R_3 C_3} \quad (7\text{-}111)$$

The resulting equations are

$$
\begin{aligned}
R_2 &= \frac{|G(j\omega_{co})|R_1}{\sqrt{K}} \\
C_1 &= \frac{\sqrt{K}}{\omega_{co}R_2} = \frac{\sqrt{K}}{2\pi f_{co}R_2} \\
C_2 &= \frac{1}{\omega_{co}R_2\sqrt{K}} = \frac{1}{2\pi f_{co}R_2\sqrt{K}} \\
C_3 &= \frac{\sqrt{K}}{\omega_{co}R_1} = \frac{\sqrt{K}}{2\pi f_{co}R_1} \\
R_3 &= \frac{1}{\omega_{co}\sqrt{K}C_3} = \frac{1}{2\pi f_{co}\sqrt{K}C_3}
\end{aligned}
\qquad (7\text{-}112)
$$

EXAMPLE 7-10

Design of a Type 3 Compensated Error Amplifier

For the buck converter shown in Fig.7-29a,

$V_s = 10$ V	with an output of 5 V
$f = 100$ kHz	
$L = 100$ μH	with a series resistance of 0.1 Ω
$C = 100$ μF	with an equivalent series resistance of 0.1 Ω
$R = 5$ Ω	
$V_p = 3$ V	in PWM circuit

Design a type 3 compensated error amplifier that results in a stable control system. Design for a crossover frequency of 10 kHz and a phase margin of 45°. Note that all parameters are the same as in Example 7-8 except that the ESR of the capacitor is much smaller.

■ **Solution**

A PSpice ac frequency sweep shows that the output voltage is -10.5 dB at 10 kHz and the phase angle is $-144°$. The PWM circuit produces an additional gain of -9.5 dB. Therefore, the compensating error amplifier must have a gain of $10.5 + 9.5 = 20$ dB at 10 kHz. A gain of 20 dB corresponds to a gain of 10.

The required phase angle of the amplifier is determined from Eq. (7-105),

$$\theta_{comp} = \theta_{phase\ margin} - \theta_{converter} = 45° - (-144°) = 189°$$

Solving for K in Eq. (7-104) yields.

$$K = \left[\tan\left(\frac{189° + 90°}{4}\right)\right]^2 = [\tan(69.75°)]^2 = 7.35$$

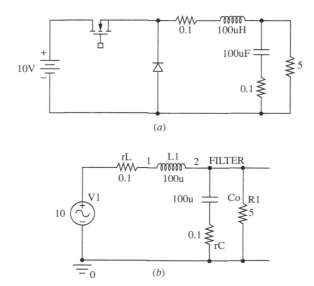

Figure 7-29 (*a*) Buck converter circuit; (*b*) The ac circuit used to determine the frequency response.

Letting $R_1 = 1$ kΩ, the other component values are computed from Eq. (7-112).

$$R_2 = \frac{|G(j\omega_{co})|R_1}{\sqrt{K}} = \frac{10(1000)}{\sqrt{7.35}} = 3.7 \text{ k}\Omega$$

$$C_1 = \frac{\sqrt{K}}{2\pi f_{co} R_2} = \frac{\sqrt{7.35}}{2\pi(10,000)(3700)} = 11.6 \text{ nF}$$

$$C_2 = \frac{1}{2\pi f_{co} R_2 \sqrt{K}} = \frac{1}{2\pi(10,000)(3700)\sqrt{7.35}} = 1.58 \text{ nF}$$

$$C_3 = \frac{\sqrt{K}}{2\pi f_{co} R_1} = \frac{\sqrt{7.35}}{2\pi(10,000)(1000)} = 43.1 \text{ nF}$$

$$R_3 = \frac{1}{2\pi f_{co} \sqrt{K} C_3} = \frac{1}{2\pi(10,000)\sqrt{7.35}(43.1)(10)^{-9}} = 136 \text{ }\Omega$$

A PSpice simulation of the converter, compensated error amplifier, and PWM circuit gives a crossover frequency of 10 kHz with a phase margin of 49°.

Note that attempting to use a type 2 compensated error amplifier for this circuit is unsuccessful because the required phase angle at the crossover frequency is greater than 180°. Comparing this converter with that of Example 7-8, the ESR of the capacitor here is smaller. Low capacitor ESR values often necessitate use of the type 3 rather than the type 2 circuit.

Table 7-1 Type 3 Compensating error amplifier zeros and poles and frequency placement

Zero or Pole	Expression	Placement
First zero	$\omega_{z_1} = \dfrac{1}{R_2 C_2}$	50% to 100% of ω_{LC}
Second zero	$\omega_{z_2} = \dfrac{1}{(R_1 + R_3)C_3} \approx \dfrac{1}{R_1 C_3}$	At ω_{LC}
First pole	$\omega_{p_1} = 0$	—
Second pole	$\omega_{p_2} = \dfrac{C_1 + C_2}{R_2 C_1 C_2} \approx \dfrac{1}{R_2 C_2}$	At the ESR zero $= 1/r_C C$
Third pole	$\omega_{p_3} = \dfrac{1}{R_3 C_3}$	At one-half the switching frequency, $2\pi f_{sw}/2$

Manual Placement of Poles and Zeros in the Type 3 Amplifier

As an alternative to the K factor method described previously, some designers place the poles and zeros of the type 3 amplifier at specified frequencies. In placing the poles and zeros, a frequency of particular interest is the resonant frequency of the LC filter in the converter. Neglecting any resistance in the inductor and capacitor,

$$\omega_{LC} = \frac{1}{\sqrt{LC}} \qquad f_{LC} = \frac{1}{2\pi\sqrt{LC}} \tag{7-113}$$

The first zero is commonly placed at 50 to 100 percent of f_{LC}, the second zero is placed at f_{LC}, the second pole is placed at the ESR zero in the filter transfer function ($1/r_C C$), and the third pole is placed at one-half the switching frequency. Table 7-1 indicates placement of the type 3 error amplifier poles and zeros.

7.14 PWM CONTROL CIRCUITS

The major elements of the feedback control of dc power supplies are available in a single integrated circuit (IC). The National Semiconductor LM2743 is one example of an integrated circuit for dc power supply control. The IC contains the error amplifier op-amp, PWM circuit, and driver circuits for the MOSFETs in a dc-dc converter using synchronous rectification. The block diagram of the IC is shown in Fig. 7-30a, and a typical application is shown in Fig. 7-30b.

7.15 THE AC LINE FILTER

In many dc power supply applications, the power source is the ac power system. The voltage and current from the ac system are often contaminated by high-frequency electrical noise. An ac line filter suppresses conductive radio-frequency interference (RFI) noise from entering or leaving the power supply.

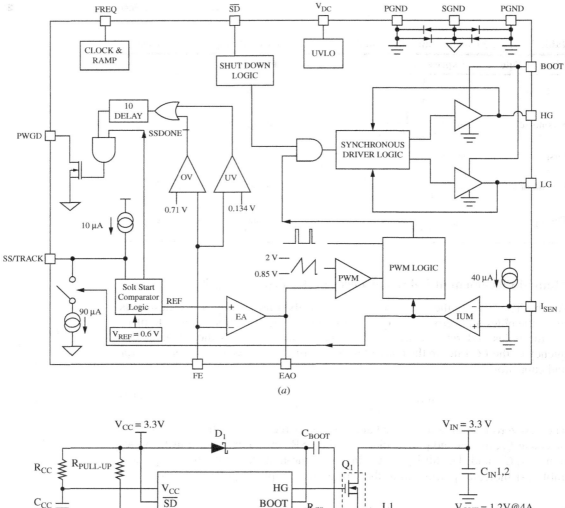

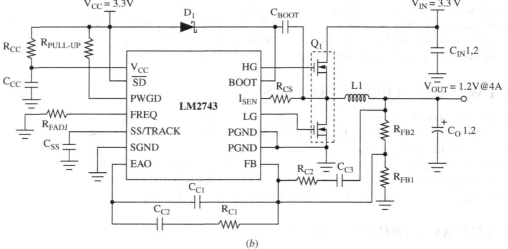

Figure 7-30 (*a*) The National Semiconductor LM2743 block diagram; (*b*) An application in a buck converter circuit (with permission from National Semiconductor Corporation[1]).

[1]Copyright © 2003 National Semiconductor Corporation, 2900 Semiconductor Drive, Santa Clara, CA 95051. All rights reserved, http://www.national.com.

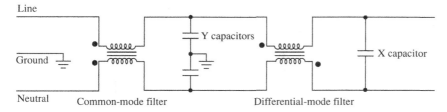

Figure 7-31 A typical ac line filter.

A single-phase ac input to a power supply has a line (or phase) wire, a neutral wire, and a ground wire. Common-mode noise consists of currents in the line and neutral conductors that are in phase and return through the ground path. Differential-mode noise consists of high-frequency currents that are 180° out of phase in the line and neutral conductors, which means that current enters from the line and returns in the neutral.

A typical ac line filter circuit is shown in Fig. 7-31. The first stage is a common-mode filter, consisting of a transformer with adjacent polarity markings and a capacitor connected from each line to ground. The capacitors in this stage are referred to as the *Y capacitors*. The second stage of the filter, consisting of a transformer with opposite polarity markings and a single capacitor connected across the ac lines, removes differential-mode noise from the ac signal. The capacitor in this stage is referred to as the *X capacitor*.

7.16 THE COMPLETE DC POWER SUPPLY

A complete dc power supply consists of an input ac line filter, a power factor correction stage, and a dc-dc converter, as illustrated in the block diagram of Fig. 7-32. The power factor correction stage is discussed in Sec. 7.11, and the dc-dc converter could be any of the converters discussed in this chapter or in Chap. 6.

Low-power applications such as cell phone chargers can be implemented with a topology like that shown in Fig. 7-33. A full-wave rectifier with a capacitor filter (Chap. 4) produces a dc voltage from the ac line voltage source, and a flyback dc-dc converter reduces the dc voltage to the appropriate level for the application. An optically coupled feedback loop preserves electrical isolation between the source and the load, and a control circuit adjusts the duty ratio of the switch for a regulated output. Integrated-circuit packages include the control

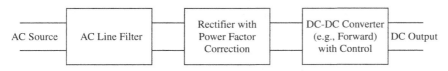

Figure 7-32 A complete power supply when the source is the ac power system.

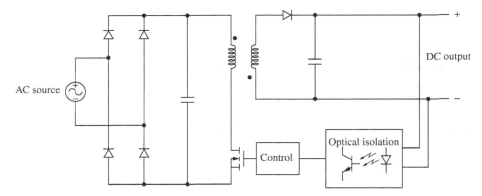

Figure 7-33 An off-line power supply for low-power applications.

function and the switching transistor. Some such integrated circuits can be powered directly from the high-voltage output of the rectifier, and others require another winding on the flyback converter to produce the IC supply voltage. This type of power supply is often called an off-line converter.

7.17 Bibliography

S. Ang and A. Oliva, *Power-Switching Converters*, 2d ed., Taylor & Francis, Boca Raton, Fla., 2005.

C. Basso, *Switch-Mode Power Supplies,* McGraw-Hill, New York, 2008.

B. K. Bose, *Power Electronics and Motor Drives: Advances and Trends*, Elsevier/Academic Press, Boston, 2006.

M. Day, "Optimizing Low-Power DC/DC Designs—External versus Internal Compensation," Texas Instruments, Incorporated, 2004.

R. W. Erickson and D. Maksimović, *Fundamentals of Power Electronics*, 2d ed., Kluwer Academic, 2001.

A. J. Forsyth and S. V. Mollov, "Modeling and Control of DC-DC converters," *Power Engineering Journal*, vol. 12, no. 5, 1998, pp. 229–236.

Y. M. Lai, *Power Electronics Handbook*, edited by M. H. Rashid, Academic Press, Calif., San Diego, 2001, Chapter 20.

LM2743 Low Voltage N-Channel MOSFET Synchronous Buck Regulator Controller, National Semiconductor, 2005.

D. Mattingly, "Designing Stable Compensation Networks for Single Phase Voltage Mode Buck Regulators," Intersil Technical Brief TB417.1, Milpitas, Calif., 2003.

N. Mohan, T. M. Undeland, and W. P. Robbins, *Power Electronics: Converters, Applications, and Design*, 3d ed., Wiley, New York, 2003.

G. Moschopoulos and P. Jain, "Single-Phase Single-Stage Power-Factor-Corrected Converter Topologies," *IEEE Transactions on Industrial Electronics*, vol. 52, no. 1, February 2005, pp. 23–35.

M. Nave, *Power Line Filter Design for Switched-Mode Power Supplies*, Van Nostrand Reinhold, Princeton, N.J., 1991.

A. I. Pressman, K. Billings, and T. Morey, *Switching Power Supply Design*, McGraw-Hill, New York, 2009.

M. H. Rashid, *Power Electronics: Circuits, Devices, and Systems,* 3d ed., Prentice-Hall, Upper Saddle River, N.J., 2004.

M. Qiao, P. Parto, and R. Amirani, "Stabilize the Buck Converter with Transconductance Amplifier," International Rectifier Application Note AN-1043, 2002.

D. Venable, "The *K* Factor: A New Mathematical Tool for Stability Analysis and Synthesis," *Proceedings Powercon 10,* 1983.

V. Vorperian, "Simplified Analysis of PWM Converters Using Model of PWM Switch," *IEEE Transactions on Aerospace and Electronic Systems,* May 1990.

"8-Pin Synchronous PWM Controller," International Rectifier Data Sheet No. PD94173 revD, 2005.

Problems

Flyback Converter

7.1 The flyback converter of Fig. 7-2 has parameters $V_s = 36$ V, $D = 0.4$, $N_1/N_2 = 2$, $R = 20\ \Omega$, $L_m = 100\ \mu$H, and $C = 50\ \mu$F, and the switching frequency is 100 kHz. Determine (*a*) the output voltage; (*b*) the average, maximum, and minimum inductor currents; and (*c*) the output voltage ripple.

7.2 The flyback converter of Fig. 7-2 has parameters $V_s = 4.5$ V, $D = 0.6$, $N_1/N_2 = 0.4$, $R = 15\ \Omega$, $L_m = 10\ \mu$H, and $C = 10\ \mu$F, and the switching frequency is 250 kHz. Determine (*a*) the output voltage; (*b*) the average, maximum, and minimum inductor currents; and (*c*) the output voltage ripple.

7.3 The flyback converter of Fig. 7-2 has an input of 44 V, an output of 3 V, a duty ratio of 0.32, and a switching frequency of 300 kHz. The load resistor is $1\ \Omega$. (*a*) Determine the transformer turns ratio. (*b*) Determine the transformer magnetizing inductance L_m such that the minimum inductor current is 40 percent of the average.

7.4 Design a flyback converter for an input of 24 V and an output of 40 W at 40 V. Specify the transformer turns ratio and magnetizing inductance, switching frequency, and capacitor to limit the ripple to less than 0.5 percent.

7.5 (*a*) What is the value of load resistance that separates continuous and discontinuous magnetizing inductance current in the flyback converter of Example 7-1? (*b*) Graph V_o/V_s as the load changes from 5 to $20\ \Omega$.

7.6 For the flyback converter operating in the discontinuous-current mode, derive an expression for the time at which the magnetizing current i_{L_m} returns to zero.

Forward Converter

7.7 The forward converter of Fig. 7-5*a* has parameters $V_s = 100$ V, $N_1/N_2 = N_1/N_3 = 1$, $L_m = 1$ mH, $L_x = 70\ \mu$H, $R = 20\ \Omega$, $C = 33\ \mu$F, and $D = 0.35$, and the switching frequency is 150 kHz. Determine (*a*) the output voltage and output voltage ripple; (*b*) the average, maximum, and minimum values of the current in L_x; (*c*) the peak current in L_m in the transformer model; and (*d*) the peak current in the switch and the physical transformer primary.

7.8 The forward converter of Fig. 7-5*a* has parameters $V_s = 170$ V, $N_1/N_2 = 10$, $N_1/N_3 = 1$, $L_m = 340\ \mu$H, $L_x = 20\ \mu$H, $R = 10\ \Omega$, $C = 10\ \mu$F, $D = 0.3$, and the switching frequency is 500 kHz. (*a*) Determine the output voltage and output voltage ripple. (*b*) Sketch the currents in L_x, L_m, each transformer winding, and V_s.

(c) Determine the power returned to the source by the tertiary (third) transformer winding from the recovered stored energy in L_m.

7.9 A forward converter has a source of 80 V and a load of 250 W at 50 V. The output filter has $L_x = 100$ μH and $C = 150$ μF. The switching frequency is 100 kHz. (a) Select a duty ratio and transformer turns ratios N_1/N_2 and N_1/N_3 to provide the required output voltage. Verify continuous current in L_x. (b) Determine the output voltage ripple.

7.10 The forward converter of Fig. 7-5a has parameters $V_s = 100$ V, $N_1/N_2 = 5$, $N_1/N_3 = 1$, $L_m = 333$ μH, $R = 2.5$ Ω, $C = 10$ μF, and $D = 0.25$, and the switching frequency is 375 kHz. (a) Determine the output voltage and output voltage ripple. (b) Sketch the currents i_{L_x}, I_1, i_2, i_3, i_{L_m}, and i_s. Determine the power returned to the source by the tertiary (third) transformer winding from the recovery storage energy in L_m.

7.11 A forward converter has parameters $V_s = 125$ V, $V_o = 50$ V, and $R = 25$ Ω, and the switching frequency is 250 kHz. Determine (a) the transformer turns ratio N_1/N_2 such that the duty ratio is 0.3, (b) the inductance L_x such that the minimum current in L_x is 40 percent of the average current, and (c) the capacitance required to limit the output ripple voltage to 0.5 percent.

7.12 Design a forward converter to meet these specifications: $V_s = 170$ V, $V_o = 48$ V, output power-150 W, and the output voltage ripple must be less than 1 percent. Specify the transformer turns ratios, the duty ratio of the switch, the switching frequency, the value of L_x to provide continuous current, and the output capacitance.

7.13 Design a forward converter to produce an output voltage of 30 V when the input dc voltage is unregulated and varies from 150 to 175 V. The output power varies from 20 to 50 W. The duty ratio of the switch is varied to compensate for the fluctuations in the source to regulate the output at 30 V. Specify the switching frequency and range of required duty ratio of the switch, the turns ratios of the transformer, the value of L_x, and the capacitance required to limit the output ripple to less than 0.2 percent. Your design must work for all operating conditions.

7.14 The current waveforms in Fig. 7-6 for the forward converter show the transformer currents based on the transformer model of Fig. 7-1d. Sketch the currents that exist in the three windings of the physical three-winding transformer. Assume that $N_1/N_2 = N_1/N_3 = 1$.

Push-Pull Converter

7.15 The push-pull converter of Fig. 7-8a has the following parameters: $V_s = 50$ V, $N_p/N_s = 2$, $L_x = 60$ μH, $C = 39$ μF, $R = 8$ Ω, $f = 150$ kHz, and $D = 0.35$. Determine (a) the output voltage, (b) the maximum and minimun inductor currents, and (c) the output voltage ripple.

7.16 For the push-pull converter in Prob. 7-12, sketch the current in L_x, D_1, D_2, Sw_1, Sw_2, and the source.

7.17 The push-pull converter of Fig. 7-8a has a transformer with a magnetizing inductance $L_m = 2$ mH which is placed across winding P_1 in the model. Sketch the current in L_m for the circuit parameters given in Prob. 7-11.

7.18 For the push-pull converter of Fig. 7-8a, (a) sketch the voltage waveform v_{L_x}, and (b) derive the expression for output voltage [Eq. (7-44)] on the basis that the average inductor voltage is zero.

Current-Fed Converter

7.19 The current-fed converter of Fig. 7-11a has an input voltage of 24 V and a turns ratio $N_p/N_s = 2$. The load resistance is 10 Ω, and the duty ratio of each switch is 0.65. Determine the output voltage and the input current. Assume that the input inductor is very large. Determine the maximum voltage across each switch.

7.20 The current-fed converter of Fig. 7-11a has an input voltage of 30 V and supplies a load of 40 W at 50 V. Specify a transformer turns ratio and a switch duty ratio. Determine the average current in the inductor.

7.21 The output voltage for the current-fed converter of Fig. 7-11a was derived on the basis of the average inductor voltage being zero. Derive the output voltage [Eq. (7-56)] on the basis that the power supplied by the source must equal the power absorbed by the load for an ideal converter.

PSpice

7.22 Run a PSpice simulation for the flyback converter in Example 7-2. Use the voltage-controlled switch Sbreak with Ron = 0.2 Ω, and use the default diode model Dbreak. Display the output for voltage for steady-state conditions. Compare output voltage and output voltage ripple to the results from Example 7-2. Display the transformer primary and secondary current, and determine the average value of each. Comment on the results.

7.23 Run a PSpice simulation for the forward converter of Example 7-4. Use the voltage-controlled switch Sbreak with Ron = 0.2 Ω and use the default diode model Dbreak. Let the capacitance be 20 μF. Display the steady-state currents in L_x and each of the transformer windings. Comment on the results.

Control

7.24 Design a type 2 compensated error amplifier (Fig. 7-23a) that will give a phase angle at crossover $\theta_{co} = -210°$ and a gain of 20 dB for a crossover frequency of 12 kHz.

7.25 A buck converter has a filter transfer function that has a magnitude of -15 dB and phase angle of $-105°$ at 5 kHz. The gain of the PWM circuit is -9.5 dB. Design a type 2 compensated error amplifier (Fig. 7-23a) that will give a phase margin of at least 45° for a crossover frequency of 5 kHz.

7.26 A buck converter has $L = 50$ μH, $C = 20$ μF, $r_c = 0.5$ Ω, and a load resistance $R = 4$ Ω. The PWM converter has $V_p = 3$ V. A type 2 error amplifier has $R_1 = 1$ kΩ, $R_2 = 5.3$ kΩ, $C_1 = 11.4$ nF, and $C_2 = 1.26$ nF. Use PSpice to determine the phase margin of the control loop (as in Example 7-8) and comment on the stability. Run a PSpice control loop simulation as in Example 7-10.

7.27 A buck converter has $L = 200$ μH with a series resistance $r_L = 0.2$ Ω, $C = 100$ μF with $r_c = 0.5$ Ω, and a load $R = 4$ Ω. The PWM converter has $V_p = 3$ V. (a) Use PSpice to determine the magnitude and phase angle of the filter and load at 10 kHz. (b) Design a type 2 compensated error amplifier (Fig. 7-23a) that will give a phase margin of at least 45° at a crossover frequency of 10 kHz. Verify your results with a PSpice simulation of a step change in load resistance from 4 to 2 Ω as in Example 7-10. Let $V_s = 20$ V and $V_{ref} = 8$ V.

7.28 A buck converter has $L = 200$ μH with a series resistance $r_L = 0.1\ \Omega$, $C = 200$ μF with $r_c = 0.4\ \Omega$, and a load $R = 5\ \Omega$. The PWM converter has $V_p = 3$ V. (a) Use PSpice to determine the magnitude and phase angle of the filter and load at 8 kHz. (b) Design a type 2 compensated error amplifier (Fig. 7-23a) that will give a phase margin of at least 45° at a crossover frequency of 10 kHz. Verify your results with a PSpice simulation of a step change in load resistance as in Fig. 7-27. Let $V_s = 20$ V and $V_{\text{ref}} = 8$ V.

7.29 For the type 3 compensated error amplifier of Fig. 7-28a, determine the K factor for an error amplifier phase angle of 195°. For a gain of 15 dB at a crossover frequency of 15 kHz, determine the resistance and capacitance values for the amplifier.

7.30 The frequency response of a buck converter shows that the output voltage is -8 dB, and the phase angle is $-140°$ at 15 kHz. The ramp function in the PWM control circuit has a peak value of 3 V. Use the K factor method to determine values of the resistors and capacitors for the type 3 error amplifier of Fig. 7-28a for a crossover frequency of 15 kHz.

7.31 The buck converter circuit of Fig. 7-29 has $L = 40$ μH, $r_L = 0.1\ \Omega$, $C_o = 500$ μF, $r_C = 30$ mΩ, and $R_L = 3\ \Omega$. The ramp function in the PWM control circuit has a peak value of 3 V. Use the K factor method to design a type 3 compensated error amplifier for a stable control system with a crossover frequency of 10 kHz. Specify the resistor and capacitor values in the error amplifier.

CHAPTER 8

Inverters
Converting dc to ac

8.1 INTRODUCTION

Inverters are circuits that convert dc to ac. More precisely, inverters transfer power from a dc source to an ac load. The controlled full-wave bridge converters in Chap. 4 can function as inverters in some instances, but an ac source must pre-exist in those cases. In other applications, the objective is to create an ac voltage when only a dc voltage source is available. The focus of this chapter is on inverters that produce an ac output from a dc input. Inverters are used in applications such as adjustable-speed ac motor drives, uninterruptible power supplies (UPS), and running ac appliances from an automobile battery.

8.2 THE FULL-BRIDGE CONVERTER

The full-bridge converter of Fig 8-1a is the basic circuit used to convert dc to ac. The full-bridge converter was introduced as part of a dc power supply circuit in Chap. 7. In this application, an ac output is synthesized from a dc input by closing and opening the switches in an appropriate sequence. The output voltage v_o can be $+V_{dc}$, $-V_{dc}$, or zero, depending on which switches are closed. Figure 8-1b to e shows the equivalent circuits for switch combinations.

Switches Closed	Output Voltage v_o
S_1 and S_2	$+V_{dc}$
S_3 and S_4	$-V_{dc}$
S_1 and S_3	0
S_2 and S_4	0

332 CHAPTER 8 Inverters

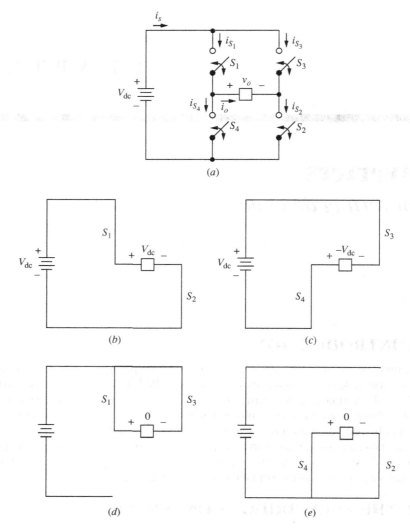

Figure 8-1 (a) Full-bridge converter; (b) S_1 and S_2 closed; (c) S_3 and S_4 closed; (d) S_1 and S_3 closed; (e) S_2 and S_4 closed.

Note that S_1 and S_4 should not be closed at the same time, nor should S_2 and S_3. Otherwise, a short circuit would exist across the dc source. Real switches do not turn on or off instantaneously, as was discussed in Chap. 6. Therefore, switching transition times must be accommodated in the control of the switches. Overlap of switch "on" times will result in a short circuit, sometimes called a *shoot-through* fault, across the dc voltage source. The time allowed for switching is called *blanking* time.

8.3 THE SQUARE-WAVE INVERTER

The simplest switching scheme for the full-bridge converter produces a square wave output voltage. The switches connect the load to $+V_{dc}$ when S_1 and S_2 are closed or to $-V_{dc}$ when S_3 and S_4 are closed. The periodic switching of the load voltage between $+V_{dc}$ and $-V_{dc}$ produces a square wave voltage across the load. Although this alternating output is nonsinusoidal, it may be an adequate ac waveform for some applications.

The current waveform in the load depends on the load components. For the resistive load, the current waveform matches the shape of the output voltage. An inductive load will have a current that has more of a sinusoidal quality than the voltage because of the filtering property of the inductance. An inductive load presents some considerations in designing the switches in the full-bridge circuit because the switch currents must be bidirectional.

For a series RL load and a square wave output voltage, assume switches S_1 and S_2 in Fig. 8-1a close at $t = 0$. The voltage across the load is $+V_{dc}$, and current begins to increase in the load and in S_1 and S_2. The current is expressed as the sum of the forced and natural responses

$$i_o(t) = i_f(t) + i_n(t)$$
$$= \frac{V_{dc}}{R} + Ae^{-t/\tau} \quad \text{for } 0 \le t \le T/2 \quad (8\text{-}1)$$

where A is a constant evaluated from the initial condition and $\tau = L/R$. At $t = T/2$, S_1 and S_2 open, and S_3 and S_4 close. The voltage across the RL load becomes $-V_{dc}$, and the current has the form

$$i_o(t) = \frac{-V_{dc}}{R} + Be^{-(t-T/2)/\tau} \quad \text{for } T/2 \le t \le T \quad (8\text{-}2)$$

where the constant B is evaluated from the initial condition.

When the circuit is first energized and the initial inductor current is zero, a transient occurs before the load current reaches a steady-state condition. At steady state, i_o is periodic and symmetric about zero, as illustrated in Fig. 8-2. Let the initial condition for the current described in Eq. (8-1) be I_{min}, and let the initial condition for the current described in Eq. (8-2) be I_{max}.
Evaluating Eq. (8-1) at $t = 0$,

$$i_o(0) = \frac{V_{dc}}{R} + Ae^0 = I_{min}$$

or

$$A = I_{min} - \frac{V_{dc}}{R} \quad (8\text{-}3)$$

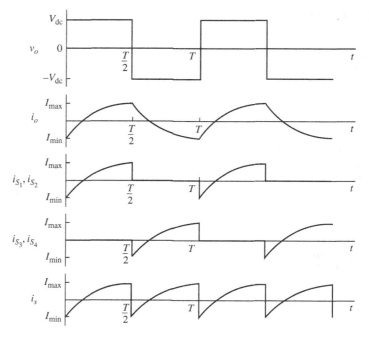

Figure 8-2 Square wave output voltage and steady-state current waveform for an *RL* load.

Likewise, Eq. (8-2) is evaluated at $t = T/2$.

$$i_o(T/2) = \frac{-V_{dc}}{R} + Be^0 = I_{max}$$

or

$$B = I_{max} + \frac{V_{dc}}{R} \tag{8-4}$$

In steady state, the current waveforms described by Eqs. (8-1) and (8-2) then become

$$i_o(t) = \begin{cases} \dfrac{V_{dc}}{R} + \left(I_{min} - \dfrac{V_{dc}}{R}\right)e^{-t/\tau} & \text{for } 0 < t < \dfrac{T}{2} \\ \dfrac{-V_{dc}}{R} + \left(I_{max} + \dfrac{V_{dc}}{R}\right) - e^{(t-T/2)/\tau} & \text{for } \dfrac{T}{2} < t < T \end{cases} \tag{8-5}$$

8.3 The Square-Wave Inverter

An expression is obtained for I_{max} by evaluating the first part of Eq. (8-5) at $t = T/2$

$$i(T/2) = I_{max} = \frac{V_{dc}}{R} + \left(I_{min} - \frac{V_{dc}}{R}\right)e^{-(T/2\tau)} \quad (8\text{-}6)$$

and by symmetry,

$$I_{min} = -I_{max} \quad (8\text{-}7)$$

Substituting $-I_{max}$ for I_{min} in Eq. (8-6) and solving for I_{max},

$$\boxed{I_{max} = -I_{min} = \frac{V_{dc}}{R}\left(\frac{1 - e^{-T/2\tau}}{1 + e^{-T/2\tau}}\right)} \quad (8\text{-}8)$$

Thus, Eqs. (8-5) and (8-8) describe the current in an RL load in the steady state when a square wave voltage is applied. Figure 8-2 shows the resulting currents in the load, source, and switches.

Power absorbed by the load can be determined from $I_{rms}^2 R$, where rms load current is determined from the defining equation from Chap. 2. The integration may be simplified by taking advantage of the symmetry of the waveform. Since the square each of the current half-periods is identical, only the first half-period needs to be evaluated:

$$I_{rms} = \sqrt{\frac{1}{T}\int_0^T i^2(t)\,d(t)} = \sqrt{\frac{2}{T}\int_0^{T/2}\left[\frac{V_{dc}}{R} + \left(I_{min} - \frac{V_{dc}}{R}\right)e^{-t/\tau}\right]^2 dt} \quad (8\text{-}9)$$

If the switches are ideal, the power supplied by the source must be the same as absorbed by the load. Power from a dc source is determined from

$$P_{dc} = V_{dc}I_s \quad (8\text{-}10)$$

as was derived in Chap. 2.

EXAMPLE 8-1

Square-Wave Inverter with RL Load

The full-bridge inverter of Fig. 8-1 has a switching sequence that produces a square wave voltage across a series RL load. The switching frequency is 60 Hz, $V_{dc} = 100$ V, $R = 10\,\Omega$, and $L = 25$ mH. Determine (a) an expression for load current, (b) the power absorbed by the load, and (c) the average current in the dc source.

■ **Solution**

(a) From the parameters given,

$T = 1/f = 1/60 = 0.0167$ s

$\tau = L/R = 0.025/10 = 0.0025$ s

$T/2\tau = 3.33$

Equation (8-8) is used to determine the maximum and minimum current.

$$I_{max} = -I_{min} = \frac{100}{10}\left(\frac{1-e^{-3.33}}{1+e^{-3.33}}\right) = 9.31 \text{ A}$$

Equation (8-5) is then evaluated to give load current.

$$i_o(t) = \frac{100}{10} + \left(-9.31 - \frac{100}{10}\right)e^{-t/0.0025}$$

$$= 10 - 19.31e^{-t/0.0025} \quad 0 \leq t \leq \frac{1}{120}$$

$$i_o(t) = -\frac{100}{10} + \left(9.31 + \frac{100}{10}\right)e^{-(t-0.0167/2)/0.0025}$$

$$= -10 + 19.31e^{-(t-0.00835)0.0025} \quad \frac{1}{120} \leq t \leq \frac{1}{60}$$

(b) Power is computed from $I_{rms}^2 R$, where I_{rms} is computed from Eq. (8-9).

$$I_{rms} = \sqrt{\frac{1}{120}\int_0^{1/120}[(10-19.31)e^{-t/0.0025}]^2 dt} = 6.64 \text{ A}$$

Power absorbed by the load is

$$P = I_{rms}^2 R = (6.64)^2(10) = 441 \text{ W}$$

(c) Average source current can also be computed by equating source and load power, assuming a lossless converter. Using Eq. (8-10),

$$I_s = \frac{P_{dc}}{V_{dc}} = \frac{441}{100} = 4.41 \text{ A}$$

Average power could also be computed from the average of the current expression in part (a).

The switch currents in Fig. 8-2 show that the switches in the full-bridge circuit must be capable of carrying both positive and negative currents for *RL* loads. However, real electronic devices may conduct current in one direction only. This problem is solved by placing feedback diodes in parallel (anitparallel) with each switch. During the time interval when the current in the switch must be negative, the feedback diode carries the current. The diodes are reverse-biased when current is positive in the switch. Figure 8-3a shows the full-bridge inverter with switches implemented as insulated gate bipolar transistors (IGBTs) with feedback diodes. Transistor and diode currents for a square wave voltage and an *RL* load are indicated in Fig 8-3b. Power semiconductor modules often include feedback diodes with the switches.

8.4 Fourier Series Analysis

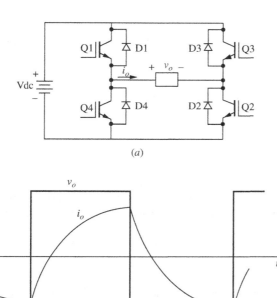

Figure 8-3 (a) Full-bridge inverter using IGBTs; (b) Steady-state current for an *RL* load.

When IGBTs Q_1 and Q_2 are turned off in Fig. 8-3a, the load current must be continuous and will transfer to diodes D_3 and D_4, making the output voltage $-V_{dc}$, effectively turning on the switch paths 3 and 4 before Q_3 and Q_4 are turned on. IGBTs Q_3 and Q_4 must be turned on before the load current decays to zero.

8.4 FOURIER SERIES ANALYSIS

The Fourier series method is often the most practical way to analyze load current and to compute power absorbed in a load, especially when the load is more complex than a simple resistive or *RL* load. A useful approach for inverter analysis is to express the output voltage and load current in terms of a Fourier series. With no dc component in the output,

$$v_o(t) = \sum_{n=1}^{\infty} V_n \sin(n\omega_0 t + \theta_n) \tag{8-11}$$

and

$$i_o(t) = \sum_{n=1}^{\infty} I_n \sin(n\omega_0 t + \phi_n) \tag{8-12}$$

Power absorbed by a load with a series resistance is determined from $I_{\text{rms}}^2 R$, where the rms current can be determined from the rms currents at each of the components in the Fourier series by

$$I_{\text{rms}} = \sqrt{\sum_{n=1}^{\infty} I_{n,\text{rms}}^2} = \sqrt{\sum_{n=1}^{\infty} \left(\frac{I_n}{\sqrt{2}}\right)^2} \qquad (8\text{-}13)$$

where

$$I_n = \frac{V_n}{Z_n} \qquad (8\text{-}14)$$

and Z_n is the load impedance at harmonic n.

Equivalently, the power absorbed in the load resistor can be determined for each frequency in the Fourier series. Total power can be determined from

$$P = \sum_{n=1}^{\infty} P_n = \sum_{n=1}^{\infty} I_{n,\text{rms}}^2 R \qquad (8\text{-}15)$$

where $I_{n,\text{rms}}$ is $I_n/\sqrt{2}$.

In the case of the square wave, the Fourier series contains the odd harmonics and can be represented as

$$v_o(t) = \sum_{n \, \text{odd}} \frac{4 V_{\text{dc}}}{n \pi} \sin n \omega_0 t \qquad (8\text{-}16)$$

EXAMPLE 8-2

Fourier Series Solution for the Square-Wave Inverter

For the inverter in Example 8-1 ($V_{\text{dc}} = 100$ V, $R = 10 \, \Omega$, $L = 25$ mH, $f = 60$ Hz), determine the amplitudes of the Fourier series terms for the square wave load voltage, the amplitudes of the Fourier series terms for load current, and the power absorbed by the load.

■ Solution

The load voltage is represented as the Fourier series in Eq. (8-16). The amplitude of each voltage term is

$$V_n = \frac{4 V_{\text{dc}}}{n \pi} = \frac{4(400)}{n \pi}$$

The amplitude of each current term is determined from Eq. (8-14),

$$I_n = \frac{V_n}{Z_n} = \frac{V_n}{\sqrt{R^2 + (n \omega_0 L)^2}} = \frac{4(400)/n\pi}{\sqrt{10^2 + [n(2\pi 60)(0.025)]^2}}$$

Power at each frequency is determined from Eq. (8-15).

$$P_n = I_{n,\text{rms}}^2 R = \left(\frac{I_n}{\sqrt{2}}\right)^2 R$$

Table 8-1 Fourier Series Quantities for Example 8-2

n	f_n (Hz)	V_n (V)	Z_n (Ω)	I_n (A)	P_n (W)
1	60	127.3	13.7	9.27	429.3
3	180	42.4	30.0	1.42	10.0
5	300	25.5	48.2	0.53	1.40
7	420	18.2	66.7	0.27	0.37
9	540	14.1	85.4	0.17	0.14

Table 8-1 summarizes the Fourier series quantities for the circuit of Example 8-1. As the harmonic number n increases, the amplitude of the Fourier voltage component decreases and the magnitude of the corresponding impedance increases, both resulting in small currents for higher-order harmonics. Therefore, only the first few terms of the series are of practical interest. Note how the current and power terms become vanishingly small for all but the first few frequencies.

Power absorbed by the load is computed from Eq. (8-15).

$$P = \sum P_n = 429.3 + 10.0 + 1.40 + 0.37 + 0.14 + \cdots \approx 441 \text{ W}$$

which agrees with the result in Example 8-1.

8.5 TOTAL HARMONIC DISTORTION

Since the objective of the inverter is to use a dc voltage source to supply a load requiring ac, it is useful to describe the quality of the ac output voltage or current. The quality of a nonsinusoidal wave can be expressed in terms of total harmonic distortion (THD), defined in Chap. 2. Assuming no dc component in the output,

$$\text{THD} = \frac{\sqrt{\sum_{n=2}^{\infty}(V_{n,\text{rms}})^2}}{V_{1,\text{rms}}} = \frac{\sqrt{V_{\text{rms}}^2 - V_{1,\text{rms}}^2}}{V_{1,\text{rms}}} \qquad (8\text{-}17)$$

The THD of current is determined by substituting current for voltage in the above equation. The THD of load current is often of greater interest than that of output voltage. This definition for THD is based on the Fourier series, so there is some benefit in using the Fourier series method for analysis when the THD must be determined. Other measures of distortion such as distortion factor, as presented in Chap. 2, can also be applied to describe the output waveform for inverters.

EXAMPLE 8-3

THD for a Square-Wave Inverter

Determine the total harmonic distortion of the load voltage and the load current for the square-wave inverter in Examples 8-1 and 8-2.

CHAPTER 8 Inverters

■ **Solution**

Use the Fourier series for the square wave in Eq. (8-16) and the definition of THD in Eq. (8-17). The rms value of the square wave voltage is the same as the peak value, and the fundamental frequency component is the first term in Eq. (8-16),

$$V_{\text{rms}} = V_{\text{dc}}$$

$$V_{1,\text{rms}} = \frac{V_1}{\sqrt{2}} = \frac{4V_{\text{dc}}}{\sqrt{2}\pi}$$

Using Eq. (8-17) to compute the total harmonic distortion for voltage,

$$\text{THD}_V = \frac{\sqrt{V_{\text{rms}}^2 - V_{1,\text{rms}}^2}}{V_{1,\text{rms}}} = \frac{\sqrt{V_{\text{dc}}^2 - (4V_{\text{dc}}/\sqrt{2}\pi)^2}}{4V_{\text{dc}}/\sqrt{2}\pi} = 0.483 = 48.3\%$$

The THD of the current is computed using the truncated Fourier series which was determined in Example 8-2.

$$\text{THD}_I = \frac{\sqrt{\sum_{n=2}^{\infty}(I_{n,\text{rms}})^2}}{I_{1,\text{rms}}}$$

$$\approx \frac{\sqrt{(1.42/\sqrt{2})^2 + (0.53/\sqrt{2})^2 + (0.27/\sqrt{2})^2 + (0.17/\sqrt{2})^2}}{9.27/\sqrt{2}} = 0.167 = 16.7\%$$

8.6 PSPICE SIMULATION OF SQUARE-WAVE INVERTERS

Computer simulation of inverter circuits can include various levels of circuit detail. If only the current waveform in the load is desired, it is sufficient to provide a source that will produce the appropriate voltage that would be expected on the inverter output. For example, a full-bridge inverter producing a square wave output might be replaced with a square wave voltage source using the VPULSE source. This simplified simulation will predict the behavior of the current in the load but will give no direct information about the switches. Also, this approach assumes that the switching operation correctly produces the desired output.

EXAMPLE 8-4

PSpice Simulation for Example 8-1

For a series (*RL*) load in a full-bridge inverter circuit with a square wave output, the dc supply is 100 V, $R = 10\ \Omega$, $L = 25$ mH, and the switching frequency is 60 Hz (Example 8-1). (*a*) Assuming ideal switches, use PSpice to determine the maximum and minimum

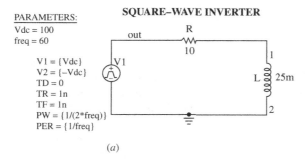

Figure 8-4 (a) Square-wave inverter simulation using an ideal source; (b) Square-wave inverter using switches and diodes.

current in the load in the steady state. (b) Determine the power absorbed by the load. (c) Determine the total harmonic distortion of the load current.

■ Solution 1

Since individual switch currents are not of concern in this problem, a square wave voltage source (VPULSE), as shown in Fig. 8-4a, across the load can simulate the converter output.

Set up a simulation profile for a transient analysis having a run time of 50 ms (three periods), and start saving data after 16.67 ms (one period) so the output represents steady-state current.

Fourier analysis is performed under *Simulation Settings, Output File Options, Perform Fourier Analysis, Center Frequency: 60 Hz, Number of Harmonics: 15, Output Variables: V(OUT) I(R)*.

(a) When in Probe, enter the expression I(R) to obtain a display of the current in the load resistor. The first period contains the start-up transient, but steady-state current like that in Fig. 8-2 is displayed thereafter. The maximum and minimum steady-state current values are approximately 9.31 and −9.31 A, which can be obtained precisely by using the cursor option.

(b) Average power can be obtained from Probe by displaying load current, making sure that the data represent the steady-state condition and entering the expression AVG(W(R)) or AVG(V(OUT)*I(R)). This shows that the resistor absorbs approximately 441 W. The rms current is determined by entering RMS(I(R)), resulting in 6.64 A, as read from the end of the trace. These results agree with the analysis in Example 8-1.

(c) The THD is obtained from the Fourier series for I(R) in the output file as 16.7 percent, agreeing with the Fourier analysis in Examples 8-2 and 8-3. Note that the THD for the square wave in the output file is 45 percent, which is lower than the 48.3 percent computed in Example 8-3. The THD in PSpice is based on the truncated Fourier series through $n = 15$. The magnitudes of higher-order harmonics are not insignificant for the square wave, and omitting them underestimates the THD. The higher-order current harmonics are small, so there is little error in omitting them from the analysis. The number of harmonics in the output file can be increased if desired.

■ Solution 2

The inverter is simulated using the full-bridge circuit of Fig. 8-4b. (This requires the full version of PSpice.) The result of this simulation gives information about the currents and voltages for the switching devices. Voltage-controlled switches (Sbreak) and the default diode (Dbreak) are used. Diodes are included in the switch model to make the switches unidirectional. The model for Sbreak is changed so $R_{on} = 0.01\ \Omega$, and the model for Dbreak is changed so $n = 0.01$, approximating an ideal diode. The output voltage is between nodes out+ and out−. Models for the switches and diodes can be changed to determine the behavior of the circuit using realistic switching devices.

8.7 AMPLITUDE AND HARMONIC CONTROL

The amplitude of the fundamental frequency for a square wave output from of the full-bridge inverter is determined by the dc input voltage [Eq. (8-16)]. A controlled output can be produced by modifying the switching scheme. An output voltage of the form shown in Fig. 8-5a has intervals when the output is zero as well as $+V_{dc}$ and $-V_{dc}$. This output voltage can be controlled by adjusting the interval α on each side of the pulse where the output is zero. The rms value of the voltage waveform in Fig 8-5a is

$$V_{rms} = \sqrt{\frac{1}{\pi}\int_{\alpha}^{\pi-\alpha} V_{dc}^2\, d(\omega t)} = V_{dc}\sqrt{1 - \frac{2\alpha}{\pi}} \qquad (8\text{-}18)$$

8.7 Amplitude and Harmonic Control   **343**

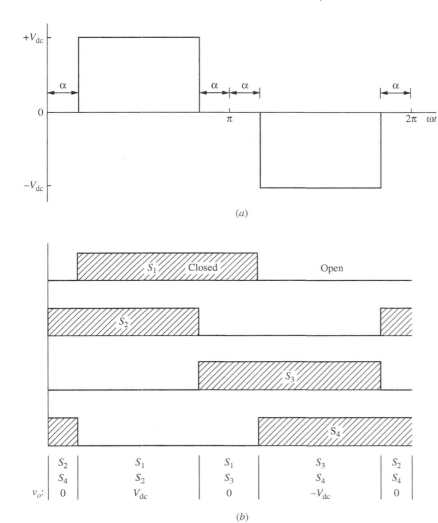

Figure 8-5 (*a*) Inverter output for amplitude and harmonic control; (*b*) Switching sequence for the full-bridge inverter of Fig. 8-1*a*.

The Fourier series of the waveform is expressed as

$$v_o(t) = \sum_{n\ \text{odd}} V_n \sin(n\omega_0 t) \qquad (8\text{-}19)$$

Taking advantage of half-wave symmetry, the amplitudes are

$$V_n = \frac{2}{\pi} \int_{\alpha}^{\pi-\alpha} V_{dc} \sin(n\omega_0 t)\, d(\omega_0 t) = \frac{4V_{dc}}{n\pi} \cos(n\alpha) \qquad (8\text{-}20)$$

344 CHAPTER 8 Inverters

where α is the angle of zero voltage on each end of the pulse. The amplitude of each frequency of the output is a function of α. In particular, the amplitude of the fundamental frequency ($n = 1$) is controllable by adjusting α:

$$V_1 = \left(\frac{4V_{dc}}{\pi}\right) \cos \alpha \qquad (8\text{-}21)$$

Harmonic content can also be controlled by adjusting α. If $\alpha = 30°$, for example, $V_3 = 0$. This is significant because the third harmonic can be eliminated from the output voltage and current. Other harmonics can be eliminated by choosing a value of α which makes the cosine term in Eq. (8-20) to go to zero. Harmonic n is eliminated if

$$\alpha = \frac{90°}{n} \qquad (8\text{-}22)$$

The switching scheme required to produce an output like Fig. 8-5a must provide intervals when the output voltage is zero, as well as $\pm V_{dc}$. The switching sequence of Fig. 8-5b is a way to implement the required output waveform.

Amplitude control and harmonic reduction may not be compatible. For example, establishing α at 30° to eliminate the third harmonic fixes the amplitude of the output fundamental frequency at $V_1 = (4V_{dc}/\pi) \cos 30° = 1.1V_{dc}$ and removes further controllability. To control both amplitude and harmonics using this switching scheme, it is necessary to be able to control the dc input voltage to the inverter. A dc-dc converter (Chap. 6 and 7) placed between the dc source and the inverter can provide a controlled dc input to the inverter.

A graphical representation of the integration in the Fourier series coefficient of Eq. (8-20) gives some insight into harmonic elimination. Recall from Chap. 2 that the Fourier coefficients are determined from the integral of the product of the waveform and a sinusoid. Figure 8-6a shows the output waveform for $\alpha = 30°$ and the sinusoid of $\omega = 3\omega_o$. The product of these two waveforms has an area of zero, showing that the third harmonic is zero. Figure 8-6b shows the waveform for $\alpha = 18°$ and the sinusoid of $\omega = 5\omega_o$, showing that the fifth harmonic is eliminated for this value of α.

Other switching schemes can eliminate multiple harmonics. For example, the output waveform shown in Fig. 8-6c eliminates both the third and fifth harmonics, as indicated by the areas of both being zero.

EXAMPLE 8-5

Harmonic Control of the Full-Bridge Inverter Output

Design an inverter that will supply the series RL load of the previous examples ($R = 10 \, \Omega$ and $L = 25$ mH) with a fundamental-frequency current amplitude of 9.27 A, but with a THD of less than 10 percent. A variable dc source is available.

■ **Solution**

A square-wave inverter produces a THD for current of 16.7 percent (Example 8-3), which does not meet the specification. The dominant harmonic current is for $n = 3$, so a switching

8.7 Amplitude and Harmonic Control 345

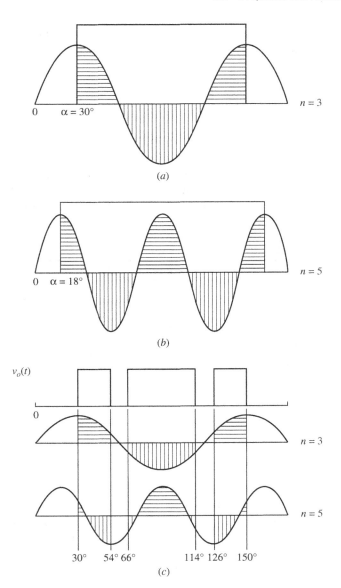

Figure 8-6 Harmonic elimination; (*a*) Third harmonic; (*b*) Fifth harmonic; (*c*) Third and fifth harmonics.

scheme to eliminate the third harmonic will reduce the THD. The required voltage amplitude at the fundamental frequency is

$$V_1 = I_1 Z_1 = I_1 \sqrt{R^2 + (\omega_0 L)^2} = 9.27\sqrt{10^2 + [2\pi 60(0.025)]^2} = 127 \text{ V}$$

Using the switching scheme of Fig. 8-5b, Eq. (8-21) describes the amplitude of the fundamental-frequency voltage,

$$V_1 = \left(\frac{4V_{dc}}{\pi}\right) \cos \alpha$$

Solving for the required dc input with $\alpha = 30°$,

$$V_{dc} = \frac{V_1 \pi}{4 \cos \alpha} = \frac{127\pi}{4 \cos 30°} = 116 \text{ V}$$

Other harmonic voltages are described by Eq. (8-20), and currents for these harmonics are determined from voltage amplitude and load impedance using the same technique as for the square-wave inverter of Example 8-2. The results are summarized in Table 8-2.

Table 8-2 Fourier Series Quantities for Example 8-5

n	f_n (Hz)	V_n (V)	Z_n (Ω)	I_n (A)
1	60	127	13.7	9.27
3	180	0	30.0	0
5	300	25.5	48.2	0.53
7	420	18.2	66.7	0.27
9	540	0	85.4	0
11	660	11.6	104	0.11

The THD of the load current is then

$$\text{THD}_I = \frac{\sqrt{\sum_{n=2}^{\infty} I_{n,\text{rms}}^2}}{I_{1,\text{rms}}} \approx \frac{\sqrt{(0.53/\sqrt{2})^2 + (0.27/\sqrt{2})^2 + (0.11/\sqrt{2})^2}}{9.27/\sqrt{2}} = 0.066 = 6.6\%$$

which more than satisfies the design specifications.

A PSpice circuit for the full-bridge inverter with harmonic and amplitude control is shown in Fig. 8-7a. The user must enter the parameters alpha, output fundamental frequency, dc input voltage to the bridge, and load. The Probe output for voltage and current is shown in Fig. 8-7b. The current is scaled by a factor of 10 to show its relationship to the voltage waveform. The THD of the load current is obtained from the Fourier analysis in the output file as 6.6 percent.

8.8 THE HALF-BRIDGE INVERTER

The half-bridge converter of Fig. 8-8 can be used as an inverter. This circuit was introduced in Chap. 7 as applied to dc power supply circuits. In this circuit, the number of switches is reduced to 2 by dividing the dc source voltage into two parts with the capacitors. Each capacitor will be the same value and will have

8.8 The Half-Bridge Inverter

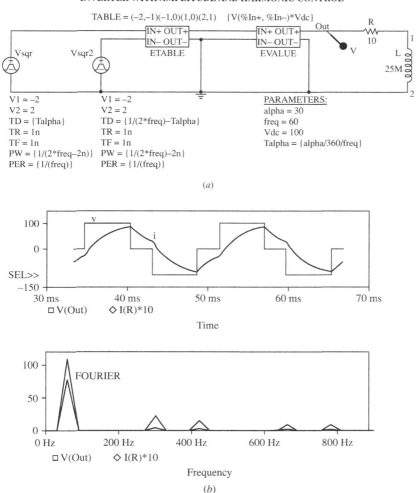

Figure 8-7 (a) A PSpice circuit for Example 8-5 to produce the voltage waveform in Fig. 8-5a; (b) Probe output for showing harmonic elimination.

voltage $V_{dc}/2$ across it. When S_1 is closed, the load voltage is $-V_{dc}/2$. When S_2 is closed, the load voltage is $+V_{dc}/2$. Thus, a square wave output or a bipolar pulse-width-modulated output, as described in Sec. 8.10, can be produced.

The voltage across an open switch is twice the load voltage, or V_{dc}. As with the full-bridge inverter, blanking time for the switches is required to prevent a short circuit across the source, and feedback diodes are required to provide continuity of current for inductive loads.

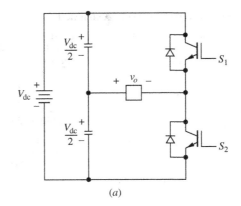

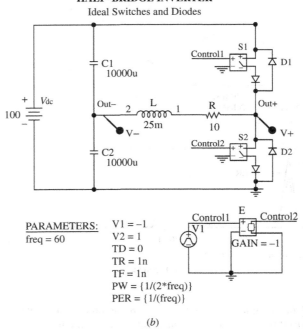

Figure 8-8 (a) A half-bridge inverter using IGBTs. The output is $\pm V_{dc}$; (b) A PSpice implementation using voltage-controlled switches and diodes.

8.9 MULTILEVEL INVERTERS

The H bridge inverter previously illustrated in Figs. 8-1 and 8-3 produces output voltages of V_{dc}, 0, and $-V_{dc}$. The basic H bridge switching concept can be expanded to other circuits that can produce additional output voltage levels.

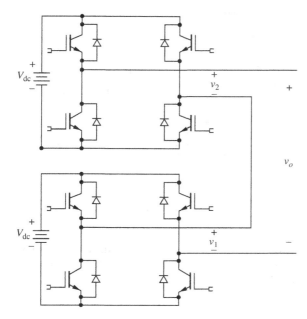

Figure 8-9 An inverter with two dc sources, each with an H bridge implemented with IGBTs.

These multilevel-output voltages are more sinelike in quality and thus reduce harmonic content. The multilevel inverter is suitable for applications including adjustable-speed motor drives and interfacing renewable energy sources such as photovoltaics to the electric power grid.

Multilevel Converters with Independent DC Sources

One multilevel inverter method uses independent dc sources, each with an H bridge. A circuit with two dc voltage sources is shown in Fig. 8-9. The output of each of the H bridges is $+V_{dc}$, $-V_{dc}$, or 0, as was illustrated in Fig. 8-1. The total instantaneous voltage v_o on the output of the multilevel converter is any combination of individual bridge voltages. Thus, for a two-source inverter, v_o can be any of the five levels $+2V_{dc}$, V_{dc}, 0, $-V_{dc}$, or $-2V_{dc}$.

Each H bridge operates with a switching scheme like that of Fig. 8.5 in Sec. 8.7, which was used for amplitude or harmonic control. Each bridge operates at a different delay angle α, resulting in bridge and total output voltages like those shown in Fig. 8-10.

The Fourier series for the total output voltage v_o for the two-source circuit contains only the odd-numbered harmonics and is

$$v_o(t) = \frac{4V_{dc}}{\pi} \sum_{n=1,3,5,7,\ldots}^{\infty} [\cos(n\alpha_1) + \cos(n\alpha_2)] \frac{\sin(n\omega_0 t)}{n} \qquad (8\text{-}23)$$

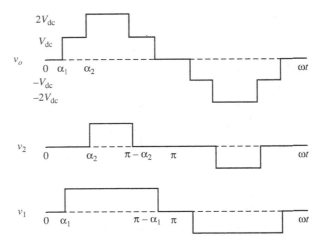

Figure 8-10 Voltage output of each of the H bridges and the total voltage for the two-source multilevel inverter of Fig. 8-9.

The Fourier coefficients for this series are

$$V_n = \frac{4V_{dc}}{n\pi}[\cos(n\alpha_1) + \cos(n\alpha_2)] \quad (8\text{-}24)$$

The modulation index M_i is the ratio of the amplitude of the fundamental frequency component of v_o to the amplitude of the fundamental frequency component of a square wave of amplitude $2V_{dc}$, which is $2(4V_{dc}/\pi)$.

$$M_i = \frac{V_1}{2(4V_{dc}/\pi)} = \frac{\cos\alpha_1 + \cos\alpha_2}{2} \quad (8\text{-}25)$$

Some harmonics can be eliminated from the output voltage waveform with the proper selection of α_1 and α_2 in Eq. (8-24). For the two-source converter, harmonic m can be eliminated by using delay angles such that

$$\cos(m\alpha_1) + \cos(m\alpha_2) = 0 \quad (8\text{-}26)$$

To eliminate the mth harmonic and also meet a specified modulation index for the two-source inverter requires the simultaneous solution to Eq. (8-26) and the additional equation derived from Eq. (8-25),

$$\cos(\alpha_1) + \cos(\alpha_2) = 2M_i \quad (8\text{-}27)$$

To solve Eqs. (8-26) and (8-27) simultaneously requires an iterative numerical method such as the Newton-Raphson method.

EXAMPLE 8-6

A Two-Source Multilevel Inverter

For the two-source multilevel inverter of Fig. 8-9 with $V_{dc} = 100$ V: (a) Determine the Fourier coefficients through $n = 9$ and the modulation index for $\alpha_1 = 20°$ and $\alpha_2 = 40°$. (b) Determine α_1 and α_2 such that the third harmonic ($n = 3$) is eliminated and $M_i = 0.8$.

■ **Solution**
Using Eq. 8-24 to evaluate the Fourier coefficients,

$$V_n = \frac{4V_{dc}}{n\pi}[\cos(n\alpha_1) + \cos(n\alpha_2)] = \frac{4(100)}{n\pi}[\cos(n20°) + \cos(n40°)]$$

resulting in $V_1 = 217$, $V_3 = 0$, $V_5 = -28.4$, $V_7 = -10.8$, and $V_9 = 0$. Note that the third and ninth harmonics are eliminated. The even harmonics are not present.

The modulation index M_i is evaluated from Eq. (8-25).

$$M_i = \frac{\cos\alpha_1 + \cos\alpha_2}{2} = \frac{\cos 20° + \cos 40°}{2} = 0.853$$

The amplitude of the fundamental frequency voltage is therefore 85.3 percent of that of a square wave of ±200 V.

(b) To achieve simultaneous elimination of the third harmonic and a modulation index of $M_i = 0.8$ requires the solution to the equations

$$\cos(3\alpha_1) + \cos(3\alpha_2) = 0$$

and

$$\cos(\alpha_1) + \cos(\alpha_2) = 2M_i = 1.6$$

Using an iterative method, $\alpha_1 = 7.6°$ and $\alpha_2 = 52.4°$.

The preceding concept can be extended to a multilevel converter having several dc sources. For k separate sources connected in cascade, there are $2k+1$ possible voltage levels. As more dc sources and H bridges are added, the total output voltage has more steps, producing a staircase waveform that more closely approaches a sinusoid. For a five-source system as shown in Fig. 8-11, there are 11 possible output voltage levels, as illustrated in Fig. 8-12.

The Fourier series for a staircase waveform such as that in Fig 8-12 for k separate dc sources each equal to V_{dc} is

$$v_o(t) = \frac{4V_{dc}}{\pi} \sum_{n=1,3,5,7,\ldots}^{\infty} [\cos(n\alpha_1) + \cos(n\alpha_2) + \cdots + \cos(n\alpha_k)]\frac{\sin(n\omega_0 t)}{n} \quad (8\text{-}28)$$

The magnitudes of the Fourier coefficients are thus

$$V_n = \frac{4V_{dc}}{n\pi}[\cos(n\alpha_1) + \cos(n\alpha_2) + \cdots + \cos(n\alpha_k)]$$
$$\text{for } n = 1, 3, 5, 7, \ldots \quad (8\text{-}29)$$

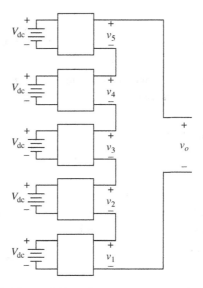

Figure 8-11 A five-source cascade multilevel converter.

The modulation index M_i for k dc sources each equal to V_{dc} is

$$M_i = \frac{V_1}{4kV_{dc}/\pi} = \frac{\cos(\alpha_1) + \cos(\alpha_2) + \cdots + \cos(\alpha_k)}{k} \qquad (8\text{-}30)$$

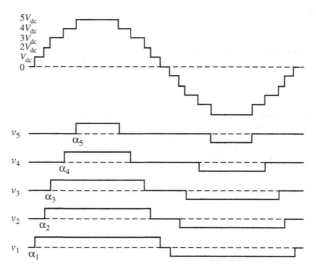

Figure 8-12 Voltages at each H bridge in Fig. 8-11 and the total output voltage.

Specific harmonics can be eliminated from the output voltage. To eliminate the mth harmonic, the delay angles must satisfy the equation

$$\cos(m\alpha_1) + \cos(m\alpha_2) + \cdots + \cos(m\alpha_k) = 0 \quad (8\text{-}31)$$

For k dc sources, $k-1$ harmonics can be eliminated while establishing a particular M_i.

EXAMPLE 8-7

A Five-Source Multilevel Inverter

Determine the delay angles required for a five-source cascade multilevel converter that will eliminate harmonics 5, 7, 11, and 13 and will have a modulation index $M_i = 0.8$.

■ **Solution**

The delay angles must satisfy these simultaneous equations:

$$\cos(5\alpha_1) + \cos(5\alpha_2) + \cos(5\alpha_3) + \cos(5\alpha_4) + \cos(5\alpha_5) = 0$$
$$\cos(7\alpha_1) + \cos(7\alpha_2) + \cos(7\alpha_3) + \cos(7\alpha_4) + \cos(7\alpha_5) = 0$$
$$\cos(11\alpha_1) + \cos(11\alpha_2) + \cos(11\alpha_3) + \cos(11\alpha_4) + \cos(11\alpha_5) = 0$$
$$\cos(13\alpha_1) + \cos(13\alpha_2) + \cos(13\alpha_3) + \cos(13\alpha_4) + \cos(13\alpha_5) = 0$$
$$\cos(\alpha_1) + \cos(\alpha_2) + \cos(\alpha_3) + \cos(\alpha_4) + \cos(\alpha_5) = 5M_i = 5(0.8) = 4$$

An iteration method such as the Newton-Raphson method must be used to solve these equations. The result is $\alpha_1 = 6.57°$, $\alpha_2 = 18.94°$, $\alpha_3 = 27.18°$, $\alpha_4 = 45.14°$, and $\alpha_5 = 62.24°$. See the references in the Bibliography for information on the technique.

Equalizing Average Source Power with Pattern Swapping

In the two-source inverter of Fig. 8-9 using the switching scheme of Fig. 8-10, the source and H bridge producing the voltage v_1 supplies more average power (and energy) than the source and H bridge producing v_2 due to longer pulse widths in both the positive and negative half cycles. If the dc sources are batteries, one battery will discharge faster than the other. A technique known as pattern swapping or duty swapping equalizes the average power supplied by each dc source.

The principle of pattern swapping is to have each dc source conduct for an equal amount of time on average. An alternate switching scheme for the two-source circuit is shown in Fig. 8-13. In this scheme, the first source conducts for a longer time in the first half-cycle while the second source conducts for more time in the second half-cycle. Thus, over one complete period, the sources conduct equally, and average power from each source is the same.

For the five-source converter in Fig. 8-11, a switching scheme to equalize average power is shown in Fig. 8-14. Note that five half cycles are required to equalize power.

A variation of the H bridge multilevel inverter is to have the dc sources be of different values. The output voltage would be a staircase waveform, but not in equal

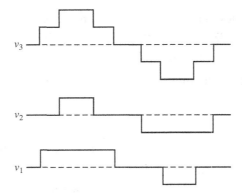

Figure 8-13 Pattern swapping to equalize average power in each source for the two-source inverter of Fig. 8-9.

voltage increments. The Fourier series of the output voltage would have different-valued harmonic amplitudes which may be an advantage in some applications.

Because independent voltage sources are needed, the multiple-source implementation of multilevel converters is best suited in applications where batteries, fuel cells, or photovoltaics are the sources.

Diode-Clamped Multilevel Inverters

A multilevel converter circuit that has the advantage of using a single dc source rather than multiple sources is the diode-clamped multilevel converter shown in Fig. 8-15a. In this circuit, the dc voltage source is connected to a pair of series capacitors, each charged to $V_{dc}/2$. The following analysis shows how the output voltage can have the levels of V_{dc}, $V_{dc}/2$, 0, $-V_{dc}/2$, or $-V_{dc}$.

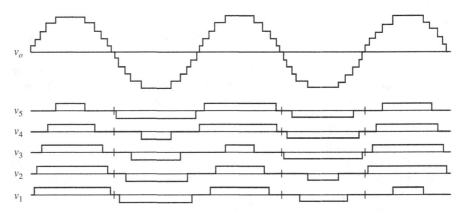

Figure 8-14 Pattern swapping to equalize average source power for the five-source multilevel inverter of Fig. 8-11.

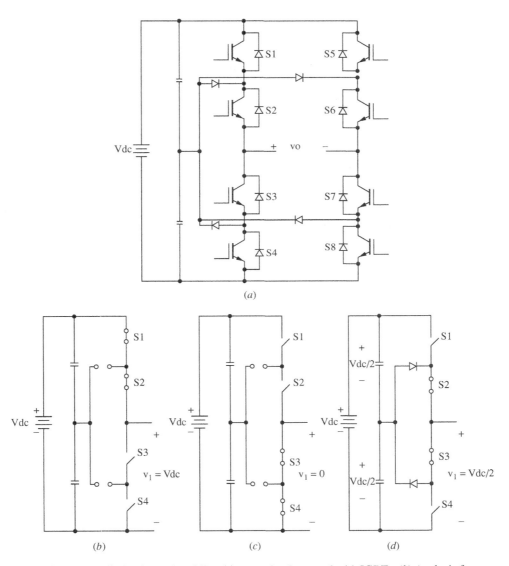

Figure 8-15 (a) A diode-clamped multilevel inverter implemented with IGBTs. (b) Analysis for one-half of the circuit for $v_1 = V_{dc}$, (c) for $v_1 = 0$, and (d) for $v_1 = \frac{1}{2}V_{dc}$.

For the analysis, consider only the left half of the bridge, as shown in Fig. 8-15b, c, and d. With S_1 and S_2 closed and S_3 and S_4 open, $V_1 = V_{dc}$ (Fig. 8-15b). The diodes are off for this condition. With S_1 and S_2 open and S_3 and S_4 closed, $V_1 = 0$ (Fig. 8-15c). The diodes are off for this condition also. To produce a voltage of $V_{dc}/2$, S_2 and S_3 are closed, and S_1 and S_4 are open (Fig. 8-15d). The voltage v_1 is that of the lower capacitor, at voltage $V_{dc}/2$, connected through the antiparallel diode path that can carry load current in either direction. Note that for each of these circuits, two switches are open, and the voltage of the source divides between the two, thus reducing the voltage stress across each switch compared to the H bridge circuit of Fig. 8-1.

Using a similar analysis, the right half of the bridge can also produce the voltages V_{dc}, 0, and $V_{dc}/2$. The output voltage is the difference of the voltages between each half bridge, resulting in the five levels

$$v_o \in \left\{ V_{dc}, \frac{1}{2}V_{dc}, 0, -\frac{1}{2}V_{dc}, -V_{dc} \right\} \tag{8-32}$$

with multiple ways to achieve some of them. The switch control can establish delay angles α_1 and α_2 to produce an output voltage like that in Fig. 8-10 for the cascaded H bridge, except that the maximum value is V_{dc} instead of $2V_{dc}$.

More output voltage levels are achieved with additional capacitors and switches. Figure 8-16 shows the dc source divided across three series capacitors.

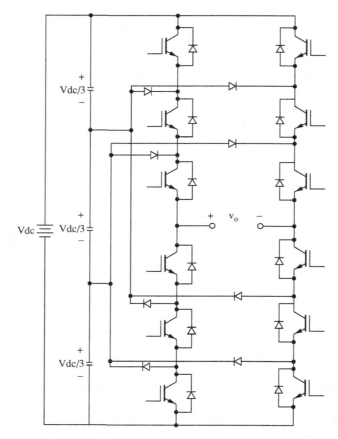

Figure 8-16 A diode-clamped multilevel inverter that produces four voltage levels on each side of the bridge and seven output voltage levels.

The voltage across each capacitor is $\frac{1}{3}$ V, producing the four voltage levels on each side of the bridge of $V_{dc}, \frac{2}{3}V_{dc}, \frac{1}{3}V_{dc}$, and 0. The output voltage can then have the seven levels

$$v_o \in \left\{ V_{dc}, \frac{2}{3}V_{dc}, \frac{1}{3}V_{dc}, 0, -\frac{1}{3}V_{dc}, -\frac{2}{3}V_{dc}, -V_{dc} \right\} \qquad (8\text{-}33)$$

8.10 PULSE-WIDTH-MODULATED OUTPUT

Pulse-width modulation (PWM) provides a way to decrease the total harmonic distortion of load current. A PWM inverter output, with some filtering, can generally meet THD requirements more easily than the square wave switching scheme. The unfiltered PWM output will have a relatively high THD, but the harmonics will be at much higher frequencies than for a square wave, making filtering easier.

In PWM, the amplitude of the output voltage can be controlled with the modulating waveforms. *Reduced filter requirements to decrease harmonics and the control of the output voltage amplitude are two distinct advantages of PWM.* Disadvantages include more complex control circuits for the switches and increased losses due to more frequent switching.

Control of the switches for sinusoidal PWM output requires (1) a reference signal, sometimes called a modulating or control signal, which is a sinusoid in this case and (2) a carrier signal, which is a triangular wave that controls the switching frequency. Bipolar and unipolar switching schemes are discussed next.

Bipolar Switching

Figure 8-17 illustrates the principle of sinusoidal bipolar pulse-width modulation. Figure 8-17a shows a sinusoidal reference signal and a triangular carrier signal. When the instantaneous value of the sine reference is larger than the triangular carrier, the output is at $+V_{dc}$, and when the reference is less than the carrier, the output is at $-V_{dc}$:

$$\begin{aligned} v_o &= +V_{dc} \quad \text{for } v_{sine} > v_{tri} \\ v_o &= -V_{dc} \quad \text{for } v_{sine} < v_{tri} \end{aligned} \qquad (8\text{-}34)$$

This version of PWM is *bipolar* because the output alternates between plus and minus the dc supply voltage.

The switching scheme that will implement bipolar switching using the full-bridge inverter of Fig. 8-1 is determined by comparing the instantaneous reference and carrier signals:

$$S_1 \text{ and } S_2 \text{ are on when } v_{sine} > v_{tri} \quad (v_o = +V_{dc})$$
$$S_3 \text{ and } S_4 \text{ are on when } v_{sine} < v_{tri}. \quad (v_o = -V_{dc})$$

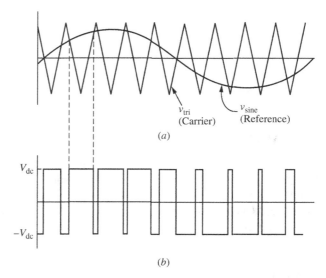

Figure 8-17 Bipolar pulse-width modulation. (*a*) Sinusoidal reference and triangular carrier; (*b*) Output is $+V_{dc}$ when $v_{sine} > v_{tri}$ and is $-V_{dc}$ when $v_{sine} < v_{tri}$.

Unipolar Switching

In a unipolar switching scheme for pulse-width modulation, the output is switched either from high to zero or from low to zero, rather than between high and low as in bipolar switching. One unipolar switching scheme has switch controls in Fig. 8-1 as follows:

S_1 is on when $v_{sine} > v_{tri}$
S_2 is on when $-v_{sine} < v_{tri}$
S_3 is on when $-v_{sine} > v_{tri}$
S_4 is on when $v_{sine} < v_{tri}$

Note that switch pairs (S_1, S_4) and (S_2, S_3) are complementary—when one switch in a pair is closed, the other is open. The voltages v_a and v_b in Fig. 8-18*a* alternate between $+V_{dc}$ and zero. The output voltage $v_o = v_{ab} = v_a - v_b$ is as shown in Fig. 8-18*d*.

Another unipolar switching scheme has only one pair of switches operating at the carrier frequency while the other pair operates at the reference frequency, thus having two high-frequency switches and two low-frequency switches. In this switching scheme,

S_1 is on when $v_{sine} > v_{tri}$ (high frequency)
S_4 is on when $v_{sine} < v_{tri}$ (high frequency)

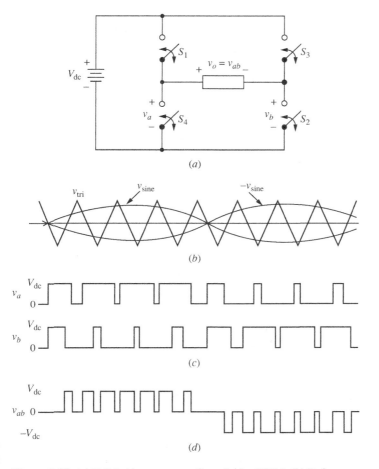

Figure 8-18 (*a*) Full-bridge converter for unipolar PWM; (*b*) Reference and carrier signals; (*c*) Bridge voltages v_a and v_b; (*d*) Output voltage.

S_2 is on when $v_{sine} > 0$ (low frequency)
S_3 is on when $v_{sine} < 0$ (low frequency)

where the sine and triangular waves are as shown in Fig. 8-19a. Alternatively, S_2 and S_3 could be the high-frequency switches, and S_1 and S_4 could be the low-frequency switches.

8.11 PWM DEFINITIONS AND CONSIDERATIONS

At this point, some definitions and considerations relevant when using PWM should be stated.

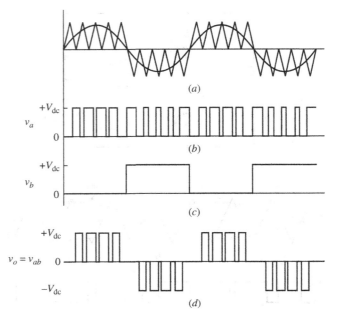

Figure 8-19 Unipolar PWM with high- and low-frequency switches. (*a*) Reference and control signals; (*b*) v_a (Fig. 8-18*a*); (*c*) v_b; (*d*) output $v_a - v_b$.

1. *Frequency modulation ratio m_f.* The Fourier series of the PWM output voltage has a fundamental frequency which is the same as the reference signal. Harmonic frequencies exist at and around multiples of the switching frequency. The magnitudes of some harmonics are quite large, sometimes larger than the fundamental. However, because these harmonics are located at high frequencies, a simple low-pass filter can be quite effective in removing them. Details of the harmonics in PWM are given in the next section. The *frequency modulation ratio m_f* is defined as the ratio of the frequencies of the carrier and reference signals,

$$m_f = \frac{f_{\text{carrier}}}{f_{\text{reference}}} = \frac{f_{\text{tri}}}{f_{\text{sine}}} \qquad (8\text{-}35)$$

Increasing the carrier frequency (increasing m_f) increases the frequencies at which the harmonics occur. A disadvantage of high switching frequencies is higher losses in the switches used to implement the inverter.

2. *Amplitude modulation ratio m_a.* The *amplitude modulation ratio m_a* is defined as the ratio of the amplitudes of the reference and carrier signals:

$$m_a = \frac{V_{m,\text{reference}}}{V_{m,\text{carrier}}} = \frac{V_{m,\text{sine}}}{V_{m,\text{tri}}} \qquad (8\text{-}36)$$

If $m_a \leq 1$, the amplitude of the fundamental frequency of the output voltage V_1 is linearly proportional to m_a. That is,

$$V_1 = m_a V_{dc} \qquad (8\text{-}37)$$

The amplitude of the fundamental frequency of the PWM output is thus controlled by m_a. This is significant in the case of an unregulated dc supply voltage because the value of m_a can be adjusted to compensate for variations in the dc supply voltage, producing a constant-amplitude output. Alternatively, m_a can be varied to change the amplitude of the output. If m_a is greater than 1, the amplitude of the output increases with m_a, but not linearly.

3. *Switches.* The switches in the full-bridge circuit must be capable of carrying current in either direction for pulse-width modulation just as they did for square wave operation. Feedback diodes across the switching devices are necessary, as was done in the inverter in Fig. 8-3a. Another consequence of real switches is that they do not turn on or off instantly. Therefore, it is necessary to allow for switching times in the control of the switches just as it was for the square-wave inverter.

4. *Reference voltage.* The sinusoidal reference voltage must be generated within the control circuit of the inverter or taken from an outside reference. It may seem as though the function of the inverter bridge is unnecessary because a sinusoidal voltage must be present before the bridge can operate to produce a sinusoidal output. However, there is very little power required from the reference signal. The power supplied to the load is provided by the dc power source, and this is the intended purpose of the inverter. The reference signal is not restricted to a sinusoid, and other waveshapes can function as the reference signal.

8.12 PWM HARMONICS

Bipolar Switching

The Fourier series of the bipolar PWM output illustrated in Fig. 8-17 is determined by examining each pulse. The triangular waveform is synchronized to the reference as shown in Fig 8-17a, and m_f is chosen to be an odd integer. The PWM output then exhibits odd symmetry, and the Fourier series can then be expressed

$$v_o(t) = \sum_{n=1}^{\infty} V_n \sin(n\omega_0 t) \qquad (8\text{-}38)$$

For the kth pulse of the PWM output in Fig. 8-20, the Fourier coefficient is

$$V_{nk} = \frac{2}{\pi} \int_0^T v(t) \sin(n\omega_0 t)\, d(\omega_0 t)$$

$$= \frac{2}{\pi} \left[\int_{\alpha_k}^{\alpha_k + \delta_k} V_{dc} \sin(n\omega_0 t)\, d(\omega_0 t) + \int_{\alpha_k + \delta_k}^{\alpha_{k+1}} -(V_{dc}) \sin(n\omega_0 t)\, (d(\omega_0 t)) \right]$$

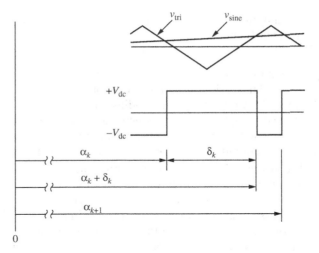

Figure 8-20 Single PWM pulse for determining Fourier series for bipolar PWM.

Performing the integration,

$$V_{nk} = \frac{2V_{dc}}{n\pi} [\cos n\alpha_k + \cos n\alpha_{k+1} - 2 \cos n(\alpha_k + \delta_k)] \qquad (8\text{-}39)$$

Each Fourier coefficient V_n for the PWM waveform is the sum of V_{nk} for the p pulses over one period,

$$V_n = \sum_{k=1}^{p} V_{nk} \qquad (8\text{-}40)$$

The normalized frequency spectrum for bipolar switching for $m_a = 1$ is shown in Fig. 8-21. The harmonic amplitudes are a function of m_a because the

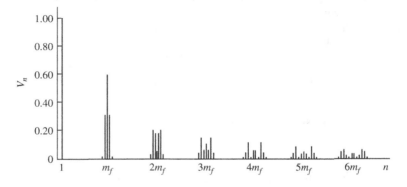

Figure 8-21 Frequency spectrum for bipolar PWM with $m_a = 1$.

8.12 PWM Harmonics

Table 8-3 Normalized Fourier Coefficients V_n/V_{dc} for Bipolar PWM

	$m_a=1$	0.9	0.8	0.7	0.6	0.5	0.4	0.3	0.2	0.1
$n=1$	1.00	0.90	0.80	0.70	0.60	0.50	0.40	0.30	0.20	0.10
$n=m_f$	0.60	0.71	0.82	0.92	1.01	1.08	1.15	1.20	1.24	1.27
$n=mf\pm2$	0.32	0.27	0.22	0.17	0.13	0.09	0.06	0.03	0.02	0.00

width of each pulse depends on the relative amplitudes of the sine and triangular waves. The first harmonic frequencies in the output spectrum are at and around m_f. Table 8-3 indicates the first harmonics in the output for bipolar PWM. The Fourier coefficients are not a function of m_f if m_f is large (≥ 9).

EXAMPLE 8-8

A PWM Inverter

The full-bridge inverter is used to produce a 60-Hz voltage across a series RL load using bipolar PWM. The dc input to the bridge is 100 V, the amplitude modulation ratio m_a is 0.8, and the frequency modulation ratio m_f is 21 [$f_{tri} = (21)(60) = 1260$ Hz]. The load has a resistance of $R = 10 \ \Omega$ and series inductance $L = 20$ mH. Determine (a) the amplitude of the 60-Hz component of the output voltage and load current, (b) the power absorbed by the load resistor, and (c) the THD of the load current.

■ **Solution**

(a) Using Eq. (8-38) and Table 8-3, the amplitude of the 60-Hz fundamental frequency is

$$V_1 = m_a V_{dc} = (0.8)(100) = 80 \text{ V}$$

The current amplitudes are determined using phasor analysis:

$$I_n = \frac{V_n}{Z_n} = \frac{V_n}{\sqrt{R^2 + (n\omega_0 L)^2}} \quad (8\text{-}41)$$

For the fundamental frequency,

$$I_1 = \frac{80}{\sqrt{10^2 + [(1)(2\pi60)(0.02)]^2}} = 6.39 \text{ A}$$

(b) With $m_f = 21$, the first harmonics are at $n = 21, 19,$ and 23. Using Table 8-3,

$$V_{21} = (0.82)(100) = 82 \text{ V}$$
$$V_{19} = V_{23} = (0.22)(100) = 22 \text{ V}$$

Current at each of the harmonics is determined from Eq. (8-41).
Power at each frequency is determined from

$$P_n = (I_{n,\text{rms}})^2 R = \left(\frac{I_n}{\sqrt{2}}\right)^2 R$$

Table 8-4 Fourier Series Quantities for the PWM Inverter of Example 8-8

n	$f_{n\,(Hz)}$	V_n (V)	Z_n (Ω)	I_n (A)	$I_{n,\mathrm{rms}}$ (A)	P_n (W)
1	60	80.0	12.5	6.39	4.52	204.0
19	1140	22.0	143.6	0.15	0.11	0.1
21	1260	81.8	158.7	0.52	0.36	1.3
23	1380	22.0	173.7	0.13	0.09	0.1

The resulting voltage amplitudes, currents, and powers at these frequencies are summarized in Table 8-4.

Power absorbed by the load resistor is

$$P = \sum P_n \approx 204.0 + 0.1 + 1.3 + 0.1 = 205.5 \text{ W}$$

Higher-order harmonics contribute little power and can be neglected.

(c) The THD of the load current is determined using Eq. (8-17) with the rms current of the harmonics approximated by the first few terms indicated in Table 8-4.

$$\text{THD}_I = \frac{\sqrt{\sum_{n=2}^{\infty} I_{n,\mathrm{rms}}^2}}{I_{1,\mathrm{rms}}} \approx \frac{\sqrt{(0.11)^2 + (0.36)^2 + (0.09)^2}}{4.52} = 0.087 = 8.7\%$$

By using the truncated Fourier series in Table 8-4, the THD will be underestimated. However, since the impedance of the load increases and the amplitudes of the harmonics generally decrease as n increases, the above approximation should be acceptable. (Including currents through $n = 100$ gives a THD of 9.1 percent.)

EXAMPLE 8-9

PWM Inverter Design

Design a bipolar PWM inverter that will produce a 75-V rms 60-Hz output from a 150-V dc source. The load is a series RL combination with $R = 12\ \Omega$ and $L = 60$ mH. Select the switching frequency such that the current THD is less than 10 percent.

■ Solution

The required amplitude modulation ratio is determined from Eq. (8-38),

$$m_a = \frac{V_1}{V_{dc}} = \frac{75\sqrt{2}}{150} = 0.707$$

The current amplitude at 60 Hz is

$$I_1 = \frac{V_1}{Z_1} = \frac{75\sqrt{2}}{\sqrt{12^2 + [(2\pi 60)(0.06)]^2}} = 4.14 \text{ A}$$

The rms value of the harmonic current has a limit imposed by the required THD,

$$\sqrt{\sum_{n=2}^{\infty}(I_{n,\text{rms}})^2} < 0.1\, I_{1,\text{rms}} = 0.1\left(\frac{4.14}{\sqrt{2}}\right) = 0.293 \text{ A}$$

The term that will produce the dominant harmonic current is at the switching frequency. As an approximation, assume that the harmonic content of the load current is the same as the dominant harmonic at the carrier frequency:

$$\sqrt{\sum_{n=2}^{\infty}(I_{n,\text{rms}})^2} \approx I_{mf,\text{rms}} = \frac{I_{mf}}{\sqrt{2}}$$

The amplitude of the current harmonic at the carrier frequency is then approximated as

$$I_{mf} < (0.1)(4.14) = 0.414 \text{ A}$$

Table 8-3 indicates that the normalized voltage harmonic for $n = m_f$ and for $m_a = 0.7$ is 0.92. The voltage amplitude for $n = m_f$ is then

$$V_{mf} = 0.92\, V_{dc} = (0.92)(150) = 138 \text{ V}$$

The minimum load impedance at the carrier frequency is then

$$Z_{mf} = \frac{V_{mf}}{I_{mf}} = \frac{138}{0.414} = 333 \text{ }\Omega$$

Because the impedance at the carrier frequency must be much larger than the 12-Ω load resistance, assume the impedance at the carrier frequency is entirely inductive reactance,

$$Z_{mf} \approx \omega L = m_f \omega_0 L$$

For the load impedance to be greater than 333 Ω,

$$m_f \omega_0 L > 333$$

$$m_f > \frac{333}{(377)(0.06)} = 14.7$$

Selecting m_f to be at least 15 would marginally meet the design specifications. However, the estimate of the harmonic content used in the calculations will be low, so a higher carrier frequency is a more prudent selection. Let $m_f = 17$, which is the next odd integer. The carrier frequency is then

$$f_{\text{tri}} = m_f f_{\text{ref}} = (17)(60) = 1020 \text{ Hz}$$

Further increasing m_f would reduce the current THD, but at the expense of larger switching losses. A PSpice simulation, as discussed later in this chapter, can be used to verify that the design meets the specifications.

Unipolar Switching

With the unipolar switching scheme in Fig. 8-18, some harmonics that were in the spectrum for the bipolar scheme are absent. The harmonics in the output begin at around $2m_f$, and m_f is chosen to be an even integer. Figure 8-22 shows the frequency spectrum for unipolar switching with $m_a = 1$.

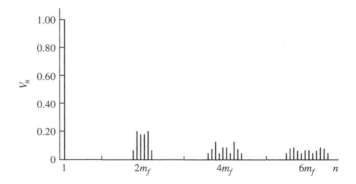

Figure 8-22 Frequency spectrum for unipolar PWM with $m_a = 1$.

Table 8-5 indicates the first harmonics in the output for unipolar PWM.

The unipolar PWM scheme using high- and low-frequency switches shown in Fig. 8-19 will have similar results as indicated above, but the harmonics will begin at around m_f rather than $2m_f$.

Table 8-5 Normalized Fourier Coefficients V_n/V_{dc} for Unipolar PWM in Fig. 8-18

	$m_a = 1$	0.9	0.8	0.7	0.6	0.5	0.4	0.3	0.2	0.1
$n=1$	1.00	0.90	0.80	0.70	0.60	0.50	0.40	0.30	0.20	0.10
$n=2m_f \pm 1$	0.18	0.25	0.31	0.35	0.37	0.36	0.33	0.27	0.19	0.10
$n=2m_f \pm 3$	0.21	0.18	0.14	0.10	0.07	0.04	0.02	0.01	0.00	0.00

8.13 CLASS D AUDIO AMPLIFIERS

The reference signal for the PWM control circuit can be an audio signal, and the full-bridge circuit could be used as a PWM audio amplifier. A PWM audio amplifier is referred to as a *class D amplifier*. The triangular wave carrier signal for this application is typically 250 kHz to provide adequate sampling, and the PWM waveform is low-pass filtered to recover the audio signal and deliver power to a speaker. The spectrum of the PWM output signal is dynamic in this case.

Class D amplifiers are much more efficient than other types of audio power amplifiers. The class AB amplifier, the traditional circuit for audio applications, has a maximum theoretical efficiency of 78.5 percent for a sine wave of maximum undistorted output. In practice, with real audio signals, class AB efficiency is much lower, on the order of 20 percent. The theoretical efficiency of the class D amplifier is 100 percent because the transistors are used as switches. Because transistor switching and filtering are imperfect, practical class D amplifiers are about 75 percent efficient.

Class D audio amplifiers are becoming more prevalent in consumer electronics applications where greater efficiency results in reduced size and increased battery life. In high-power applications such as at rock concerts, class D amplifiers are used to reduce the size of the amplifier and for reduced heat requirements in the equipment.

8.14 SIMULATION OF PULSE-WIDTH-MODULATED INVERTERS

Bipolar PWM

PSpice can be used to simulate the PWM inverter switching schemes presented previously in this chapter. As with other power electronics circuits, the level of circuit detail depends on the objective of the simulation. If only the voltages and currents in the load are desired, a PWM source may be created without modeling the individual switches in the bridge circuit. Figure 8-23 shows two ways to produce a bipolar PWM voltage. The first uses an ABM2 block, and the second uses

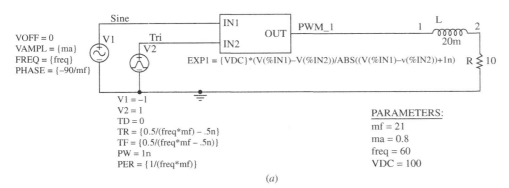

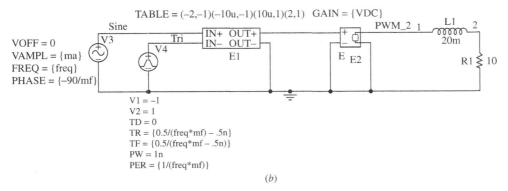

Figure 8-23 PSpice functional circuits for producing a bipolar PWM voltage using (*a*) an ABM block and (*b*) an ETABLE voltage source.

a voltage-controlled voltage source ETABLE. Both methods compare a sine wave to a triangular wave. Either method allows the behavior of a specific load to a PWM input to be investigated.

If the load contains an inductance and/or capacitance, there will be an initial transient in the load current. Since the steady-state load current is usually of interest, one or more periods of the load current must be allowed to run before meaningful output is obtained. One way to achieve this in PSpice is to delay output in the transient command, and another way is to restrict the data to steady-state results in Probe. The reference signal is synchronized with the carrier signal as in Fig. 8-17a. When the triangular carrier voltage has negative slope going through zero, the sinusoidal reference voltage must have positive slope going through zero. The triangular waveform starts at the positive peak with negative slope. The phase angle of the reference sinusoid is adjusted to make the zero crossing correspond to that of the triangular wave by using a phase angle of $-90°/m_f$. The following example illustrates a PSpice simulation of a bipolar PWM application.

EXAMPLE 8-10

PSpice Simulation of PWM

Use PSpice to analyze the PWM inverter circuit of Example 8-8.

■ **Solution**

Using either PWM circuit in Fig. 8-23, the Probe output will be the waveforms shown in Fig. 8-24a. The current is scaled by a factor of 10 to show more clearly its relationship

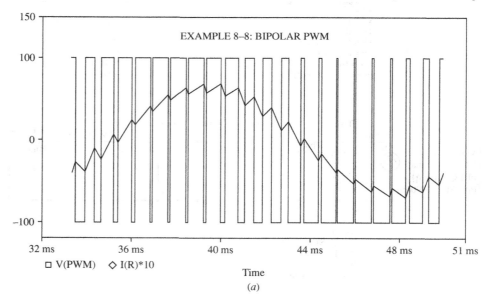

Figure 8-24 (a) Probe output for Example 8-10 showing PWM voltage and load current (current is scaled for illustration); (b) Frequency spectra for voltage and current.

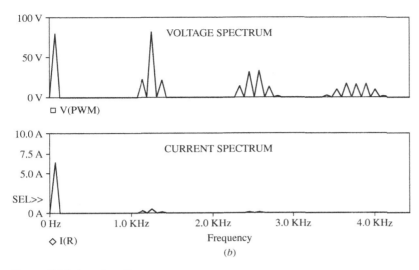

Figure 8-24 (*continued*)

with output voltage. Note the sinelike quality of the current. The Fourier coefficients of voltage and current are determined by using the Fourier option under the x axis menu or by pressing the FFT icon. Figure 8-24b shows the frequency spectra of voltage and current with the range on the x axis selected to show the lower frequencies. The cursor option is used to determine the Fourier coefficients.

Table 8-6 summarizes the results. Note the close correspondence with the results of Example 8-8.

Table 8-6 PSpice Results of Example 8-10

n	f_n (Hz)	V_n (V)	I_n (A)
1	60	79.8	6.37
19	1140	21.8	0.15
21	1260	82.0	0.52
23	1380	21.8	0.13

If the voltages and currents in the source and switches are desired, the PSpice input file must include the switches. A somewhat idealized circuit using voltage-controlled switches with feedback diodes is shown in Fig. 8-25. To simulate pulse-width modulation, the control for the switches in the inverter is the voltage difference between a triangular carrier voltage and a sine reference voltage. While this does not represent a model for real switches, this circuit is useful to simulate either bipolar or unipolar PWM. A more realistic bridge model would include devices such as BJTs, MOSFETs, or IGBTs for the switches. The model that is appropriate will depend on how completely switch performance must be investigated.

CHAPTER 8 Inverters

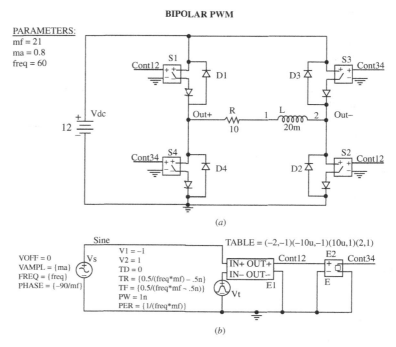

Figure 8-25 PSpice circuits for a PWM inverter (*a*) using voltage-controlled switches and diodes but requires the full PSpice version and (*b*) generating a PWM function.

Unipolar PWM

Again, unipolar PWM can be simulated using various levels of switch models. The input file shown in Fig. 8-26 utilizes dependent sources to produce a unipolar PWM output.

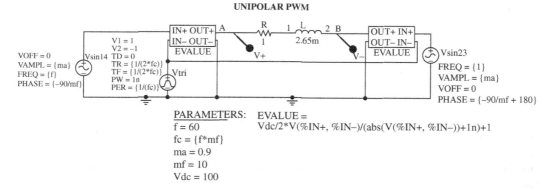

Figure 8-26 A PSpice circuit for generating a unipolar PWM voltage. The output voltage is between nodes A and B.

EXAMPLE 8-11

Pulse-Width Modulation PSpice Simulation

Pulse-width modulation is used to provide a 60-Hz voltage across a series RL load with R = 1 Ω and L = 2.65 mH. The dc supply voltage is 100 V. The amplitude of the 60-Hz voltage is to be 90 V, requiring m_a = 0.9. Use PSpice to obtain the current waveform in the load and the THD of the current waveform in the load. Use (a) bipolar PWM with m_f = 21, (b) bipolar PWM with m_f = 41, and (c) unipolar PWM with m_f = 10.

■ **Solution**

(a) The PSpice circuit for bipolar PWM (Fig. 8-25b) is run with m_a = 0.9 and m_f = 21. The voltage across the load and the current in the load resistor are shown in Fig. 8-27a. The currents for the 60-Hz fundamental and the lowest-order harmonics are obtained from the Fourier option under x axis in Probe. The harmonic amplitudes correspond to the peaks, and the cursor option determines precise values. The rms current can be obtained from Probe by entering the expression RMS(I(R)). The total harmonic distortion based on the truncated Fourier series is computed from Eq. (8-17). Results are in the table in this example.

(b) The PSpice circuit is modified for m_f = 41. The voltage and current waveforms are shown in Fig. 8-27b. The resulting harmonic currents are obtained from the Fourier option in Probe.

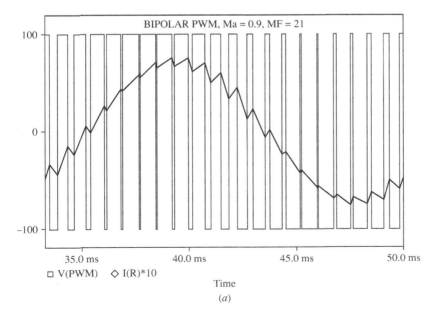

Figure 8-27 Voltage and current for Example 8-11 for (a) bipolar PWM with m_f = 21, (b) bipolar PWM with m_f = 41, (c) Unipolar PWM with m_f = 10.

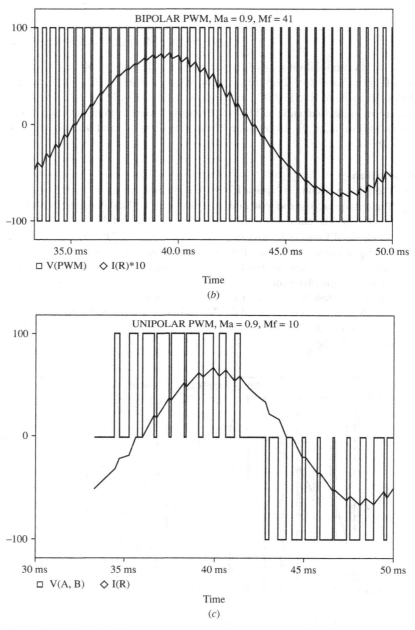

Figure 8-27 (*continued*)

(c) The PSpice input file for unipolar switching in Fig. 8-26 is run with the parameter $m_f = 10$. The output voltage and current are shown in Fig. 8-27c. The results of the three simulations for this example are shown in the following table.

	Bipolar $m_f = 21$		Bipolar $m_f = 41$		Unipolar $m_f = 10$	
	f_n	I_n	f_n	I_n	f_n	I_n
	60	63.6	60	64.0	60	62.9
	1140	1.41	2340	0.69	1020	1.0
	1260	3.39	2460	1.7	1140	1.4
	1380	1.15	2580	0.62	1260	1.24
					1380	0.76
I_{rms}		45.1		45.0		44.5
THD		6.1%		3.2%		3.6%

Note that the THD is relatively low in each of these PWM switching schemes, and increasing the switching frequency (increasing m_f) decreases the harmonic currents in this type of a load.

8.15 THREE-PHASE INVERTERS

The Six-Step Inverter

Figure 8-28a shows a circuit that produces a three-phase ac output from a dc input. A major application of this circuit is speed control of induction motors, where the output frequency is varied. The switches are closed and opened in the sequence shown in Fig. 8-28b.

Each switch has a duty ratio of 50 percent (not allowing for blanking time), and a switching action takes place every $T/6$ time interval, or 60° angle interval. Note that switches S_1 and S_4 close and open opposite of each other, as do switch pairs (S_2, S_5) and (S_3, S_6). As with the single-phase inverter, these switch pairs must coordinate so they are not closed at the same time, which would result in a short circuit across the source. With this scheme, the instantaneous voltages v_{A0}, v_{B0}, and v_{C0} are $+V_{dc}$ or zero, and line-to-line output voltages v_{AB}, v_{BC}, and v_{CA} are $+V_{dc}$, 0, or $-V_{dc}$. The switching sequence in Fig. 8-28b produces the output voltages shown in Fig. 8-28c.

The three-phase load connected to this output voltage may be connected in delta or ungrounded neutral wye. For a wye-connected load, which is the more common load connection, the voltage across each phase of the load is a line-to-neutral voltage, shown in Fig. 8-28d. Because of the six steps in the output waveforms for the line-to-neutral voltage resulting from the six switching transitions per period, this circuit with this switching scheme is called a *six-step inverter*.

The Fourier series for the output voltage has a fundamental frequency equal to the switching frequency. Harmonic frequencies are of order $6k \pm 1$ for $k = 1$, 2, ... ($n = 5, 7, 11, 13 \ldots$). The third harmonic and multiples of the third do not

Figure 8-28 (*a*) Three-phase inverter; (*b*) Switching sequence for six-step output; (*c*) Line-to-line output voltages; (*d*) Line-to-neutral voltages for an ungrounded Y-connected load; (*e*) Current in phase *A* for an *RL* load.

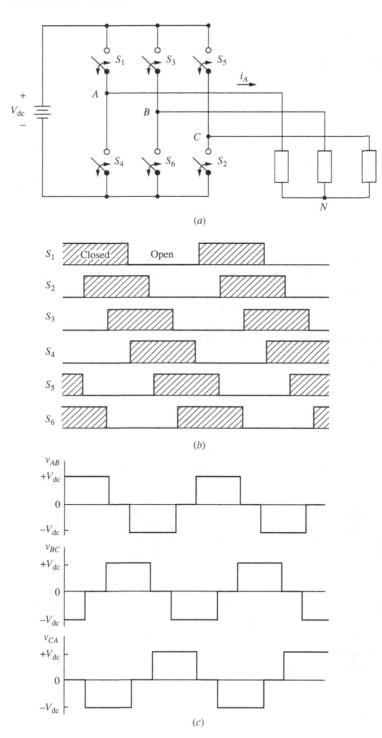

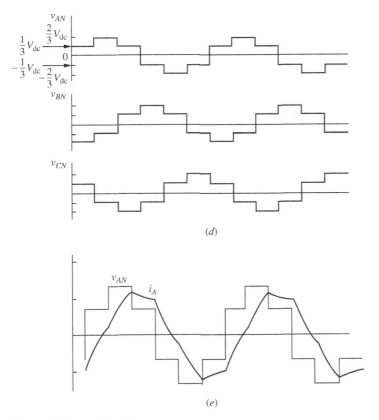

Figure 8-28 (*continued*)

exist, and even harmonics do not exist. For an input voltage of V_{dc}, the output for an ungrounded wye-connected load has the following Fourier coefficients:

$$V_{n,L-L} = \left| \frac{4V_{dc}}{n\pi} \cos\left(n\frac{\pi}{6}\right) \right|$$

$$V_{n,L-N} = \left| \frac{2V_{dc}}{3n\pi} \left[2 + \cos\left(n\frac{\pi}{3}\right) - \cos\left(n\frac{2\pi}{3}\right) \right] \right| \qquad n = 1, 5, 7, 11, 13, \ldots \quad (8\text{-}42)$$

The THD of both the line-to-line and line-to-neutral voltages can be shown to be 31 percent from Eq. (8-17). The THD of the currents in load-dependent 15 are smaller for an *RL* load. An example of the line-to-neutral voltage and line current for an *RL* wye-connected load is shown in Fig. 8-28*e*.

The output frequency can be controlled by changing the switching frequency. The magnitude of the output voltage depends on the value of the dc supply voltage. To control the output voltage of the six-step inverter, the dc input voltage must be adjusted.

EXAMPLE 8-12

Six-Step Three-Phase Inverter

For the six-step three-phase inverter of Fig. 8-28a, the dc input is 100 V and the fundamental output frequency is 60 Hz. The load is wye-connected with each phase of the load a series RL connection with $R = 10\ \Omega$ and $L = 20$ mH. Determine the total harmonic distortion of the load current.

■ Solution

The amplitude of load current at each frequency is

$$I_n = \frac{V_{n,L-N}}{Z_n} = \frac{V_{n,L-N}}{\sqrt{R^2 + (n\omega_0 L)^2}} = \frac{V_{n,L-N}}{\sqrt{10^2 + [n(2\pi 60)(0.02)]^2}}$$

where $V_{n,L-N}$ is determined from Eq. (8-42). Table 8-7 summarizes the results of the Fourier series computation.

Table 8.7 Fourier Components for the Six-Step Inverter of Example 8-12

n	$V_{n,L\text{-}N}$ (V)	Z_n (Ω)	I_n (A)	$I_{n,\text{rms}}$ (A)
1	63.6	12.5	5.08	3.59
5	12.73	39.0	0.33	0.23
7	9.09	53.7	0.17	0.12
11	5.79	83.5	0.07	0.05
13	4.90	98.5	0.05	0.04

The THD of the load current is computed from Eq. (8-17) as

$$\text{THD}_I = \frac{\sqrt{\sum_{n=2}^{\infty} I_{n,\text{rms}}^2}}{I_{1,\text{rms}}} \approx \frac{\sqrt{(0.23)^2 + (0.12)^2 + (0.05)^2 + (0.04)^2}}{3.59} = 0.07 = 7\%$$

PWM Three-Phase Inverters

Pulse-width modulation can be used for three-phase inverters as well as for single-phase inverters. The advantages of PWM switching are the same as for the single-phase case: reduced filter requirements for harmonic reduction and the controllability of the amplitude of the fundamental frequency.

PWM switching for the three-phase inverter is similar to that of the single-phase inverter. Basically, each switch is controlled by comparing a sinusoidal reference wave with a triangular carrier wave. The fundamental frequency of the output is the same as that of the reference wave, and the amplitude of the output is determined by the relative amplitudes of the reference and carrier waves.

As in the case of the six-step three-phase inverter, switches in Fig. 8-28a are controlled in pairs (S_1, S_4), (S_2, S_5), and (S_3, S_6). When one switch in a pair is closed, the other is open. Each pair of switches requires a separate sinusoidal reference wave. The three reference sinusoids are 120° apart to produce a balanced three-phase output. Figure 8-29a shows a triangular carrier and the three reference waves. Switch controls are such that

S_1 is on when $v_a > v_{tri}$
S_2 is on when $v_c > v_{tri}$
S_3 is on when $v_b > v_{tri}$
S_4 is on when $v_a < v_{tri}$
S_5 is on when $v_c < v_{tri}$
S_6 is on when $v_b < v_{tri}$

Harmonics will be minimized if the carrier frequency is chosen to be an odd triple multiple of the reference frequency, that is, 3, 9, 15, . . . times the reference. Figure 8-29b shows the line-to-line output voltages for a PWM three-phase inverter.

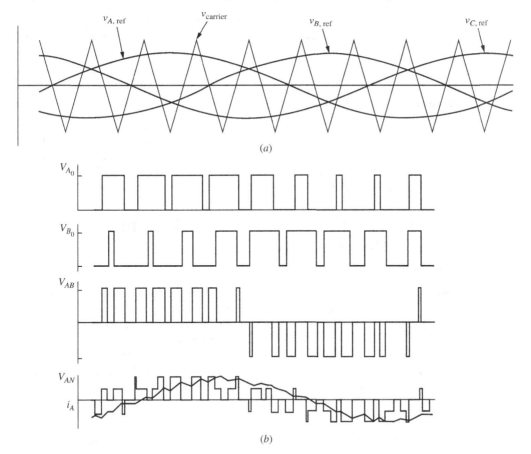

Figure 8-29 (a) Carrier and reference waves for PWM operation with $m_f = 9$ and $m_a = 0.7$ for the three-phase inverter of Fig. 8-28a; (b) Output waveforms—current is for an RL load.

The Fourier coefficients for the line-to-line voltages for the three-phase PWM switching scheme are related to those of single-phase bipolar PWM (V_n in Table 8-3) by

$$V_{n3} = \sqrt{A_{n3}^2 + B_{n3}^2} \qquad (8\text{-}43)$$

where

$$A_{n3} = V_n \sin\left(\frac{n\pi}{2}\right)\sin\left(\frac{n\pi}{3}\right)$$
$$B_{n3} = V_n \cos\left(\frac{n\pi}{2}\right)\sin\left(\frac{n\pi}{3}\right) \qquad (8\text{-}44)$$

Significant Fourier coefficients are listed in Table 8-8.

Table 8-8 Normalized Amplitudes V_{n3}/V_{dc} for Line-to-Line Three-Phase PWM Voltages

	$m_a=1$	0.9	0.8	0.7	0.6	0.5	0.4	0.3	0.2	0.1
$n=1$	0.866	0.779	0.693	0.606	0.520	0.433	0.346	0.260	0.173	0.087
$m_f=2$	0.275	0.232	0.190	0.150	0.114	0.081	0.053	0.030	0.013	0.003
$2m_f=1$	0.157	0.221	0.272	0.307	0.321	0.313	0.282	0.232	0.165	0.086

Multilevel Three-Phase Inverters

Each of the multilevel inverters described in Sec. 8.9 can be expanded to three-phase applications. Figure 8-30 shows a three-phase diode-clamped multilevel inverter circuit. This circuit can be operated to have a stepped-level output similar to the six-step converter, or, as is most often the case, it can be operated to have a pulse-width-modulated output.

8.16 PSPICE SIMULATION OF THREE-PHASE INVERTERS

Six-Step Three-Phase Inverters

PSpice circuits that will simulate a six-step three-phase inverter are shown in Fig. 8-31. The first circuit is for a complete switching scheme described in Fig. 8-28. Voltage-controlled switches with feedback diodes are used for switching. (The full version of PSpice is required for this circuit.) The second circuit is for generating the appropriate output voltages for the converter so load currents can be analyzed. The output nodes of the inverter are nodes A, B, and C. The parameters shown are those in Example 8-12.

PWM Three-Phase Inverters

The circuit in Fig. 8-32 produces the voltages of the PWM three-phase inverter without showing the switching details. Dependent sources compare sine waves to a triangular carrier wave, as in Example 8-8 for the single-phase case.

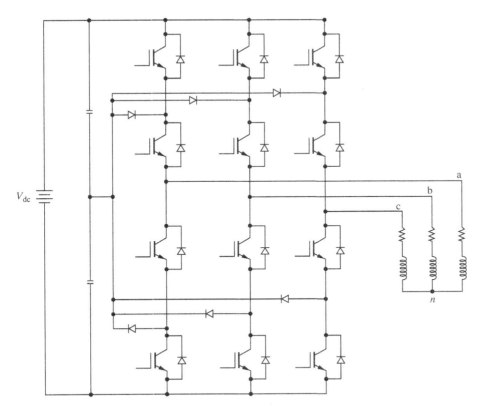

Figure 8-30 A three-phase diode-clamped multilevel inverter.

8.17 INDUCTION MOTOR SPEED CONTROL

The speed of an induction motor can be controlled by adjusting the frequency of the applied voltage. The synchronous speed ω_s of an induction motor is related to the number of poles p and the applied electrical frequency ω by

$$\omega_s = \frac{2\omega}{p} \qquad (8\text{-}45)$$

Slip s, is defined in terms of the rotor speed ω_r

$$s = \frac{\omega_s - \omega_r}{\omega_s} \qquad (8\text{-}46)$$

and torque is proportional to slip.

If the applied electrical frequency is changed, the motor speed will change proportionally. However, if the applied voltage is held constant when the frequency is lowered, the magnetic flux in the air gap will increase to the point of

380 CHAPTER 8 Inverters

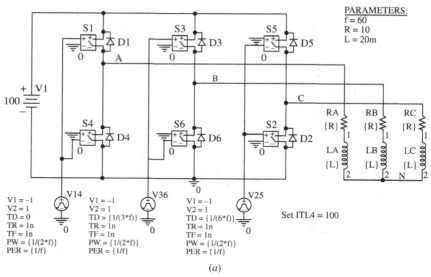

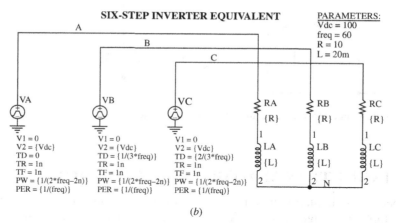

Figure 8-31 (a) A six-step inverter using switches and diodes (requires the full PSpice version); (b) A PSpice circuit for generating three-phase six-step converter voltages.

saturation. It is desirable to keep the air-gap flux constant and equal to its rated value. This is accomplished by varying the applied voltage proportionally with frequency. The ratio of applied voltage to applied frequency should be constant.

$$\frac{V}{f} = \text{constant} \qquad (8\text{-}47)$$

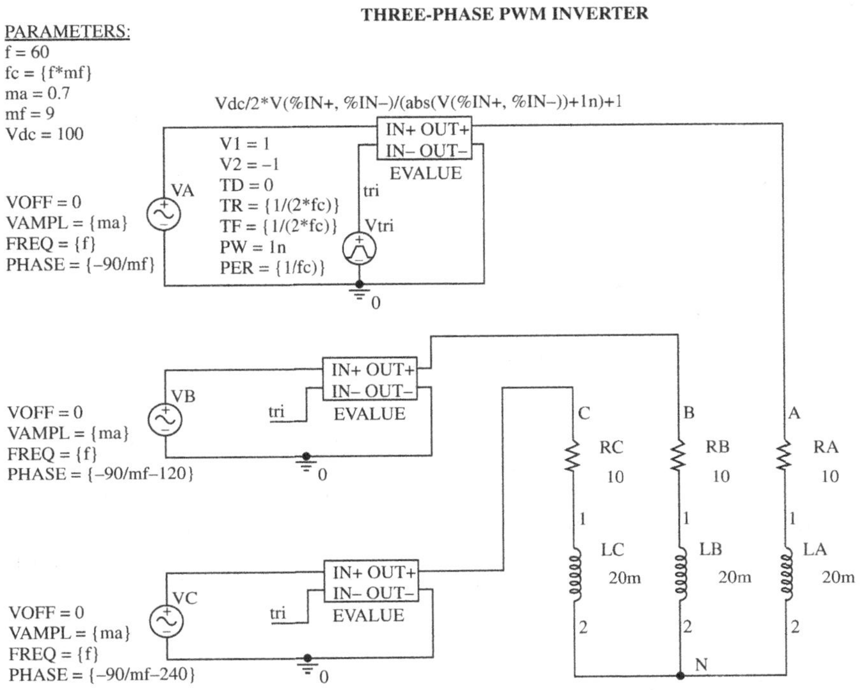

THREE-PHASE PWM INVERTER

PARAMETERS:
f = 60
fc = {f*mf}
ma = 0.7
mf = 9
Vdc = 100

VOFF = 0
VAMPL = {ma}
FREQ = {f}
PHASE = {−90/mf}

VOFF = 0
VAMPL = {ma}
FREQ = {f}
PHASE = {−90/mf−120}

VOFF = 0
VAMPL = {ma}
FREQ = {f}
PHASE = {−90/mf−240}

Vdc/2*V(%IN+, %IN−)/(abs(V(%IN+, %IN−))+1n)+1

V1 = 1
V2 = −1
TD = 0
TR = {1/(2*fc)}
TF = {1/(2*fc)}
PW = 1n
PER = {1/fc}

RC 10, RB 10, RA 10
LC 20m, LB 20m, LA 20m

Figure 8-32 A PSpice functional circuit for generating three-phase PWM voltages.

The term *volts/hertz control* is often used for this situation. The induction motor torque-speed curves of Fig. 8-33 are for different frequencies and constant volts/hertz.

The six-step inverter can be used for this application if the dc input is adjustable. In the configuration of Fig. 8-34, an adjustable dc voltage is produced from a controlled rectifier, and an inverter produces an ac voltage at the desired

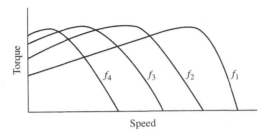

Figure 8-33 Induction motor torque-speed curves for constant volts/hertz variable-speed control.

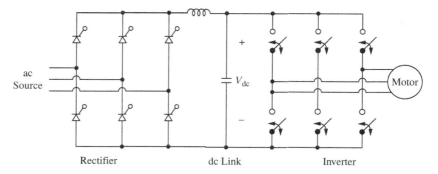

Figure 8-34 AC-AC converter with a dc link.

frequency. If the dc source is not controllable, a dc-dc converter may be inserted between the dc source and the inverter.

The PWM inverter is useful in a constant volts/hertz application because the amplitude of the output voltage can be adjusted by changing the amplitude modulation ratio m_a. The dc input to the inverter can come from an uncontrolled source in this case. The configuration in Fig. 8-34 is classified as an ac-ac converter with a dc link between the two ac voltages.

8.18 Summary

- The full- or half-bridge converters can be used to synthesize an ac output from a dc input.
- A simple switching scheme produces a square wave voltage output, which has a Fourier series that contains the odd harmonic frequencies of amplitudes

$$V_n = \frac{4V_{dc}}{n\pi}$$

- Amplitude and harmonic control can be implemented by allowing a zero-voltage interval of angle α at each end of a pulse, resulting in Fourier coefficients

$$V_n = \left(\frac{4V_{dc}}{n\pi}\right)\cos(n\alpha)$$

- Multilevel inverters use more than one dc voltage source or split a single voltage source with a capacitor voltage divider to produce multiple voltage levels on the output of an inverter.
- Pulse-width modulation (PWM) provides amplitude control of the fundamental output frequency. Although the harmonics have large amplitudes, they occur at high frequencies and are filtered easily.
- Class D audio amplifiers use PWM techniques for high efficiency.
- The six-step inverter is the basic switching scheme for producing a three-phase ac output from a dc source.
- A PWM switching scheme can be used with a three-phase inverter to reduce the THD of the load current with modest filtering.
- Speed control of induction motors is a primary application of three-phase inverters.

8.19 Bibliography

J. Almazan, N. Vazquez, C. Hernandez, J. Alvarez, and J. Arau, "Comparison between the Buck, Boost and Buck-Boost Inverters," *International Power Electronics Congress*, Acapulco, Mexico, October 2000, pp. 341–346.

B. K. Bose, *Power Electronics and Motor Drives: Advances and Trends*, Elsevier/Academic Press, 2006.

J. N. Chiasson, L. M. Tolbert, K. J. McKenzie, and D. Zhong, "A Unified Approach to Solving the Harmonic Elimination Equations in Multilevel Converters," *IEEE Transactions on Power Electronics*, March 2004, pp. 478–490.

K. A. Corzine, "Topology and Control of Cascaded Multi-Level Converters," Ph.D. dissertation, University of Missouri, Rolla, 1997.

T. Kato, "Precise PWM Waveform Analysis of Inverter for Selected Harmonic Elimination," 1986 IEEE/IAS Annual Meeting, pp. 611–616.

N. Mohan, T. M. Undeland, and W. P. Robbins, *Power Electronics: Converters, Applications, and Design*, 3d ed., Wiley, New York, 2003.

L. G. Franquelo, "Multilevel Converters: Current Developments and Future Trends," *IEEE International Conference on Industrial Technology*, Chengdu, China, 2008.

J. R. Hauser, *Numerical Methods for Nonlinear Engineering Models*, Springer Netherlands, Dordrecht, 2009.

J. Holtz, "Pulsewidth Modulation—A Survey," *IEEE Transactions on Industrial Electronics*, vol. 39, no. 5, Dec. 1992, pp. 410–420.

S. Miaosen, F. Z. Peng, and L. M. Tolbert, "Multi-level DC/DC Power Conversion System with Multiple DC Sources," *IEEE 38th Annual Power Electronics Specialists Conference*, Orlando, Fla., 2007.

L. M. Tolbert, and F. Z. Peng, "Multilevel Converters for Large Electric Drives," *Applied Power Electronics Conference and Exposition*, anaheim, Calif., 1998.

M. H. Rashid, *Power Electronics: Circuits, Devices, and Systems*, 3d ed., Prentice-Hall, Upper Saddle River, N.J., 2004.

L. Salazar and G. Joos, "PSpice Simulation of Three-Phase Inverters by Means of Switching Functions," *IEEE Transactions on Power Electronics*, vol. 9, no. 1, Jan. 1994, pp. 35–42.

B. Wu, *High-Power Converters and AC Drives*, Wiley, New York, 2006.

X. Yuan, and I. Barbi, "Fundamentals of a New Diode Clamping Multilevel Inverter," *IEEE Transactions on Power Electronics*, vol. 15, no. 4, July 2000, pp. 711–718.

Problems

Square-Wave Inverter

8-1. The square-wave inverter of Fig. 8-1*a* has $V_{dc} = 125$ V, an output frequency of 60 Hz, and a resistive load of 12.5 Ω. Sketch the currents in the load, each switch, and the source, and determine the average and rms values of each.

8-2. A square-wave inverter has a dc source of 96 V and an output frequency of 60 Hz. The load is a series RL load with $R = 5$ Ω and $L = 100$ mH. When the load is first energized, a transient precedes the steady-state waveform described by Eq. (8-5). (*a*) Determine the peak value of the steady-state current. (*b*) Using Eq. (8-1) and assuming zero initial inductor current, determine the maximum current that occurs during the transient. (*c*) Simulate the circuit with the PSpice input file of Fig. 8.4a and compare the results with parts (*a*) and (*b*). How many periods must elapse before the current reaches steady state? How many L/R time constants elapse before steady state?

8-3. The square-wave inverter of Fig. 8-3 has a dc input of 150 V and supplies a series RL load with $R = 20\ \Omega$ and $L = 40$ mH. (*a*) Determine an expression for steady-state load current. (*b*) Sketch the load current and indicate the time intervals when each switch component (Q1, D1; ... Q4, D4) is conducting. (*c*) Determine the peak current in each switch component. (*d*) What is the maximum voltage across each switch? Assume ideal components.

8-4. A square-wave inverter has a dc source of 125 V, an output frequency of 60 Hz, and an RL series load with $R = 20\ \Omega$ and $L = 25$ mH. Determine (*a*) an expression for load current, (*b*) rms load current, and (*c*) average source current.

8-5. A square-wave inverter has an RL load with $R = 15\ \Omega$ and $L = 10$ mH. The inverter output frequency is 400 Hz. (*a*) Determine the value of the dc source required to establish a load current that has a fundamental frequency component of 8 A rms. (*b*) Determine the THD of the load current.

8-6. A square-wave inverter supplies an RL series load with $R = 25\ \Omega$ and $L = 25$ mH. The output frequency is 120 Hz. (*a*) Specify the dc source voltage such that the load current at the fundamental frequency is 2.0 A rms. (*b*) Verify your results with PSpice. Determine the THD from PSpice.

8-7. A square-wave inverter has a dc input of 100 V, an output frequency of 60 Hz, and a series RLC combination with $R = 10\ \Omega$, $L = 25$ mH, and $C = 100\ \mu$F. Use the PSpice simplified square-wave inverter circuit of Fig. 8-4*a* to determine the peak and rms value of the steady-state current. Determine the total harmonic distortion of the load current. On a printout of one period of the current, indicate the intervals where each switch component in the inverter circuit of Fig. 8-3 is conducting for this load if that circuit were used to implement the converter.

Amplitude and Harmonic Control

8-8. For the full-bridge inverter, the dc source is 125 V, the load is a series RL connection with $R = 10\ \Omega$ and $L = 20$ mH, and the switching frequency is 60 Hz. (*a*) Use the switching scheme of Fig. 8-5 and determine the value of α to produce an output with an amplitude of 90 V at the fundamental frequency. (*b*) Determine the THD of the load current.

8-9. An inverter that produces the type of output shown in Fig. 8-5*a* is used to supply an RL series load with $R = 10\ \Omega$ and $L = 35$ mH. The dc input voltage is 200 V and the output frequency is 60 Hz. (*a*) Determine the rms value of the fundamental frequency of the load current when $\alpha = 0$. (*b*) If the output fundamental frequency is lowered to 30 Hz, determine the value of α required to keep the rms current at the fundamental frequency at the same value of part (*a*).

8-10. Use the PSpice circuit of Fig. 8-7*a* to verify that (*a*) the waveform of Fig. 8-5*a* with $\alpha = 30°$ contains no third harmonic frequency and (*b*) the waveform of Fig. 8-5*a* with $\alpha = 18°$ contains no fifth harmonic.

8-11. (*a*) Determine the value of α that will eliminate the seventh harmonic from the inverter output of Fig. 8-5*a*. (*b*) Verify your answer with a PSpice simulation.

8-12. Determine the rms value of the notched waveform to eliminate the third and fifth harmonics in Fig. 8-6.

8-13. Use PSpice to verify that the notched waveform of Fig. 8-6c contains no third or fifth harmonic. What are the magnitudes of the fundamental frequency and the first four nonzero harmonics? (The piecewise linear type of source may be useful.)

Multilevel Inverters

8-14. For a multilevel inverter having three separate dc sources of 48 V each, $\alpha_1 = 15°$, $\alpha_2 = 25°$, and $\alpha_3 = 55°$. (a) Sketch the output voltage waveform. (b) Determine the Fourier coefficients through $n = 9$. (c) Determine the modulation index M_i.

8-15. For a three-source multilevel inverter, select values of α_1, α_2, and α_3 such that the third harmonic frequency ($n = 3$) in the output voltage waveform is eliminated. Determine the modulation index M_i for your selection.

8-16. The five-source multilevel inverter of Fig. 8-11 has $\alpha_1 = 16.73°$, $\alpha_2 = 26.64°$, $\alpha_3 = 46.00°$, $\alpha_4 = 60.69°$, and $\alpha_5 = 62.69°$. Determine which harmonics will be eliminated from the output voltage. Determine the amplitude of the fundamental-frequency output voltage.

8-17. The concept of the two-source multilevel inverters of Figs. 8-9 and 8-11 is extended to have three independent sources and H bridges and three delay angles α_1, α_2, and α_3. Sketch the voltages at the output of each bridge of a three-source multilevel converter such that the average power from each source is the same.

Pulse-Width-Modulated Inverters

8-18. The dc source supplying an inverter with a bipolar PWM output is 96 V. The load is an *RL* series combination with $R = 32\,\Omega$ and $L = 24$ mH. The output has a fundamental frequency of 60 Hz. (a) Specify the amplitude modulation ratio to provide a 54-V rms fundamental frequency output. (b) If the frequency modulation ratio is 17, determine the total harmonic distortion of the load current.

8-19. The dc source supplying an inverter with a bipolar PWM output is 250 V. The load is an *RL* series combination with $R = 20\,\Omega$ and $L = 50$ mH. The output has a fundamental frequency of 60 Hz. (a) Specify the amplitude modulation ratio to provide a 160-V rms fundamental frequency output. (b) If the frequency modulation ratio is 31, determine the total harmonic distortion of the load current.

8-20. Use PSpice to verify that the design in Example 8-9 meets the THD specifications.

8-21. Design an inverter that has a PWM output across an *RL* series load with $R = 10\,\Omega$ and $L = 20$ mH. The fundamental frequency of the output voltage must be 120 V rms at 60 Hz, and the total harmonic distortion of the load current must be less than 8 percent. Specify the dc input voltage, the amplitude modulation ratio m_a, and the switching frequency (carrier frequency). Verify the validity of your design with a PSpice simulation.

8-22. Design an inverter that has a PWM output across an *RL* series load with $R = 30\,\Omega$ and $L = 25$ mH. The fundamental frequency of the output voltage must be 100 V rms at 60 Hz, and the total harmonic distortion of the load current must be less than 10 percent. Specify the dc input voltage, the amplitude modulation ratio m_a, and the switching frequency (carrier frequency). Verify the validity of your design with a PSpice simulation.

8-23. Pulse-width modulation is used to provide a 60-Hz voltage across a series RL load with $R = 12\,\Omega$ and $L = 20$ mH. The dc supply voltage is 150 V. The amplitude of the 60-Hz voltage is to be 120 V. Use PSpice to obtain the current waveform in the load and the THD of the current waveform in the load. Use (a) bipolar PWM with $m_f = 21$, (b) bipolar PWM with $m_f = 41$, and (c) unipolar PWM with $m_f = 10$.

Three-Phase Inverters

8-24. A six-step three-phase inverter has a 250-V dc source and an output frequency of 60 Hz. A balanced Y-connected load consists of a series 25-Ω resistance and 20-mH inductance in each phase. Determine (a) the rms value of the 60-Hz component of load current and (b) the THD of the load current.

8-25. A six-step three-phase inverter has a 400-V dc source and an output frequency that varies from 25 to 100 Hz. The load is a Y connection with a series 10-Ω resistance and 30-mH inductance in each phase. (a) Determine the range of the rms value of the fundamental-frequency component of load current as the frequency is varied. (b) What is the effect of varying frequency on the THD of the load current and the THD of the line-to-neutral voltage?

8-26. A six-step three-phase inverter has an adjustable dc input. The load is a balanced Y connection with a series RL combination in each phase, with $R = 5\,\Omega$ and $L = 50$ mH. The output frequency is to be varied between 30 and 60 Hz.
(a) Determine the range of the dc input voltage required to maintain the fundamental-frequency component of current at 10 A rms. (b) Use PSpice to determine the THD of load current in each case. Determine the peak current and rms load current for each case.

CHAPTER 9

Resonant Converters

9.1 INTRODUCTION

Imperfect switching is a major contributor to power loss in converters, as discussed in Chap. 6. Switching devices absorb power when they turn on or off if they go through a transition when both voltage and current are nonzero. As the switching frequency increases, these transitions occur more often and the average power loss in the device increases. High switching frequencies are otherwise desirable because of the reduced size of filter components and transformers, which reduces the size and weight of the converter.

In resonant switching circuits, switching takes place when voltage and/or current is zero, thus avoiding simultaneous transitions of voltage and current and thereby eliminating switching losses. This type of switching is called *soft* switching, as opposed to *hard* switching in circuits such as the buck converter. Resonant converters include resonant switch converters, load resonant converters, and resonant dc link converters. This chapter introduces the basic concept of the resonant converter and gives a few examples.

9.2 A RESONANT SWITCH CONVERTER: ZERO-CURRENT SWITCHING

Basic Operation

One method for taking advantage of the oscillations caused by an *LC* circuit for reducing the switching losses in a dc-dc converter is shown in the circuit of Fig. 9-1a. This circuit is similar to the buck converter described in Chap. 6. The current in the output inductor L_o is assumed to be ripple-free and equal to the load

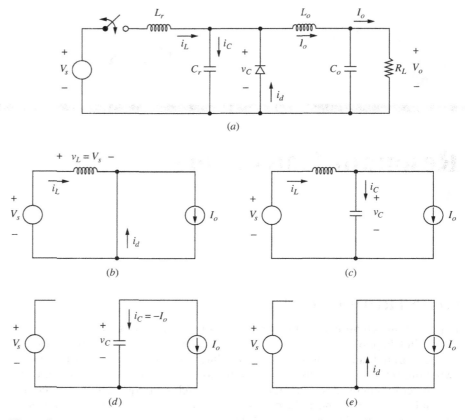

Figure 9-1 (a) A resonant converter with zero-current switching; (b) Switch closed and diode on ($0 < t < t_1$); (c) Switch closed and diode off ($t_1 < t < t_2$); (d) Switch open and diode off ($t_2 < t < t_3$); (e) Switch open and diode on ($t_3 < t < T$); (f) Waveforms; (g) Normalized output vs. switching frequency with $r = R_L/Z_0$ as a parameter. © 1992 IEEE, B.K. Bose, Modern Power Electronics: Evolution, Technology, and Applications. Reprinted with permission.

current I_o. When the switch is open, the diode is forward-biased to carry the output inductor current, and the voltage across C_r is zero. When the switch closes, the diode initially remains forward-biased to carry I_o, and the voltage across L_r is the same as the source voltage V_s (Fig. 9-1b). The current in L_r increases linearly, and the diode remains forward-biased while i_L is less than I_o. When i_L reaches I_o, the diode turns off, and the equivalent circuit is that of Fig. 9-1c. If I_o is a constant, the load appears as a current source, and the underdamped LC circuit oscillates. Consequently, i_L returns to zero and remains there, assuming the switch is unidirectional. The switch is turned off after the current reaches zero, resulting in zero-current switching and no switching power loss.

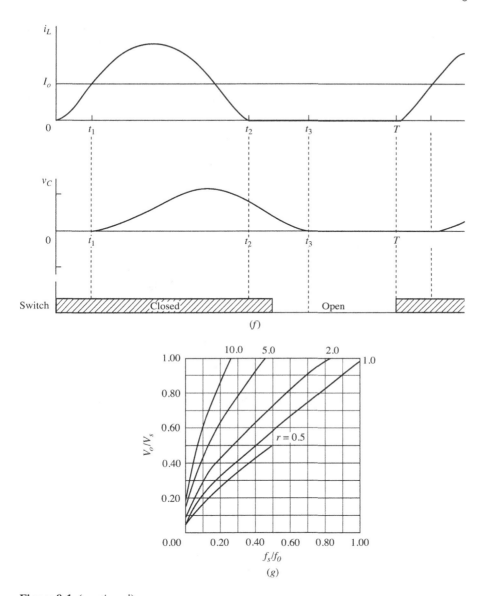

Figure 9-1 (*continued*)

After the current in the switch reaches zero, the positive capacitor voltage keeps the diode reverse-biased, so load current I_o flows through C_r, with $i_c = -I_o$ (Fig. 9-1d). If I_o is constant, the capacitor voltage decreases linearly. When the capacitor voltage reaches zero, the diode becomes forward-biased to carry I_o (Fig. 9-1e). The circuit is then back at the starting point. The analysis for each time interval is given next.

Analysis for $0 \leq t \leq t_1$ The switch is closed at $t = 0$, the diode is on, and the voltage across L_r is V_s (Fig. 9-1b). The current in L_r is initially zero and is expressed as

$$i_L(t) = \frac{1}{L_r} \int_0^t V_s \, d\lambda = \frac{V_s}{L_r} t \tag{9-1}$$

At $t = t_1$, i_L reaches I_o, and the diode turns off. Solving for t_1,

$$i_L(t_1) = I_o = \frac{V_s}{L_r} t_1 \tag{9-2}$$

or

$$\boxed{t_1 = \frac{I_o L_r}{V_s}} \tag{9-3}$$

Capacitor voltage is zero in this interval.

Analysis for $t_1 \leq t \leq t_2$ (Fig. 9-1c) When the diode turns off at $t = t_1$, the circuit is equivalent to that in Fig. 9-1c. In the circuit of Fig. 9-1c, these equations apply:

$$v_C(t) = V_s - L_r \frac{di_L(t)}{dt} \tag{9-4}$$

$$i_C(t) = i_L(t) - I_o \tag{9-5}$$

Differentiating Eq. (9-4) and using the voltage-current relationship for the capacitor,

$$\frac{dv_C(t)}{dt} = -L_r \frac{d^2 i_L(t)}{dt^2} = \frac{i_C(t)}{C_r} \tag{9-6}$$

Substituting for i_C using Eq. (9-5),

$$L_r \frac{d^2 i_L(t)}{dt^2} = \frac{I_o - i_L(t)}{C_r} \tag{9-7}$$

or

$$\frac{d^2 i_L(t)}{dt^2} + \frac{i_L(t)}{L_r C_r} = \frac{I_o}{L_r C_r} \tag{9-8}$$

The solution to the preceding equation with the initial condition $i_L(t_1) = I_o$ is

$$i_L(t) = I_o + \frac{V_s}{Z_0} \sin \omega_0 (t - t_1) \tag{9-9}$$

where Z_0 is the characteristic impedance

$$Z_0 = \sqrt{\frac{L_r}{C_r}} \tag{9-10}$$

and ω_0 is the frequency of oscillation

$$\omega_0 = \frac{1}{\sqrt{L_r C_r}} \qquad (9\text{-}11)$$

Equation (9-9) is valid until i_L reaches zero at $t = t_2$. Solving for the time interval $t_2 - t_1$ when the oscillation occurs,

$$t_2 - t_1 = \frac{1}{\omega_0} \sin^{-1}\left(\frac{-I_o Z_0}{V_s}\right) \qquad (9\text{-}12)$$

which can be expressed as

$$t_2 - t_1 = \frac{1}{\omega_0}\left[\sin^{-1}\left(\frac{I_o Z_0}{V_s}\right) + \pi\right] \qquad (9\text{-}13)$$

Solving for capacitor voltage by substituting i_L from Eq. (9-9) into Eq. (9-4) gives

$$v_C(t) = V_s\{1 - \cos[\omega_0(t - t_1)]\} \qquad (9\text{-}14)$$

which is also valid until $t = t_2$. Maximum capacitor voltage is therefore $2V_s$.

Analysis for $t_2 \leq t \leq t_3$ After the inductor current reaches zero at t_2, switch current is zero and it can be opened without power loss. The equivalent circuit is shown in Fig. 9-1d. The diode is off because $v_C > 0$. Capacitor current is $-I_o$, resulting in a linearly decreasing capacitor voltage expressed as

$$v_C(t) = \frac{1}{C_r}\int_{t_2}^{t} -I_o\, d\lambda + v_C(t_2) = \frac{I_o}{C_r}(t_2 - t) + v_C(t_2) \qquad (9\text{-}15)$$

Equation (9-15) is valid until the capacitor voltage reaches zero and the diode turns on. Letting the time at which the capacitor voltage reaches zero be t_3, Eq. (9-15) gives an expression for the time interval $t_3 - t_2$:

$$t_3 - t_2 = \frac{C_r v_C(t_2)}{I_o} = \frac{C_r V_s\{1 - \cos[\omega_0(t_2 - t_1)]\}}{I_o} \qquad (9\text{-}16)$$

where $v_C(t_2)$ is obtained from Eq. (9-14).

Analysis for $t_3 \leq t \leq T$ In this time interval, i_L is zero, the switch is open, the diode is on to carry I_o, and $v_C = 0$ (Fig. 9-1e). The duration of this interval is the difference between the switching period T and the other time intervals, which are determined from other circuit parameters.

Output Voltage

Output voltage can be determined from energy balance. Energy supplied by the source is equal to energy absorbed by the load during a switching period. Energy supplied by the source in one period is

$$W_s = \int_0^T p_s(t)\,d(t) = V_s \int_0^T i_L(t)\,dt \tag{9-17}$$

Energy absorbed by the load is

$$W_o = \int_0^T p_o(t)\,dt = V_o I_o T = \frac{V_o I_o}{f_s} \tag{9-18}$$

where f_s is the switching frequency. From Eqs. (9-1) and (9-9),

$$\int_0^T i_L(t)\,dt = \int_0^{t_1} \frac{V_s t}{L_r}\,dt + \int_{t_1}^{t_2} \left\{ I_o + \frac{V_s}{Z_0} \sin[\omega_0(t - t_1)] \right\} dt \tag{9-19}$$

Using $W_s = W_o$ and solving for V_o using Eqs. (9-17) to (9-19),

$$V_o = V_s f_s \left(\frac{t_1}{2} + (t_2 - t_1) + \frac{V_s C_r}{I_o}\{1 - \cos[\omega_0(t_2 - t_1)]\} \right) \tag{9-20}$$

Using Eq. (9-16), output voltage can be expressed in terms of the time intervals for each circuit condition:

$$\boxed{V_o = V_s f_s \left[\frac{t_1}{2} + (t_2 - t_1) + (t_3 - t_2) \right]} \tag{9-21}$$

where the time intervals are determined from Eqs. (9-3), (9-13), and (9-16).

Equation (9-21) shows that the output voltage is a function of the switching frequency. Increasing f_s increases V_o. The switching period must be greater than t_3, and output voltage is less than input voltage, as is the case for the buck converter of Chap. 6. Note that the time intervals are a function of output current I_o, so output voltage for this circuit is load-dependent. When the load is changed, the switching frequency must be adjusted to maintain a constant output voltage. Figure 9-1g shows the relationship between output voltage and switching frequency. The quantity $r = R_L/Z_0$ is used as a parameter where R_L is the load resistance and Z_0 is defined in Eq. (9-10).

A diode placed in antiparallel with the switch in Fig. 9-1a creates a resonant switch converter which includes negative inductor current. For that circuit, V_o/V_s is nearly a linear function of switching frequency independent of load (that is, $V_o/V_s = f_s/f_0$).

9.2 A Resonant Switch Converter: Zero-Current Switching

The resonant switch converter with zero-current switching has theoretically zero switching losses. However, junction capacitance in switching devices stores energy which is dissipated in the device, resulting in small losses.

Note that output voltage is the average of the capacitor voltage v_c, yielding an alternate method of deriving Eq. (9-21).

EXAMPLE 9-1

Resonant Switch DC-DC Converter: Zero-Current Switching

In the circuit of Fig. 9-1a,

$$V_s = 12 \text{ V}$$
$$C_r = 0.1 \text{ μF}$$
$$L_r = 10 \text{ μH}$$
$$I_o = 1 \text{ A}$$
$$f_s = 100 \text{ kHz}$$

(a) Determine the output voltage of the converter. (b) Determine the peak current in L_r and the peak voltage across C_r. (c) What is the required switching frequency to produce an output voltage of 6 V for the same load current? (d) Determine the maximum switching frequency. (e) If the load resistance is changed to 20 Ω, determine the switching frequency required to produce an output voltage of 8 V.

■ **Solution**

(a) Using the given circuit parameters,

$$\omega_0 = \frac{1}{\sqrt{L_r C_r}} = \frac{1}{\sqrt{10(10)^{-6}(0.1)(10)^{-6}}} = 10^6 \text{ rad/s}$$

$$Z_0 = \sqrt{\frac{L_r}{C_r}} = \sqrt{\frac{10(10)^{-6}}{0.1(10)^{-6}}} = 10 \text{ Ω}$$

Output voltage is determined from Eq. (9-21). The time t_1 is determined from Eq. (9-3):

$$t_1 = \frac{I_o L_r}{V_s} = \frac{(1)(10)(10)^{-6}}{12} = 0.833 \text{ μs}$$

From Eq. (9-13),

$$t_2 - t_1 = \frac{1}{\omega_0}\left[\sin^{-1}\left(\frac{I_o Z_0}{V_s}\right) + \pi\right] = \frac{1}{10^6}\left[\sin^{-1}\frac{(1)(10)}{12} + \pi\right] = 4.13 \text{ μs}$$

From Eq. (9-16),

$$t_3 - t_2 = \frac{C_r V_s}{I_o}\{1 - \cos[\omega_0(t_2 - t_1)]\}$$

$$= \frac{(0.1)(10)^{-6}(12)}{1}\{1 - \cos[10^6 (4.13)(10)^{-6}]\} = 1.86 \text{ μs}$$

Output voltage from Eq. (9-21) is then

$$V_o = V_s f_s \left[\frac{t_1}{2} + (t_2 - t_1) + (t_3 - t_2) \right]$$

$$= (12)(100)(10^5)\left(\frac{0.833}{2} + 4.13 + 1.86\right)(10^{-6}) = 7.69 \text{ V}$$

(b) Peak current in L_r is determined from Eq. (9-9).

$$I_{L,\text{peak}} = I_o + \frac{V_s}{Z_0} = 1 + \frac{12}{10} = 2.2 \text{ A}$$

Peak voltage across C_r is determined from Eq. (9-14):

$$V_{C,\text{peak}} = 2V_s = 2(12) = 24 \text{ V}$$

(c) Since output voltage is proportional to frequency [Eq. (9-21)] if I_o remains unchanged, the required switching frequency for a 6-V output is

$$f_s = 100 \text{ kHz}\left(\frac{6 \text{ V}}{7.69 \text{ V}}\right) = 78 \text{ kHz}$$

(d) Maximum switching frequency for this circuit occurs when the interval $T - t_3$ is zero. Time $t_3 = t_1 + (t_2 - t_1) + (t_3 - t_2) = (0.833 + 4.13 + 1.86)$ μs $= 6.82$ μs, resulting in

$$f_{s,\text{max}} = \frac{1}{T_{\text{min}}} = \frac{1}{t_3} = \frac{1}{(6.82)(10^{-6})} = 146 \text{ kHz}$$

(e) The graph of Fig. 9-1g can be used to estimate the required switching frequency to obtain an output of 8 V with the load at 20 Ω. With $V_o/V_s = 8/12 = 0.67$, the curve for the parameter $r = R_L/Z_0 = 20/10 = 2$ gives $f_s/f_0 \approx 0.45$. The switching frequency is $f_s = 0.45 f_0 = 0.45(\omega_0/2\pi) = 0.45(10)^6/2\pi = 71.7$ kHz. The method used in part (a) of this problem can be used to verify the results. Note that I_o is now $V_o/R_L = 8/20 = 0.4$ A.

9.3 A RESONANT SWITCH CONVERTER: ZERO-VOLTAGE SWITCHING

Basic Operation

The circuit of Fig. 9-2a shows a method for using the oscillations of an LC circuit for switching at zero voltage. The analysis assumes that the output filter produces a ripple-free current I_o in L_o. Beginning with the switch closed, the current in the switch and in L_r is I_o, the currents in D_1 and D_s are zero, and the voltage across C_r and the switch is zero.

9.3 A Resonant Switch Converter: Zero-Voltage Switching

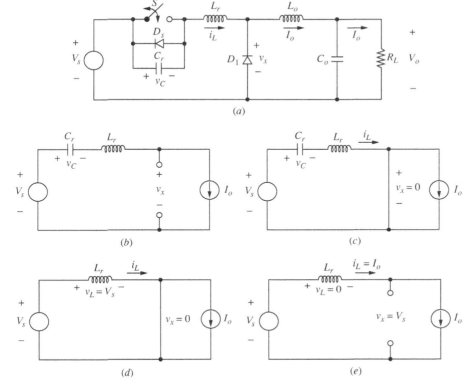

Figure 9-2 (*a*) A resonant converter with zero-voltage switching; (*b*) Switch open and D_1 off $(0 < t < t_1)$; (*c*) Switch open and D_1 on $(t_1 < t < t_2)$. (*d*) Switch closed and D_1 on $(t_2 < t < t_3)$; (*e*) Switch closed and D_1 off $(t_3 < t < T)$; (*f*) Waveforms (*g*) Normalized output vs. switching frequency with $r = R_L/Z_0$ as a parameter. © 1992 IEEE, B.K. Bose, Modern Power Electronics: Evolution, Technology, and Applications. Reprinted with permission.

The switch is opened (with zero voltage across it), and $i_L = I_o$ flows through the capacitor C_r, causing v_C to increase linearly (Fig. 9-2b). When v_C reaches the source voltage V_s, the diode D_1 becomes forward-biased, in effect forming a series circuit with V_s, C_r, and L_r as shown in Fig. 9-2c. At this time, i_L and v_C in this underdamped series circuit begin to oscillate.

When v_C returns to zero, diode D_s turns on to carry i_L, which is negative (Fig. 9-2d). The voltage across L_r is V_s, causing i_L to increase linearly. The switch should be closed just after D_s turns on for zero-voltage turn-on. When i_L becomes positive, D_s turns off and i_L is carried by the switch. When i_L reaches I_o, D_1 turns off, and circuit conditions are back at the starting point. The analysis for each circuit condition is given next.

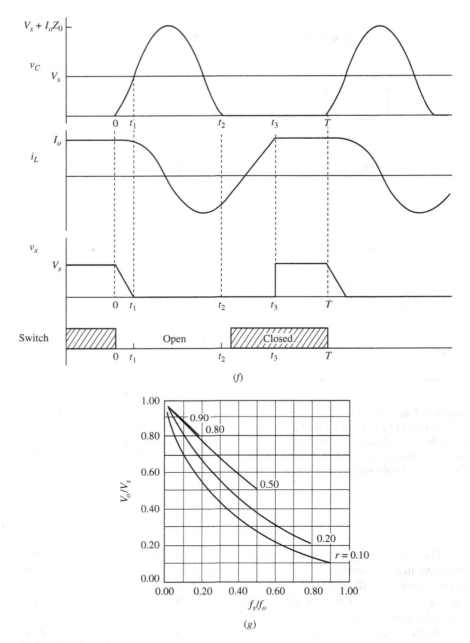

Figure 9-2 (*continued*)

9.3 A Resonant Switch Converter: Zero-Voltage Switching

Analysis for $0 \leq t \leq t_1$ The switch is opened at $t = 0$. The capacitor current is then I_o (Fig. 9-2b), causing the capacitor voltage, initially zero, to increase linearly. The voltage across C_r is

$$v_C(t) = \frac{1}{C_r} \int_0^t I_o \, d\lambda = \frac{I_o}{C_r} t \qquad (9\text{-}22)$$

The voltage across L_r is zero because inductor current is I_o, which is assumed to be constant. The voltage at the filter input v_x is

$$v_x(t) = V_s - v_C(t) = V_s - \frac{I_o}{C_r} t \qquad (9\text{-}23)$$

which is a linearly decreasing function beginning at V_s. At $t = t_1$, $v_x = 0$ and the diode turns on. Solving the preceding equation for t_1,

$$\boxed{t_1 = \frac{V_s C_r}{I_o}} \qquad (9\text{-}24)$$

Equation (9-23) can then be expressed as

$$v_x(t) = V_s \left(1 - \frac{t}{t_1}\right) \qquad (9\text{-}25)$$

Analysis for $t_1 \leq t \leq t_2$ Diode D_1 is forward-biased and has 0 V across it, and the equivalent circuit is shown in Fig. 9-2c. Kirchhoff's voltage law is expressed as

$$L_r \frac{di_L(t)}{dt} + v_C(t) = V_s \qquad (9\text{-}26)$$

Differentiating,

$$L_r \frac{d^2 i_L(t)}{dt^2} + \frac{dv_C(t)}{dt} = 0 \qquad (9\text{-}27)$$

Capacitor current is related to voltage by

$$\frac{dv_C(t)}{dt} = \frac{i_C(t)}{C_r} \qquad (9\text{-}28)$$

Since inductor and capacitor currents are the same in this time interval, Eq. (9-27) can be expressed as

$$\frac{d^2 i_L(t)}{dt^2} + \frac{i_L(t)}{L_r C_r} = 0 \qquad (9\text{-}29)$$

Solving the preceding equation for i_L by using the initial condition $i_L(t_1) = I_o$,

$$i_L(t) = I_o \cos[\omega_0(t - t_1)] \tag{9-30}$$

where

$$\omega_0 = \frac{1}{\sqrt{L_r C_r}} \tag{9-31}$$

Capacitor voltage is expressed as

$$v_C(t) = \frac{1}{C_r}\int_{t_1}^{t} i_C(\lambda)\, d\lambda + v_C(t_1) = \frac{1}{C_r}\int_{t_1}^{t} I_o \cos[\omega_0(\lambda - t_1)]\, d\lambda + V_s$$

which simplifies to

$$v_C(t) = V_s + I_o Z_0 \sin[\omega_0(t - t_1)] \tag{9-32}$$

where

$$Z_0 = \sqrt{\frac{L_r}{C_r}} \tag{9-33}$$

Note that the peak capacitor voltage is

$$V_{C,\text{peak}} = V_s + I_o Z_0 = V_s + I_o \sqrt{\frac{L_r}{C_r}} \tag{9-34}$$

which is also the maximum reverse voltage across diode D_s and is larger than the source voltage.

With diode D_1 forward-biased,

$$v_x = 0 \tag{9-35}$$

The diode D_s across C_r prevents v_C from going negative, so Eq. (9-32) is valid for $v_C > 0$. Solving Eq. (9-32) for the time $t = t_2$ when v_C returns to zero,

$$t_2 = \frac{1}{\omega_0}\left[\sin^{-1}\left(\frac{-V_s}{I_o Z_0}\right)\right] + t_1$$

which can be expressed as

$$\boxed{t_2 = \frac{1}{\omega_0}\left[\sin^{-1}\left(\frac{V_s}{I_o Z_0}\right) + \pi\right] + t_1} \tag{9-36}$$

At $t = t_2$, diode D_s turns on.

Analysis for $t_2 \leq t \leq t_3$ (Fig. 9-2d) After t_2, both diodes are forward-biased (Fig. 9-2d), the voltage across L_r is V_s, and i_L increases linearly until it reaches I_o at t_3. The switch is reclosed just after t_2 when $v_C = 0$ (zero-voltage turn-on) and the diode is on to carry a negative i_L. The current i_L in the interval from t_2 to t_3 is expressed as

$$i_L(t) = \frac{1}{L_r}\int_{t_2}^{t} V_s \, d\lambda + i_L(t_2) = \frac{V_s}{L_r}(t - t_2) + I_o \cos[\omega_0(t_2 - t_1)] \quad (9\text{-}37)$$

where $i_L(t_2)$ is from Eq. (9-30). Current at t_3 is I_o:

$$i_L(t_3) = I_o = \frac{V_s}{L_r}(t_3 - t_2) + I_o \cos[\omega_0(t_2 - t_1)] \quad (9\text{-}38)$$

Solving for t_3,

$$t_3 = \left(\frac{L_r I_o}{V_s}\right)\{1 - \cos[\omega_0(t_2 - t_1)]\} + t_2 \quad (9\text{-}39)$$

Voltage v_x is zero in this interval:

$$v_x = 0 \quad (9\text{-}40)$$

At $t = t_3$, diode D_1 turns on.

Analysis for $t_3 \leq t \leq T$ In this interval, the switch is closed, both diodes are off, the current in the switch is I_o, and

$$v_x = V_s \quad (9\text{-}41)$$

The circuit remains in this condition until the switch is reopened. The time interval $T - t_3$ is determined by the switching frequency of the circuit. All other time intervals are determined by other circuit parameters.

Output Voltage

The voltage $v_x(t)$ at the input of the output filter is shown in Fig. 9-2f. Summarizing Eqs. (9-25), (9-35), (9-40), and (9-41),

$$v_x(t) = \begin{cases} V_s\left(1 - \dfrac{t}{t_1}\right) & 0 < t < t_1 \\ 0 & t_1 < t < t_3 \\ V_s & t_3 < t < T \end{cases} \quad (9\text{-}42)$$

The output voltage is the average of $v_x(t)$. Output voltage is

$$V_o = \frac{1}{T}\int_0^T v_x\, dt = \frac{1}{T}\left[\int_0^{t_1} V_s\left(1 - \frac{t}{t_1}\right) dt + \int_{t_3}^T V_s\, dt\right] \quad (9\text{-}43)$$

$$= \frac{V_s}{T}\left[\frac{t_1}{2} + (T - t_3)\right]$$

Using $f_s = 1/T$,

$$\boxed{V_o = V_s\left[1 - f_s\left(t_3 - \frac{t_1}{2}\right)\right]} \quad (9\text{-}44)$$

Times t_1, and t_3 in the preceding equation are determined from the circuit parameters as described by Eqs. (9-24), (9-36), and (9-39). *The output voltage is controlled by changing the switching frequency.* The time interval when the switch is open is fixed, and the time interval when the switch is closed is varied. Times t_1 and t_3 are determined in part by the load current I_o, so output voltage is a function of load. Increasing the switching frequency decreases the time interval $T - t_3$ and thus reduces the output voltage. Normalized output voltage vs. switching frequency with the parameter $r = R_L/Z_0$ is shown in the graph in Fig. 9-2g. Output voltage is less than input voltage, as was the case for the buck converter in Chap. 6.

EXAMPLE 9-2

Resonant Switch Converter: Zero-Voltage Switching

In the circuit of Fig. 9-2a,

$$V_s = 20 \text{ V}$$
$$L_r = 1 \text{ } \mu\text{H}$$
$$C_r = 0.047 \text{ } \mu\text{F}$$
$$I_o = 5 \text{ A}$$

(a) Determine the switching frequency such that the output voltage is 10 V.
(b) Determine the peak voltage across D_s when it is reverse-biased.

■ **Solution**
(a) From the circuit parameters,

$$\omega_0 = \frac{1}{\sqrt{(10)^{-6}(0.047)(10)^{-6}}} = 4.61(10^6) \text{ rad/s}$$

$$Z_0 = \sqrt{\frac{L_r}{C_r}} = \sqrt{\frac{10^{-6}}{0.047(10^{-6})}} = 4.61 \text{ } \Omega$$

Using Eq. (9-24) to solve for t_1,

$$t_1 = \frac{V_s C_r}{I_o} = \frac{(20)(0.047)(10^{-6})}{5} = 0.188 \ \mu s$$

From Eq. (9-36),

$$t_2 = \frac{1}{\omega_0}\left[\sin^{-1}\left(\frac{V_s}{I_o\sqrt{L_r/C_r}}\right) + \pi\right] + t_1$$

$$= \frac{1}{4.61(10^6)}\left[\sin^{-1}\frac{20}{(5)(4.61)} + \pi\right] + 0.188 \ \mu s = 1.10 \ \mu s$$

From Eq. (9-39),

$$t_3 = \left(\frac{L_r I_o}{V_s}\right)\{1 - \cos[\omega_0(t_2 - t_1)]\} + t_2$$

$$= \left(\frac{10^{-6}(5)}{20}\right)\{1 - \cos[(4.61)(10^6)(1.10 - 0.188)(10^{-6})]\} + 1.10 \ \mu s = 1.47 \ \mu s$$

Equation (9-44) is used to determine the proper switching frequency,

$$V_o = V_s\left[1 - f_s\left(t_3 - \frac{t_1}{2}\right)\right]$$

$$10 = 20\left[1 - f_s\left(1.47 - \frac{0.188}{2}\right)(10^{-6})\right]$$

$$f_s = 363 \text{ kHz}$$

(b) Peak reverse voltage across D_s is the same as peak capacitor voltage. From Eq. (9-25),

$$V_{D_s,\text{peak}} = V_{C,\text{peak}} = V_o + I_o\sqrt{\frac{L_r}{C_r}} = 20 + (5)(4.61) = 33 \text{ V}$$

9.4 THE SERIES RESONANT INVERTER

The series resonant inverter (dc-to-ac converter) of Fig. 9-3a is one application of resonant converters. In a series resonant inverter, an inductor and a capacitor are placed in series with a load resistor. The switches produce a square wave voltage, and the inductor-capacitor combination is selected such that the resonant frequency is the same as the switching frequency.

402 **CHAPTER 9** Resonant Converters

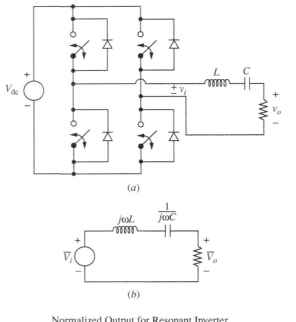

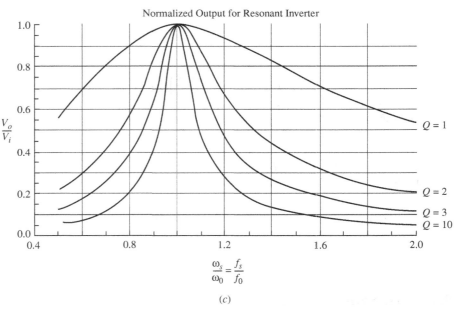

Figure 9-3 (a) A series resonant inverter; (b) Phasor equivalent of a series *RLC* Circuit; (c) Normalized frequency response.

The analysis begins by considering the frequency response of the *RLC* circuit of Fig. 9-3*b*. The input and output voltage amplitudes are related by

$$\frac{V_o}{V_i} = \frac{R}{\sqrt{R^2 + (\omega L - (1/\omega C))^2}} = \frac{1}{\sqrt{1 + ((\omega L/R) - (1/\omega RC))^2}} \quad (9\text{-}45)$$

Resonance is at the frequency

$$\omega_0 = \frac{1}{\sqrt{LC}} \quad (9\text{-}46)$$

or

$$f_0 = \frac{1}{2\pi\sqrt{LC}} \quad (9\text{-}47)$$

At resonance, the impedances of the inductance and capacitance cancel, and the load appears as a resistance. If the bridge output is a square wave at frequency f_0, the *LC* combination acts as a filter, passing the fundamental frequency and attenuating the harmonics. If the third and higher harmonics of the square wave bridge output are effectively removed, the voltage across the load resistor is essentially a sinusoid at the square wave's fundamental frequency.

The amplitude of the fundamental frequency of a square wave voltage of $\pm V_{dc}$ is

$$V_1 = \frac{4V_{dc}}{\pi} \quad (9\text{-}48)$$

The frequency response of the filter could be expressed in terms of bandwidth, which is also characterized by the quality factor Q.

$$Q = \frac{\omega_0 L}{R} = \frac{1}{\omega_0 RC} \quad (9\text{-}49)$$

Equation (9-45) can be expressed in terms of ω_0 and Q:

$$\frac{V_o}{V_i} = \frac{1}{\sqrt{1 + Q^2((\omega/\omega_0) - (\omega_0/\omega))^2}} \quad (9\text{-}50)$$

The normalized frequency response with Q as a parameter is shown in Fig. 9-3*c*. The total harmonic distortion (THD, as defined in Chap. 2) of the voltage across the load resistor is reduced by increasing the Q of the filter. Increasing inductance and reducing capacitance increase Q.

Switching Losses

An important feature of the resonant inverter is that switch losses are reduced over that of the inverters discussed in Chap. 8. If switching is at the resonant frequency and the Q of the circuit is high, the switches operate when the load current is at or near zero. This is significant because the power absorbed by the switches is less than in the nonresonant inverter.

Amplitude Control

If the frequency of the load voltage is not critical, the amplitude of the fundamental frequency across the load resistor can be controlled by shifting the switching frequency off of resonance. Power absorbed by the load resistor is thus controlled by the switching frequency. Induction heating is an application.

The switching frequency should be shifted higher than resonance rather than lower when controlling the output. Higher switching frequencies moves the harmonics of the square wave higher, increasing the filter's effectiveness in removing them. Conversely, shifting the frequency lower than resonance moves the harmonics, particularly the third harmonic, closer to resonance and increases their amplitudes in the output.

EXAMPLE 9-3

A Resonant Inverter

A 10-Ω resistive load requires a 1000-Hz, 50-V rms sinusoidal voltage. The THD of the load voltage must be no more than 5 percent. An adjustable dc source is available. (*a*) Design an inverter for this application. (*b*) Determine the maximum voltage across the capacitor. (*c*) Verify the design with a PSpice simulation.

■ Solution

(*a*) The full-bridge converter of Fig. 9-3*a* with 1000-Hz square wave switching and series resonant *LC* filter is selected for this design. The amplitude of a 50-V rms sinusoidal voltage is $\sqrt{2}(50) = 70.7$ V. The required dc input voltage is determined from Eq. (9-48).

$$70.7 = \frac{4V_{dc}}{\pi}$$

$$V_{dc} = 55.5 \text{ V}$$

The resonant frequency of the filter must be 1000 Hz, establishing the *LC* product. The *Q* of the filter and the THD limit are used to determine the values of *L* and *C*. The third harmonic of the square wave is the largest and will be the least attenuated by the filter. Estimating the THD from the third harmonic,

$$\text{THD} = \frac{\sqrt{\sum_{n \neq 1} V_n^2}}{V_1} \approx \frac{V_3}{V_1} \quad (9\text{-}51)$$

where V_1 and V_3 are the amplitudes of the fundamental and third harmonic frequencies, respectively, across the load. Using the foregoing approximation, the amplitude of the third harmonic of the load voltage must be at most

$$V_3 < (\text{THD})(V_1) = (0.05)(70.7) = 3.54 \text{ V}$$

9.4 The Series Resonant Inverter

For the square wave, $V_3 = V_1/3 = 70.7/3$. Using Eq. (9-50), Q is determined from the magnitude of the third harmonic output with the third harmonic input, 70.7/3, at $\omega = 3\omega_0$.

$$\frac{V_{o,3}}{V_{i,3}} = \frac{3.54}{70.7/3} = \sqrt{\frac{1}{1 + Q^2((3\omega_0/\omega_0) - (\omega_0/3\omega_0))^2}}$$

Solving the preceding equation for Q results in $Q = 2.47$. Using Eq. (9-49),

$$L = \frac{QR}{\omega_0} = \frac{(2.47)(10)}{2\pi(1000)} = 3.93 \text{ mH}$$

$$C = \frac{1}{Q\omega_0 R} = \frac{1}{(2.47)(2\pi)(1000)(10)} = 6.44 \text{ }\mu\text{F}$$

Power delivered to the load resistor at the fundamental frequency is $V_{rms}^2/R = 50^2/10 = 250$ W. Power delivered to the load at the third harmonic is $(2.5^2)/10 = 0.63$ W, showing that power at the harmonic frequencies is negligible.

(b) Voltage across the capacitor is estimated from phasor analysis at the fundamental frequency:

$$V_C = \left|\frac{I}{j\omega_0 C}\right| = \frac{V_1/R}{\omega_0 C} = \frac{70.7/10}{(2\pi)(1000)(6.44)(10^{-6})} = 175 \text{ V}$$

At resonance, the inductor has the same impedance magnitude as the capacitor, so its voltage is also 175 V. The inductor and capacitor voltages would be larger if Q were increased. Note that these voltages are larger than the output or source voltage.

(c) One method of doing a PSpice simulation is to use a square wave voltage as the input to the RLC circuit. This assumes that the switching is ideal, but it is a good starting point to verify that the design meets the specifications. The circuit is shown in Fig. 9-4a.

Output begins after three periods (3 ms) to allow steady-state conditions to be reached. The Probe output showing input and output voltages is seen in Fig. 9-4b, and a Fourier analysis (FFT) from Probe is shown in Fig. 9-4c. The amplitudes of the fundamental frequency and third harmonic are as predicted in part (a). The Fourier analysis for the output voltage is as follows:

```
FOURIER COMPONENTS OF TRANSIENT RESPONSE V(OUTPUT)
  DC COMPONENT = -2.770561E-02

HARMONIC   FREQUENCY    FOURIER      NORMALIZED    PHASE       NORMALIZED
  NO         (HZ)       COMPONENT    COMPONENT     (DEG)       PHASE (DEG)

  1        1.000E+03    7.056E+01    1.000E+00     1.079E-01    0.000E+00
  2        2.000E+03    3.404E-02    4.825E-04     3.771E+01    3.749E+01
  3        3.000E+03    3.528E+00    5.000E-02    -8.113E+01   -8.145E+01
  4        4.000E+03    1.134E-02    1.608E-04    -5.983E+00   -6.414E+00
  5        5.000E+03    1.186E+00    1.681E-02    -8.480E+01   -8.533E+01
  6        6.000E+03    8.246E-03    1.169E-04    -2.894E+01   -2.959E+01
  7        7.000E+03    5.943E-01    8.423E-03    -8.609E+01   -8.684E+01
  8        8.000E+03    7.232E-03    1.025E-04    -4.302E+01   -4.388E+01
  9        9.000E+03    3.572E-01    5.062E-03    -8.671E+01   -8.768E+01

TOTAL HARMONIC DISTORTION =  5.365782E+00 PERCENT
```

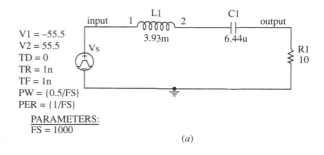

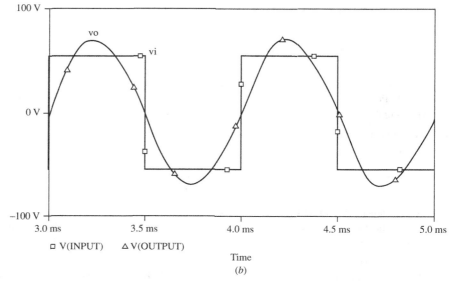

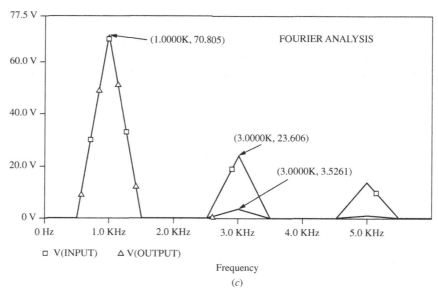

Figure 9-4 (*a*) PSpice circuit for Example 9-3; (*b*) Input and output voltages; (*c*) Fourier analysis.

The output file shows that the THD is 5.37 percent, slightly larger than the 5 percent specification. Frequencies larger than the third harmonic were neglected in the design and have a small effect on the THD. A slight increase in L and corresponding decrease in C would increase the Q of the circuit and reduce the THD to compensate for the approximation. Note that switching occurs when the current is close to zero.

9.5 THE SERIES RESONANT DC-DC CONVERTER

Basic Operation

The upper switching frequency limit on dc-dc converters in Chaps. 6 and 7 is largely due to the switching losses, which increase with frequency. A method for using resonance to reduce the switching losses in dc-dc converters is to start with a resonant inverter to produce an ac signal and then rectify the output to obtain a dc voltage. Figure 9-5a shows a half-bridge inverter with a full-wave rectifier and a capacitor output filter across the load resistor R_L. The two capacitors on the input are large and serve to split the voltage of the source. The input capacitors are not part of the resonant circuit. The basic operation of the circuit is to use the switches to produce a square wave voltage for v_a. The series combination of L_r and C_r forms a filter for the current i_L. The current i_L oscillates and is rectified and filtered to produce a dc voltage output. Converter operation is dependent on the relationship between the switching frequency and the resonant frequency of the filter.

Operation for $\omega_s > \omega_o$

For the first analysis, assume that the switching frequency ω_s is slightly larger than the resonant frequency ω_o of the series LC combination. If the switching frequency is around the resonant frequency of the LC filter, i_L is approximately sinusoidal with frequency equal to the switching frequency.

Figure 9-5b shows the square wave input voltage v_a, the current i_L, the switch current i_{S_1}, and the input to the rectifier bridge v_b. The current in the switches is turned on at zero voltage to eliminate turn-on losses, but the switches are turned off at nonzero current, so turnoff losses could exist. However, capacitors could be placed across the switches to act as lossless snubbers (see Chap. 10) to prevent turn off losses.

The series resonant dc-dc converter is analyzed by considering the fundamental frequency of the Fourier series for the voltages and currents. The input voltage to the filter v_a is a square wave of $\pm V_s/2$. If the output voltage is assumed to be a constant V_o, then the input voltage to the bridge v_b is V_o when i_L is positive and is $-V_o$ when i_L is negative because of the condition of the rectifier diodes for each of these cases. The amplitudes of the fundamental frequencies of the square waves v_a and v_b are

$$V_{a_1} = \frac{4(V_s/2)}{\pi} = \frac{2V_s}{\pi} \tag{9-52}$$

$$V_{b_1} = \frac{4V_o}{\pi} \tag{9-53}$$

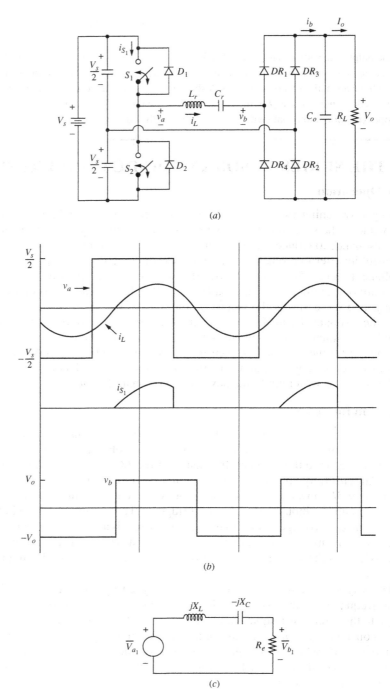

Figure 9-5 (*a*) A series resonant dc-dc converter using a half-bridge inverter; (*b*) Voltage and current waveforms for $\omega_s < \omega_o$; (*c*) Equivalent ac circuit for series resonant dc-dc converter; (*d*) Normalized frequency response.

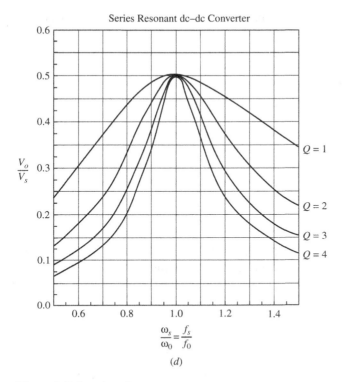

Figure 9-5 (*continued*)

The current at the output of the bridge i_b is the full-wave-rectified form of i_L. The average value of i_b is output current I_o. If i_L is approximated as a sine wave of amplitude I_{L_1}, the average value of i_b is

$$I_b = I_o = \frac{2I_{L_1}}{\pi} \tag{9-54}$$

The relationship between input and output voltages is approximated from ac circuit analysis using the fundamental frequencies of the voltage and current waveforms. Figure 9-5c shows the equivalent ac circuit. The input voltage is the fundamental of the input square wave, and the impedances are ac impedances using ω_s of the input voltage. The value of output resistance in this equivalent circuit is based on the ratio of voltage to current at the output. Using Eqs. (9-53) and (9-54),

$$R_e = \frac{V_{b_1}}{I_{L_1}} = \frac{(4V_o/\pi)}{(\pi I_o/2)} = \left(\frac{8}{\pi^2}\right)\left(\frac{V_o}{I_o}\right) = \left(\frac{8}{\pi^2}\right)(R_L) \tag{9-55}$$

The ratio of output to input voltage is determined from phasor analysis of Fig. 9-5c,

$$\frac{V_{b_1}}{V_{a_1}} = \frac{4V_o/\pi}{2V_s/\pi} = \left|\frac{R_e}{R_e + j(X_L - X_C)}\right| \quad (9\text{-}56)$$

or

$$V_o = \frac{V_s}{2}\left(\frac{1}{\sqrt{1 + [(X_L - X_C)/R_e]^2}}\right) \quad (9\text{-}57)$$

where the reactances X_L and X_C are

$$X_L = \omega_s L_r \quad (9\text{-}58)$$

$$X_C = \frac{1}{\omega_s C_r} \quad (9\text{-}59)$$

The reactances X_L and X_C depend on the switching frequency ω_s. Therefore, the output voltage can be controlled by changing the switching frequency of the converter. The sensitivity of the output to the switching frequency depends on the values of L_r and C_r. If Q is defined as

$$Q = \frac{\omega_0 L_r}{R_L} \quad (9\text{-}60)$$

V_o/V_s is plotted with Q as the parameter in Fig. 9-5d. The curves are more accurate above resonance because i_L has more of a sinusoidal quality for these frequencies. Recall that the curves are based on the approximation that the current is sinusoidal despite the square wave voltage excitation, and the results will be inexact.

EXAMPLE 9-4

Series Resonant DC-DC Converter

For the dc-dc converter of Fig. 9-5a,

$$V_s = 100 \text{ V}$$
$$L_r = 30 \text{ }\mu\text{H}$$
$$C_r = 0.08 \text{ }\mu\text{F}$$
$$R_L = 10 \text{ }\Omega$$
$$f_s = 120 \text{ kHz}$$

Determine the output voltage of the converter. Verify the result with a PSpice simulation.

■ **Solution**

The resonant frequency of the filter is

$$f_0 = \frac{1}{2\pi\sqrt{L_r C_r}} = \frac{1}{2\pi\sqrt{30(10^{-6})(0.08)(10^{-6})}} = 102.7 \text{ kHz}$$

Switching frequency is higher than resonance, and the equivalent circuit of Fig. 9-5c is used to determine the output voltage. From Eq. (9-55), the equivalent resistance is

$$R_e = \frac{8}{\pi^2}(R_L) = \frac{8}{\pi^2}(10) = 8.11\ \Omega$$

The inductive and capacitive reactances are

$$X_L = \omega_s L_r = 2\pi(120{,}000)(30)(10^{-6}) = 22.6\ \Omega$$
$$X_C = \frac{1}{\omega_s C_r} = \frac{1}{2\pi(120{,}000)(0.08)(10^{-6})} = 16.6\ \Omega$$

Using Eq. (9-57), the output voltage is

$$V_o = \frac{V_s}{2}\left(\frac{1}{\sqrt{1+[(X_L-X_C)/R_e]^2}}\right) = \frac{100}{2}\left(\frac{1}{\sqrt{1+[(22.6-16.6)/8.11]^2}}\right) = 40.1\ \text{V}$$

The output could also be approximated from the graph of Fig. 9-5d. The value of Q from Eq. (9-60) is

$$Q = \frac{\omega_0 L_r}{R_L} = \frac{2\pi(102.7)(10)^3\,30(10^{-6})}{10} = 1.94$$

Normalized switching frequency is

$$\frac{f_s}{f_0} = \frac{120\ \text{kHz}}{102.7\ \text{kHz}} = 1.17$$

Normalized output is obtained from Fig. 9-5d as approximately 0.4, making the output voltage $(0.4)(100\ \text{V}) = 40\ \text{V}$.

Simulation for this circuit could include various levels of detail. The simplest assumes that switching takes place properly, and a square wave exists at the input to the filter as shown in Fig. 9-6a. The source is then modeled as a square wave of $\pm V_s/2$ without including any details of the switches, as was done in Example 9-3. The small capacitors across the diodes aid in convergence in the transient analysis.

Figure 9-6 shows the current in L_r and the output voltage. Note that the current is not quite sinusoidal and that the output is approximately 40 V and contains some ripple. The simulation verifies the foregoing analytic solution. Note that the results of the simulation are very sensitive to the simulation parameters, include the step size of the transient analysis. A step size of 0.1 μs was used here. The diodes are made ideal by setting $n = 0.001$ in the PSpice diode model.

EXAMPLE 9-5

Series Resonant DC-DC Converter

For the series resonant dc-dc converter of Fig. 9-5a, the dc source voltage is 75 V. The desired output voltage is 25 V, and the desired switching frequency is 100 kHz. The load resistance R_L is 10 Ω. Determine L_r and C_r.

SERIES RESONANT DC–DC CONVERTER

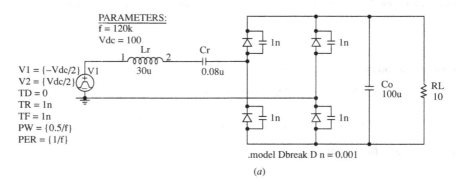

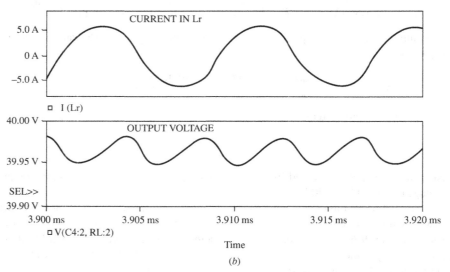

Figure 9-6 (*a*) PSpice circuit for the series resonant dc-dc converter with the source and switches replaced with a square wave. The small capacitors across the diodes aid convergence; (*b*) Probe output.

■ **Solution**
Select the resonant frequency ω_0 to be slightly less than the desired switching frequency ω_s. Let $\omega_s/\omega_0 = 1.2$,

$$\omega_0 = \frac{\omega_s}{1.2} = \frac{2\pi f_s}{1.2} = \frac{2\pi 10^5}{1.2} = 524(10^3) \text{ rad/s}$$

From the graph of Fig. 9-5*d* with $V_o/V_s = 25/75 = 0.33$ and $\omega_s/\omega_0 = 1.2$, the required Q is approximately 2.5. From Eq. (9-60),

$$L_r = \frac{QR_L}{\omega_0} = \frac{(2.5)(10)}{524(10^3)} = 47.7 \text{ }\mu\text{H}$$

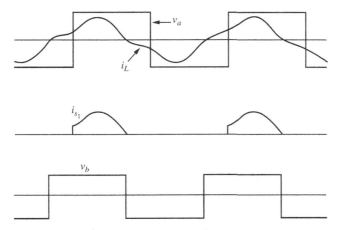

Figure 9-7 Voltage and current waveforms for the series resonant dc-dc converter, $\omega_0/2 < \omega_s < \omega_0$.

and

$$\omega_0 = \frac{1}{\sqrt{L_r C_r}} \Rightarrow C_r = \frac{1}{\omega_0^2 L_r} = \frac{1}{(524)(10^3)(47.7)(10^{-6})} = 0.0764 \ \mu\text{F}$$

Operation for $\omega_0/2 < \omega_s < \omega_0$

The series resonant dc-dc converter that has a switching frequency less than resonance but greater than $\omega_0/2$ has the current waveform for i_L as shown in Fig. 9-7. The switches turn on with positive voltage and current, resulting in turn-on switching losses. The switches turn off at zero current, resulting in no turnoff losses. Furthermore, because the switches turn off at zero current, thyristors could be used if the switching frequency is low. Analysis is done using the same technique as for $\omega_s > \omega_0$, but the harmonic content of the current waveform is now higher, and the sinusoidal approximation is not as accurate.

Operation for $\omega_s < \omega_0/2$

With this switching frequency, the current in the series LC circuit is shown in Fig. 9-8. When S_1 in Fig. 9-5a is turned on, i_L becomes positive and oscillates at frequency ω_0. When the current reaches zero at t_1 and becomes negative, diode D_1 carries the negative current. When the current again reaches zero at t_2, S_1 is off, and the current remains at zero until S_2 turns on at $T/2$. The current waveform for the second half-period is the negative of that of the first.

Switches turn on and off at zero current, resulting in zero switching losses. Since the switches turn off at zero current, thyristors could be used in low-frequency applications.

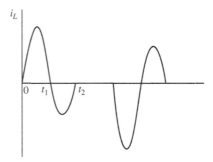

Figure 9-8 Current waveform for the series resonant dc-dc converter, $\omega_s < \omega_0$.

Current in the *LC* series combination is discontinuous for this mode of operation. In the two previously described modes of operation, the current is continuous. Since the average of the rectified inductor current must be the same as the load current, the current in the *LC* branch will have a large peak value.

PSpice simulation for discontinuous current must include unidirectional switch models because the voltage at the input to the circuit is not a square wave.

Variations on the Series Resonant DC-DC Converter

The series resonant dc-dc converter can be implemented using variations on the basic topology in Fig. 9-5a. The capacitor C_r can be incorporated into the voltage-divider capacitors in the half bridge, each being $C_r/2$. An isolation transformer can be included as part of the full-wave rectifier on the output. Figure 9-9 shows an alternate implementation of the series resonant dc-dc converter.

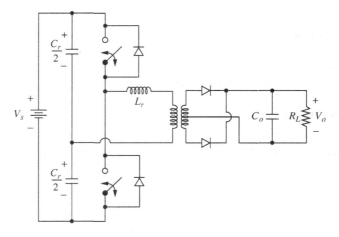

Figure 9-9 An alternate implementation of the series resonant dc-dc converter.

9.6 THE PARALLEL RESONANT DC-DC CONVERTER

The converter in Fig. 9-10a is a parallel dc-dc converter. The capacitor C_r is placed in parallel with the rectifier bridge rather than in series. An output filter inductor L_o produces essentially a constant current from the bridge output to

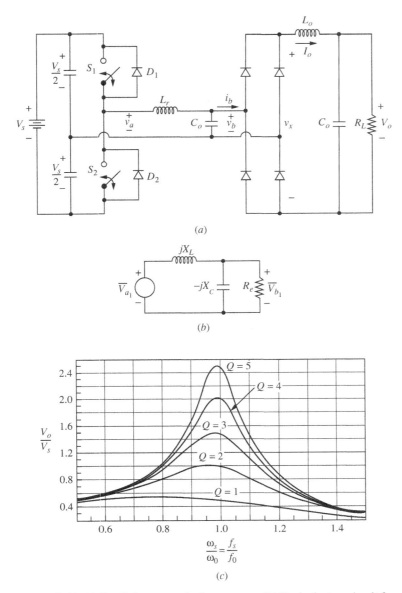

Figure 9-10 (a) Parallel resonant dc-dc converter; (b) Equivalent ac circuit for parallel resonant dc-dc converter; (c) Normalized frequency response.

the load. The switching action causes the voltage across the capacitor and bridge input to oscillate. When the capacitor voltage is positive, rectifier diodes DR_1 and DR_2 are forward-biased and carry current I_o. When the capacitor voltage is negative, DR_3 and DR_4 are forward-biased and carry current I_o. The current i_b at the input to the bridge is therefore a square wave current of $\pm I_o$. The bridge output voltage is the full-wave rectified waveform of voltage v_b. The average voltage across the output inductor L_o is zero, so the output voltage is the average of rectified v_b.

The parallel dc-dc converter can be analyzed by assuming that the voltage across the capacitor C_r is sinusoidal, taking only the fundamental frequencies of the square wave voltage input and square wave current into the bridge. The equivalent ac circuit is shown in Fig. 9-10b. The equivalent resistance for this circuit is the ratio of capacitor voltage to the fundamental frequency of the square wave current. Assuming that the capacitor voltage is sinusoidal, the average of the rectified sine wave at the bridge output (v_x) is the same as V_o,

$$V_o = V_x = \frac{2V_{x_1}}{\pi} = \frac{2V_{b_1}}{\pi} \tag{9-61}$$

where V_{b_1} is the amplitude of the fundamental frequency of v_b. The equivalent resistance is then

$$R_e = \frac{V_{b_1}}{I_{b_1}} = \frac{V_o \pi/2}{4I_o/\pi} = \frac{\pi^2}{8}\left(\frac{V_o}{I_o}\right) = \frac{\pi^2}{8} R_L \tag{9-62}$$

where I_{b_1} is the amplitude of the fundamental frequency of the square wave current i_b.

Solving for output voltage in the phasor circuit of Fig. 9-10b,

$$\frac{V_{b_1}}{V_{a_1}} = \left|\frac{1}{1 - (X_L/X_C) + j(X_L/R_e)}\right| \tag{9-63}$$

Since V_o is the average of the full-wave rectified value of v_b,

$$V_{b_1} = \frac{V_o \pi}{2} \tag{9-64}$$

V_{a_1} is the amplitude of the fundamental frequency of the input square wave:

$$V_{a_1} = \frac{4(V_s/2)}{\pi} \tag{9-65}$$

Combining Eqs. (9-64) and (9-65) with Eq. (9-63), the relationship between output and input of the converter is

$$\frac{V_o}{V_s} = \frac{4}{\pi^2}\left|\frac{1}{1 - (X_L/X_C) + j(X_L/R_e)}\right| \tag{9-66}$$

or

$$V_o = \frac{4V_s}{\pi^2\sqrt{[1 - (X_L/X_C)]^2 + (X_L/R_e)^2}} \quad (9\text{-}67)$$

V_o/V_s is plotted with Q as a parameter in Fig. 9-10c, where Q is defined as

$$Q = \frac{R_L}{\omega_0 L_r} \quad (9\text{-}68)$$

and

$$\omega_0 = \frac{1}{\sqrt{L_r C_r}} \quad (9\text{-}69)$$

The curves are more accurate for switching frequencies larger than ω_0 because of the sine-like quality of the capacitor voltage for these frequencies. Note that the output can be larger than the input for the parallel resonant dc-dc converter, but the output is limited to $V_s/2$ for the series resonant dc-dc converter.

EXAMPLE 9-6

Parallel Resonant DC-DC Converter

The circuit of Fig. 9-10a has the following parameters:

$$V_s = 100 \text{ V}$$
$$L_r = 8 \text{ } \mu\text{H}$$
$$C_r = 0.32 \text{ } \mu\text{F}$$
$$R_L = 10 \text{ } \Omega$$
$$f_s = 120 \text{ kHz}$$

Determine the output voltage of the converter. Assume the output filter components L_o and C_o produce a ripple-free output current and voltage.

■ **Solution**
From the parameters given,

$$\omega_0 = \frac{1}{\sqrt{L_r C_r}} = \frac{1}{\sqrt{8(10^{-6})0.32(10^{-6})}} = 625 \text{ krad/s}$$

$$Q = \frac{R_L}{\omega_0 L_r} = \frac{10}{625(10^3)8(10^{-6})} = 2.0$$

$$\frac{\omega_s}{\omega_0} = \frac{2\pi(120 \text{ k})}{625 \text{ k}} = 1.21$$

The normalized output can be estimated from the graph in Fig. 9-10c as 0.6, making the output approximately 60 V. The output voltage can also be obtained from Eq. (9-67). The reactances are

$$X_L = \omega_s L_r = 2\pi(120)(10^3)8(10^{-6}) = 6.03 \ \Omega$$

$$X_C = \frac{1}{\omega_s C_r} = \frac{1}{2\pi(120)(10^3)0.32(10^{-6})} = 4.14 \ \Omega$$

The equivalent resistance is

$$R_e = \frac{\pi^2}{8} R_L = \frac{\pi^2}{8}(10) = 12.3 \ \Omega$$

Equation (9-67) for output voltage becomes

$$V_o = \frac{(4)(100)}{\pi^2 \sqrt{[1-(6.03/4.14)]^2 + (6.03/12.3)^2}} = 60.7 \ \text{V}$$

9.7 THE SERIES-PARALLEL DC-DC CONVERTER

The series-parallel dc-dc converter of Fig. 9-11a has both a series and a parallel capacitor. The analysis is similar to the parallel converter discussed previously. The switches produce a square wave voltage v_a, and the voltage v_b at the input to the rectifier is ideally a sinusoid at the fundamental frequency of the input square wave. The output inductor L_o is assumed to produce a ripple-free current, causing the input current i_b to the rectifier bridge to be a square wave.

The relationship between input and output voltages is estimated from ac analysis of the circuit for the fundamental frequency of the square waves. The ac equivalent circuit is shown in Fig. 9-11b. A straightforward phasor analysis of Fig. 9-11b gives

$$\frac{V_{b_1}}{V_{a_1}} = \left| \frac{1}{1 + (X_{C_s}/X_{C_p}) - (X_L/X_{C_p}) + j(X_L/R_e - X_{C_s}/R_e)} \right| \quad (9\text{-}70)$$

where R_e is the same as for the parallel converter,

$$R_e = \frac{\pi^2}{8} R_L \quad (9\text{-}71)$$

9.7 The Series-Parallel DC-DC Converter

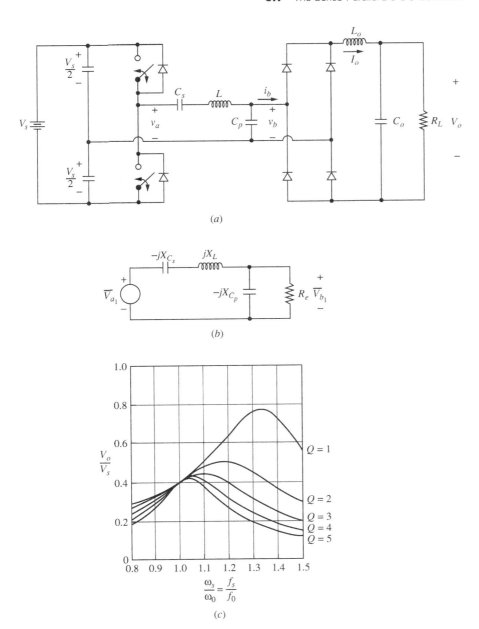

Figure 9-11 (*a*) Series-parallel resonant dc-dc converter; (*b*) Equivalent ac circuit for the series-parallel resonant dc-dc converter; (*c*) Normalized frequency response for output voltage.

and the reactances at the switching frequency are

$$X_{C_s} = \frac{1}{\omega_s C_s}$$
$$X_{C_p} = \frac{1}{\omega_s C_p} \qquad (9\text{-}72)$$
$$X_L = \omega_s L$$

Also V_{a_1} and V_{b_1} are the amplitudes of the fundamental frequencies of the waveforms at v_a and v_b. Using Eqs. (9-64) and (9-65), the relationship between input and output of the converter is

$$\frac{V_o}{V_s} = \frac{4}{\pi^2} \left| \frac{1}{1 + (X_{C_s}/X_{C_p}) - (X_L/X_{C_p}) + j(X_L/R_e - X_{C_s}/R_e)} \right| \qquad (9\text{-}73)$$

Rewriting the preceding equation in terms of ω_s,

$$\frac{V_o}{V_s} = \frac{4}{\pi^2 \sqrt{\left(1 + \dfrac{C_p}{C_s} - \omega_s^2 L C_p\right)^2 + \left(\dfrac{\omega_s L}{R_e} - \dfrac{1}{\omega_s R_e C_s}\right)^2}} \qquad (9\text{-}74)$$

Equation (9-74) for $C_s = C_p$ is plotted with Q as a parameter in Fig. 9-11c where Q is defined as

$$Q = \frac{\omega_0 L}{R_L} \qquad (9\text{-}75)$$

where

$$\omega_0 = \frac{1}{\sqrt{LC_s}} \qquad (9\text{-}76)$$

These curves are more accurate above ω_0 than below because the harmonics of the square wave are more adequately filtered, resulting in the ac analysis being more representative of the actual situation.

The series capacitor C_s can be incorporated into the voltage-divider capacitors, each equal to $C_s/2$, for the half-bridge circuit as was shown in Fig. 9-9 for the series resonant dc-dc converter.

EXAMPLE 9-7

Series-Parallel Resonant DC-DC Converter

The series-parallel resonant dc-dc converter of Fig. 9-11a has the following parameters:

$$V_s = 100 \text{ V}$$
$$C_p = C_s = 0.1 \text{ μF}$$
$$L = 100 \text{ μH}$$
$$R_L = 10 \text{ Ω}$$
$$f_s = 60 \text{ kHz}$$

The output filter components L_o and C_o are assumed to produce a ripple-free output. Determine the output voltage of the converter.

■ Solution

The resonant frequency ω_0 is determined from Eq. (9-76) as

$$\omega_0 = \frac{1}{\sqrt{LC_s}} = \frac{1}{\sqrt{(100)(10^{-6})(0.1)(10^{-6})}} = 316 \text{ krad/s}$$

$$f_0 = \frac{\omega_0}{2\pi} = 50.3 \text{ kHz}$$

The Q of the circuit is determined from Eq. (9-75) as

$$Q = \frac{\omega_0 L}{R_L} = \frac{3.16(10^3)(100)(10^{-6})}{10} = 3.16$$

The normalized switching frequency is

$$\frac{f_s}{f_0} = \frac{60(10^3)}{50.3(10^3)} = 1.19$$

From the graph of Fig. 9-11c, the normalized output is slightly less than 0.4, for an estimated output of $V_o \approx 100(0.4) = 40$ V. Equation (9-74) is evaluated, using $R_e = \pi^2 R_L/8 = 12.34$ Ω,

$$\frac{V_o}{V_s} = 0.377$$

$$V_o = V_s(0.377) = (100)(0.377) = 37.7 \text{ V}$$

9.8 RESONANT CONVERTER COMPARISON

A drawback of the series converter described previously is that the output cannot be regulated for the no-load condition. As R_L goes to infinity, Q in Eq. (9-60) goes to zero. The output voltage is then independent of frequency. However, the parallel converter is able to regulate the output at no load. In Eq. (9-68), for the parallel converter Q becomes larger as the load resistor increases, and the output remains dependent on the switching frequency.

A drawback of the parallel converter is that the current in the resonant components is relatively independent of load. The conduction losses are fixed, and the efficiency of the converter is relatively poor for light loads.

The series-parallel converter combines the advantages of the series and parallel converters. The output is controllable for no load or light load, and the light-load efficiency is relatively high.

9.9 THE RESONANT DC LINK CONVERTER

The circuit of Fig. 9-12a is the basic topology for a switching scheme for an inverter that has zero-voltage switching. The analysis proceeds like that of the resonant switch converters. During the switching interval, the load current is assumed to be essentially constant at I_o. The resistance represents losses in the circuit.

When the switch is closed, the voltage across the RL_r combination is V_s. If the time constant L_r/R is large compared to the time that the switch is closed, the current rises nearly linearly. When the switch is opened, the equivalent circuit is shown in Fig. 9-12b. Kirchhoff's voltage and current laws yield the equations

$$Ri_L(t) + L_r\frac{di_L(t)}{dt} + v_C(t) = V_s \tag{9-77}$$

$$i_C(t) = i_L(t) - I_o \tag{9-78}$$

Differentiating Eq. (9-77),

$$L_r\frac{d^2i_L(t)}{dt^2} + R\frac{di_L(t)}{dt} + \frac{dv_C(t)}{dt} = 0 \tag{9-79}$$

The derivative of the capacitor voltage is related to capacitor current by

$$\frac{dv_C(t)}{dt} = \frac{i_C(t)}{C_r} = \frac{i_L(t) - I_o}{C_r} \tag{9-80}$$

Substituting into Eq. (9-79) and rearranging,

$$\frac{d^2i_L}{dt^2} + \frac{R}{L_r}\frac{di_L(t)}{dt} + \frac{i_L(t)}{L_rC_r} = \frac{I_o}{L_rC_r} \tag{9-81}$$

If the initial conditions for inductor current and capacitor voltage are

$$i_L(0) = I_1 \qquad v_C(0) = 0, \tag{9-82}$$

the solution for current can be shown to be

$$i_L(t) = I_1 + e^{-\alpha t}\left[(I_1 - I_o)\cos(\omega t) + \frac{2V_s - R(I_1 + I_o)}{2\omega L_r}\sin(\omega t)\right] \tag{9-83}$$

9.9 The Resonant DC Link Converter

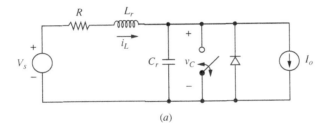

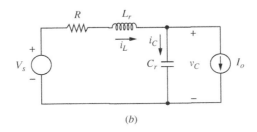

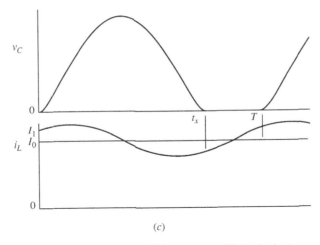

Figure 9-12 (*a*) Resonant dc link converter; (*b*) Equivalent circuit with the switch open and diode off; (*c*) Capacitor voltage and inductor current.

where

$$\alpha = \frac{R}{2L_r} \tag{9-84}$$

$$\omega_0 = \frac{1}{\sqrt{L_r C_r}} \tag{9-85}$$

$$\omega = \sqrt{\omega_0^2 - \alpha^2} \tag{9-86}$$

Capacitor voltage can be shown to be

$$v_C(t) = V_s - I_o R + e^{-\alpha t}\left((I_o R - V_s)\cos(\omega t) + \left\{\frac{R}{2\omega L_r}\left[V_s - \frac{R}{2}(I_1 + I_o)\right]\right.\right.$$

$$\left.\left. + \omega L_r(I_1 - I_o)\right\}\sin(\omega t)\right) \tag{9-87}$$

If the resistance is small, making $R \ll \omega; L_r$, Eqs. (9-83) and (9-87) become

$$i_L(t) \approx I_o + e^{-\alpha t}\left[(I_1 - I_o)\cos(\omega_0 t) + \frac{V_s}{\omega_0 L_r}\sin(\omega_0 t)\right] \tag{9-88}$$

$$v_C(t) \approx V_s + e^{-\alpha t}[-V_s\cos(\omega_0 t) + \omega_0 L_r(I_1 - I_o)\sin(\omega_0 t)] \tag{9-89}$$

When the switch is opened, the inductor current and capacitor voltage oscillate. The switch can be reclosed when the capacitor voltage returns to zero and thereby avoids switching losses. The switch should remain closed until the inductor current reaches some selected value I_1 which is above the load current I_o. This allows the capacitor voltage to return to zero for lossless switching.

An important application of this resonant switching principle is for inverter circuits. The three-phase inverter of Fig. 9-13 can have PWM switching (see Chap. 8) and can include intervals when both switches in one of the three legs are closed to cause the input voltage to the bridge to oscillate. The switches can then turn on or off when the capacitor voltage is zero.

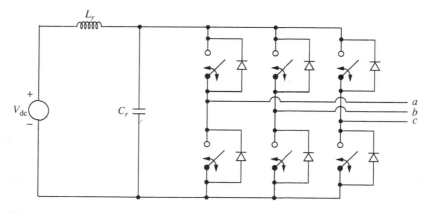

Figure 9-13 Three-phase inverter with a resonant dc link.

9.9 The Resonant DC Link Converter

EXAMPLE 9-8

Resonant DC Link

The single-switch resonant dc link converter of Fig. 9-12a has the parameters

$$V_s = 75 \text{ V}$$
$$L = 100 \text{ μH}$$
$$C = 0.1 \text{ μF}$$
$$R = 1 \text{ Ω}$$
$$I_o = 10 \text{ A}$$
$$I_1 = 12 \text{ A}$$

If the switch is opened at $t = 0$ with $i_L(0) = I_1$ and $v_C(0) = 0$, determine when the switch should be closed so the voltage across it is zero. If the switch is closed immediately after the capacitor voltage becomes zero, how long should the switch remain closed so that the inductor voltage returns to I_1?

■ Solution
From the circuit parameters,

$$\omega_0 = \frac{1}{\sqrt{LC}} = \frac{1}{\sqrt{(10^{-4})(10^{-7})}} = 316 \text{ krad/s}$$

$$\alpha = \frac{R}{2L} = \frac{1}{2(10^{-4})} = 5000$$

$$\omega = \sqrt{\omega_0^2 - \alpha^2} \approx \omega_0$$

$$\omega L_r = 316(10^3)(100)(10^{-6}) = 31.6$$

Since $\alpha \ll \omega_0$, $\omega \approx \omega_0$, and Eqs. (9-88) and (9-89) are good approximations,

$$v_C(t) \approx 75 + e^{-5000t}[-75\cos(\omega_0 t) + 31.6(12-10)\sin(\omega_0 t)]$$
$$= 75 + e^{-5000t}[-75\cos(\omega_0 t) + 63.2\sin(\omega_0 t)]$$

$$i_L(t) \approx 10 + e^{-5000t}\left[(12-10)\cos(\omega_0 t) + \frac{75}{31.6}\sin(\omega_0 t)\right]$$
$$= 10 + e^{-5000t}[2\cos(\omega_0 t) + 2.37\sin(\omega_0 t)]$$

The above equations are graphed in Fig. 9-12c. The time at which the capacitor voltage returns to zero is determined by setting v_C equal to zero and solving for t numerically, resulting in $t_x = 15.5$ μs. Current is evaluated at $t = 15.5$ μs using Eq. (9-88), resulting in $i_L(t = 15.5 \text{ μs}) = 8.07$ A.

If the switch is closed at 15.5 μs, voltage across the inductor is approximately V_s, and the current increases linearly.

$$\Delta i_L = \frac{V_s}{L}\Delta t \qquad (9\text{-}90)$$

The switch must remain closed until i_L is 12 A, requiring a time of

$$\Delta t = \frac{(\Delta i_L)(L)}{V_s} = \frac{(12 - 8.39)(100)(10^{-6})}{75} = 4.81 \ \mu s$$

9.10 Summary

Resonant converters are used to reduce switching losses in various converter topologies. Resonant converters reduce switching losses by taking advantage of voltage or current oscillations. Switches are opened and closed when the voltage or current is at or near zero. The topologies discussed in this chapter are resonant switch inverters; the series resonant inverter; the series, parallel, and series-parallel dc-dc converters; and the resonant dc link converter. Resonant converters are presently a topic of great interest in power electronics because of increased efficiency and the possibility of higher switching frequencies with associated smaller filter components. As was demonstrated in the examples, the voltage stresses on the components may be quite high for resonant converters. The sources in the Bibliography give further details on resonant converters.

9.11 Bibliography

S. Ang and A. Oliva, *Power-Switching Converters*, 2d ed., Taylor & Francis, Boca Raton, Fla., 2005.

S. Basson, and G. Moschopoulos, "Zero-Current-Switching Techniques for Buck-Type AC-DC Converters," *International Telecommunications Energy Conference,* Rome, Italy, pp. 506–513, 2007.

W. Chen, Z. Lu, and S. Ye, "A Comparative Study of the Control Type ZVT PWM Dual Switch Forward Converters: Analysis, Summary and Topology Extensions," *IEEE Applied Power Electronics Conference and Exposition (APEC)*, Washington, D.C., pp. 1404–9, 2009.

T. W. Ching. and K. U. Chan, "Review of Soft-Switching Techniques for High-Frequency Switched-Mode Power Converters," *IEEE Vehicle Power and Propulsion Conference, Austina, Tex.,* 2008.

D. M. Divan, "The Resonant DC Link Converter—A New Concept in Static Power Conversion," *IEEE Transactions. on Industry Applications*, vol. 25, no. 2, March/April 1989, pp. 317–325.

S. Freeland and R. D. Middlebrook, "A Unified Analysis of Converters with Resonant Switches," *IEEE Power Electronics Specialists Conference,* New Orleans, La., 1986, pp. 20–30.

J. Goo, J. A. Sabate, G. Hua, F. and C. Lee, "Zero-Voltage and Zero-Current-Switching Full-Bridge PWM Converter for High-Power Applications," *IEEE Transactions on Industry Applications*, vol. 1, no. 4, July 1996, pp. 622–627.

G. Hua and F. C. Lee, "Soft-Switching Techniques in PWM Converters," *Industrial Electronics Conference Proceedings*, vol. 2, pp. 637–643, 1993.

R. L. Steigerwald, R. W. DeDoncker, and M. H. Kheraluwala, "A Comparison of High-Power Dc-Dc Soft-Switched Converter Topologies," *IEEE Transactions on Industry Applications*, vol. 32, no. 5, September/October 1996, pp. 1139–1145.

Problems

Zero-current Resonant Switch Converter

9-1 In the converter of Fig. 9-1a, $V_s = 10$ V, $I_o = 5$ A, $L_r = 1$ µH, $C_r = 0.3$ µF, and $f_s = 150$ kHz. Determine the output voltage of the converter.

9-2 In the converter of Fig. 9-1a, $V_s = 18$ V, $I_o = 3$ A, $L_r = 0.5$ µH, and $C_r = 0.7$ µF. Determine the maximum switching frequency and the corresponding output voltage. Determine the switching frequency such that the output voltage is 5 V.

9-3 In the converter of Fig. 9-1a, $V_s = 36$ V, $I_o = 5$ A, $L_r = 10$ nH, $C_r = 10$ nF, and $f_s = 750$ kHz. (a) Determine the output voltage of the converter. (b) Determine the maximum inductor current and capacitor voltage. (c) Determine the switching frequency for an output of 12 V.

9-4 In the converter of Fig. 9-1a, $V_s = 50$ V, $I_o = 3$ A, $\omega_0 = 7(10^7)$ rad/s, and $V_o = 36$ V. Determine L_r and C_r such that the maximum current in L_r is 9 A. Determine the required switching frequency.

9-5 In the converter of Fig. 9-1a, $V_s = 100$ V, $L_r = 10$ µH, and $C_r = 0.01$ µF. The load current ranges from 0.5 to 3 A. Determine the range of switching frequency required to regulate the output voltage at 50 V.

9-6 In the converter of Fig. 9-1a, $V_s = 30$ V, $R_L = 5$ Ω, and $f_s = 200$ kHz. Determine values for L_r and C_r such that Z_0 is 2.5 Ω and $V_o = 15$ V.

9-7 Determine a PSpice input file to simulate the circuit of Fig. 9-1a using the parameters in Probl. 9-1. Model the load current as a current source. Use the voltage-controlled switch Sbreak for the switching device. Idealize the circuit by using $R_{on} = 0.001$ Ω in the switch model and using $n = 0.001$ in the Dbreak diode model. (a) Determine the (average) output voltage. (b) Determine the peak voltage across C_r. (c) Determine the peak, average, and rms values of the current in L_r.

Zero-voltage Resonant Switch Converter

9-8 In Example 9-2, determine the required switching frequency to produce an output voltage of 15 V. All other parameters are unchanged.

9-9 In Fig. 9-2a, $V_s = 20$, $L_r = 0.1$ µH, $C_r = 1$ nF, $I_o = 10$ A, and $f_s = 2$ MHz. Determine the output voltage and the maximum capacitor voltage and maximum inductor current.

9-10 In Fig. 9-2a, $V_s = 5$ V, $I_o = 3$ A, $L_r = 1$ µH, and $C_r = 0.01$ µF. (a) Determine the output voltage when $f_s = 500$ kHz. (b) Determine the switching frequency such that the output voltage is 2.5 V.

9-11 In Fig. 9-2a, $V_s = 12$ V, $L_r = 0.5$ µH, $C_r = 0.01$ µF, and $I_o = 10$ A. (a) Determine the output voltage when $f_s = 500$ kHz. (b) The load current I_o is expected to vary between 8 and 15 A. Determine the range of switching frequency necessary to regulate the output voltage at 5 V.

9-12 In Fig. 9-2a, $V_s = 15$ V and $I_o = 4$ A. Determine L_r and C_r such that the maximum capacitor voltage is 40 V and the resonant frequency is $1.6(10^6)$ rad/s. Determine the switching frequency to produce an output voltage of 5 V.

9-13 In Fig. 9-2a, $V_s = 30$ V, $R_L = 5\ \Omega$, and $f_s = 100$ kHz. Determine values for L_r and C_r such that Z_0 is $25\ \Omega$ and $V_o = 15$ V.

9-14 Determine a PSpice circuit to simulate the circuit of Fig. 9-2a using the parameters in Probl. 9-9. Model the load current as a current source. Use the voltage-controlled switch Sbreak for the switching device, and make it unidirectional by adding a series diode. Make the diode ideal by using $n = 0.001$ in the Dbreak model. (a) Determine the (average) output voltage. (b) Determine the peak voltage across C_r. (c) Determine the energy transferred from the source to the load in each switching period.

Resonant Inverter

9-15 The full-bridge resonant inverter of Fig. 9-3a has a 12-Ω resistive load that requires a 400-Hz, 80-V rms sinusoidal voltage. The THD of the load voltage must be no more than 5 percent. Determine the required dc input and suitable values for L and C. Determine the peak voltage across C and the peak current in L.

9-16 The full-bridge resonant inverter of Fig. 9-3a has a 8-Ω resistive load that requires a 1200-Hz, 100-V rms sinusoidal voltage. The THD of the load voltage must be no more than 10 percent. Determine the required dc input and suitable values for L and C. Simulate the inverter in PSpice and determine the THD. Adjust values of L and C if necessary so that the 10 percent THD is strictly satisfied. What is the value of current when switching takes place?

9-17 The full-bridge resonant inverter of Fig. 9-3a is required to supply 500 W to a 15-Ω load resistance. The load requires a 500-Hz ac current which has no more than 10 percent total harmonic distortion. (a) Determine the required dc input voltage. (b) Determine the values of L and C. (c) Estimate the peak voltage across C and peak current in L using the fundamental frequency. (d) Simulate the circuit in PSpice. Determine the THD, peak capacitor voltage, and peak inductor current.

Series Resonant DC-DC Converter

9-18 The series resonant dc-dc converter of Fig. 9-5a has the following operation parameters: $V_s = 10$ V, $L_r = 6\ \mu$H, $C_r = 6$ nF, $f_s = 900$ kHz, and $R_L = 10\ \Omega$. Determine the output voltage V_o.

9-19 The series resonant dc-dc converter of Fig. 9-5a has the following operation parameters: $V_s = 24$ V, $L_r = 1.2\ \mu$H, $C_r = 12$ nF, $f_s = 1.5$ MHz, and $R_L = 5\ \Omega$. Determine the output voltage V_o.

9-20 The series resonant dc-dc converter of Fig. 9-5a has an 18-V dc source and is to have a 6-V output. The load resistance is $5\ \Omega$, and the desired switching frequency is 800 kHz. Select suitable values of L_r and C_r.

9-21 The series resonant dc-dc converter of Fig. 9-5a has a 50-V dc source and is to have an 18-V output. The load resistance is $9\ \Omega$, and the desired switching frequency is 1 MHz. Select suitable values of L_r and C_r.

9-22 The series resonant dc-dc converter of Fig. 9-5a has a 40-V dc source and is to have a 15-V output. The load resistance is $5\ \Omega$, and the desired switching frequency is 800 kHz. Select suitable values of L_r and C_r. Verify your results with a PSpice simulation.

9-23 The series resonant dc-dc converter of Fig. 9-5a has a 150-V dc source and is to have a 55-V output. The load resistance is 20 Ω. Select a switching frequency and suitable values of L_r and C_r. Verify your results with a PSpice simulation.

Parallel Resonant dc-dc Converter

9-24 The parallel resonant dc-dc converter of Fig. 9-10a has the following operation parameters: $V_s = 15$ V, $R_L = 10$ Ω, $L_r = 1.3$ μH, $C_r = 0.12$ μF, and $f_s = 500$ kHz. Determine the output voltage of the converter.

9-25 The parallel resonant dc-dc converter of Fig. 9-10a has the following operation parameters: $V_s = 30$ V, $R_L = 15$ Ω, $L_r = 1.2$ μH, $C_r = 26$ nF, and $f_s = 1$ MHz. Determine the output voltage of the converter.

9-26 The parallel resonant dc-dc converter of Fig. 9-10a has $V_s = 12$ V, $R_L = 15$ Ω, and $f_s = 500$ kHz. The desired output voltage is 20 V. Determine suitable values for L_r and C_r.

9-27 The parallel resonant dc-dc converter of Fig. 9-10a has $V_s = 45$ V, $R_L = 20$ Ω, and $f_s = 900$ kHz. The desired output voltage is 36 V. Determine suitable values for L_r and C_r.

9-28 The parallel resonant dc-dc converter of Fig. 9-10a has a 50-V dc source and is to have a 60-V output. The load resistance is 25 Ω. Select a switching frequency and suitable values of L_r and C_r.

Series-parallel dc-dc Converter

9-29 The series-parallel dc-dc converter of Fig. 9-11a has the following parameters: $V_s = 100$ V, $f_s = 500$ kHz, $R_L = 10$ Ω, $L = 12$ μH, and $C_s = C_p = 12$ nF. Determine the output voltage.

9-30 The series-parallel dc-dc converter of Fig. 9-11a has $V_s = 12$ V, $f_s = 800$ kHz, and $R_L = 2$ Ω. Determine suitable values of L, C_s, and C_p such that the output voltage is 5 V. Use $C_s = C_p$.

9-31 The series-parallel dc-dc converter of Fig. 9-11a has $V_s = 20$ V and $f_s = 750$ kHz. The output voltage is to be 5 V and supply 1 A to a resistive load. Determine suitable values of L, C_s, and C_p. Use $C_s = C_p$.

9-32 The series-parallel dc-dc converter of Fig. 9-11a has $V_s = 25$ V. The output voltage is to be 10 V and supply 1 A to a resistive load. (*a*) Select a switching frequency and determine suitable values of L, C_s, and C_p. (*b*) Verify your results with a PSpice simulation.

Resonant dc Link

9-33 Create a PSpice simulation for the resonant dc link in Example 9-8. Use an ideal diode model. (*a*) Verify the results of Example 9-8. (*b*) Determine the energy supplied by the dc source during one switching period. (*c*) Determine the average power supplied by the dc source. (*d*) Determine the average power absorbed by the resistance. (*e*) How do the results change if the resistance is zero?

9-34 For the resonant link dc converter of Fig. 9-12a, $V_s = 75$ V, $I_o = 5$ A, $R = 1\ \Omega$, $L = 250\ \mu\text{H}$, and $C = 0.1\ \mu\text{F}$. If the switch is opened at $t = 0$ with $i_L(0) = I_1 = 7$ A, and $v_C(0) = 0$, determine time when the switch should be closed so the voltage across it is zero. If the switch is closed immediately after the capacitor voltage becomes zero, how long should the switch remain closed so that the inductor voltage returns to 7 A?

9-35 For the resonant link dc converter of Fig. 9-12a, $V_s = 100$ V, $I_o = 10$ A, $R = 0.5\ \Omega$, $L = 150\ \mu\text{H}$, and $C = 0.05\ \mu\text{F}$. If the switch is opened at $t = 0$ with $i_L(0) = I_1 = 12$ A, and $v_C(0) = 0$, determine time when the switch should be closed so the voltage across it is zero. If the switch is closed immediately after the capacitor voltage becomes zero, how long should the switch remain closed so that the inductor voltage returns to 12 A?

CHAPTER 10

Drive Circuits, Snubber Circuits, and Heat Sinks

10.1 INTRODUCTION

Minimizing power losses in electronic switches is an important objective when designing power electronics circuits. On-state power losses occur because the voltage across a conducting switch is not zero. Switching losses occur because a device does not make a transition from one state to the other instantaneously, and switching losses in many converters are larger than on-state losses.

Resonant converters (Chap. 9) reduce switch losses by taking advantage of natural oscillations to switch when voltage or current is zero. Switches in circuits such as the dc-dc converters of Chaps. 6 and 7 go through a transition when voltage and current are nonzero. Switch losses in those types of converters can be minimized by drive circuits designed to provide fast switching transitions. Snubber circuits are designed to alter the switching waveforms to reduce power loss and to protect the switch. Power loss in an electronic switch produces heat, and limiting device temperature is critical in the design of all converter circuits.

10.2 MOSFET AND IGBT DRIVE CIRCUITS

Low-Side Drivers

The MOSFET is a voltage-controlled device and is relatively simple to turn on and off, which gives it an advantage over a bipolar junction transistor (BJT). The on state is achieved when the gate-to-source voltage sufficiently exceeds the threshold voltage, forcing the MOSFET into the triode (also called ohmic or nonsaturation) region of operation. Typically, the MOSFET gate-to-source voltage for the on state in switching circuits is between 10 and 20 V, although some MOSFETs are designed for logic-level control voltages. The off state is achieved by a lower-than-threshold voltage. On- and off-state gate currents are essentially zero. However, the parasitic input capacitance must be charged to turn the MOSFET on and must be discharged to turn it off. Switching speeds are basically determined by how rapidly charge can be transferred to and from the gate. Insulated gate bipolar transistors (IGBTs) are similar to MOSFETs in their drive requirements, and the following discussion applies to them as well.

A MOSFET drive circuit must be capable of rapidly sourcing and sinking currents for high-speed switching. The elementary drive circuit of Fig. 10-1a will drive the transistor, but the switching time may be unacceptably high for some applications. Moreover, if the input signal is from low-voltage digital logic devices, the logic output may not be sufficient to turn on the MOSFET.

A better drive circuit is shown in Fig. 10-1b. The double emitter-follower consists of a matched NPN and PNP bipolar transistor pair. When the drive input voltage is high, Q_1 is turned on and Q_2 is off, turning the MOSFET on. When the drive input signal is low, Q_1 turns off, and Q_2 turns on and removes the charge from the gate and turns the MOSFET off. The input signal may come from open-collector TTL used for control, with the double emitter-follower used as a buffer to source and sink the required gate currents, as shown in Fig. 10-1c.

Other arrangements for MOSFET drive circuits are shown in Fig. 10-2. These are functionally equivalent to the BJT double emitter-follower of Fig. 10-1b. The upper and lower transistors are driven as complementary on off transistors, with one transistor sourcing current and the other sinking current to and from the gate of the MOSFET to turn the power MOSFET on and off. Figure 10-2a shows NPN BJT transistors, Fig. 10-2b shows N-channel MOSFETs, and Fig. 10-2c shows complementary P- and N-channel MOSFETs.

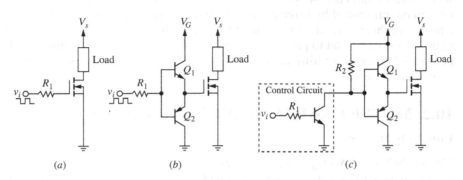

Figure 10.1 (*a*) Elementary MOSFET drive circuit; (*b*) Double emitter-follower drive circuit; (*c*) IC drive with double emitter-follower buffer.

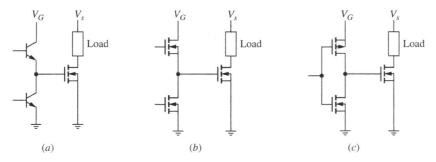

Figure 10.2 Additional MOSFET drive circuits. (*a*) NPN transistors; (*b*) N-channel MOSFETs; (*c*) P- and N-channel MOSFETs.

Example 10-1 illustrates the significance of the drive circuit on MOSFET switching speeds and power loss.

EXAMPLE 10.1

MOSFET Drive Circuit Simulation

A PSpice model for the IRF150 power MOSFET is available in the PSpice demo version in the EVAL file. (*a*) Use a PSpice simulation to determine the resulting turn-on and turnoff times and the power dissipated in the MOSFET for the circuit of Fig. 10-1*a*. Use $V_s = 80$ V and a load resistance of 10 Ω. The switch control voltage v_i is a 0- to 15-V pulse, and $R_1 = 100$ Ω. (*b*) Repeat for the circuit of Fig. 10-1*c* with $R_1 = R_2 = 1$ kΩ. The switching frequency for each case is 200 kHz, and the duty ratio of the switch control voltage is 50 percent.

■ **Solution**

(*a*) The elementary drive circuit is created for Fig. 10-1*a* using VPULSE for the switch control voltage. The resulting switching waveforms from Probe are shown in Fig. 10-3*a*. Switching transition times are roughly 1.7 and 0.5 μs for turnoff and turn-on, respectively. Average power absorbed by the MOSFET is determined from Probe by entering AVG(W(M1)), which yields a result of approximately 38 W.

(*b*) The emitter-follower drive circuit of Fig. 10-1*c* is created using 2N3904 NPN and 2N3906 PNP transistors from the evaluation library. The resulting switching waveforms are shown in Fig. 10-3*b*. The switching times are roughly 0.4 and 0.2 μs for turnoff and turn-on, and the power absorbed by the transistor is 7.8 W. Note that the emitter-follower drive circuit removes the gate charge more rapidly than the elementary drive circuit in part (*a*).

High-Side Drivers

Some converter topologies, such as the buck converter using an N-channel MOSFET, have high-side switches. The source terminal of the high-side MOSFET

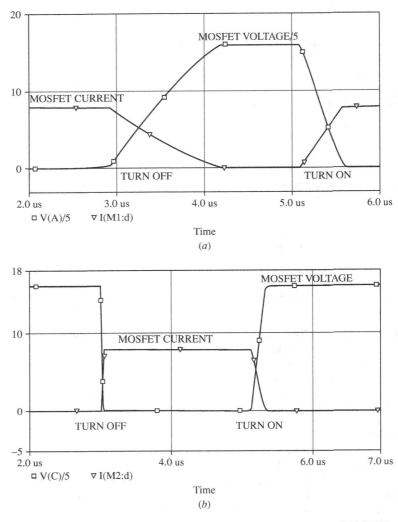

Figure 10.3 Switching waveforms for Example 10-1. (*a*) Elementary MOSFET drive circuit of Fig. 10-1*a*; (*b*) Double emitter-follower drive circuit of Fig. 10-1*b*.

is not connected to the circuit ground, as it would be in a low-side switch in a converter such as a boost converter. High-side switches require the MOSFET drive circuit to be floating with respect to the circuit ground. Drive circuits for these applications are called high-side drivers. To turn on the MOSFET, the gate-to-source voltage must be sufficiently high. When the MOSFET is on in a buck converter, for example, the voltage at the source terminal of the MOSFET is the same as the supply voltage V_s. Therefore, the gate voltage must be greater than the supply voltage.

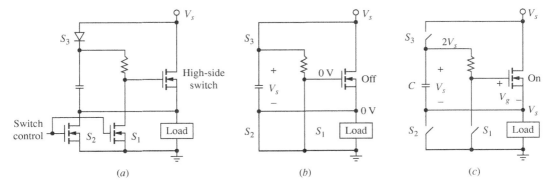

Figure 10.4 (*a*) A bootstrap circuit for driving a high-side MOSFET or IGBT; (*b*) The circuit for the switches closed, causing the capacitor to charge to V_s; (*c*) The circuit with the switches open, showing that the gate-to-source voltage is V_s.

A way to achieve a voltage higher than the source is to use a charge pump (switched-capacitor converter) as described in Chap. 6. One such high-side driver configuration is shown in Fig. 10-4*a*. The two driver MOSFETs and the diode are labeled as switches S_1, S_2, and S_3. When the control signal is high, S_1 and S_2 turn on, and the capacitor charges to V_s through the diode (Fig. 10-4*b*). When the control signal goes low, S_1 and S_2 are off, and the capacitor voltage is across the resistor and the gate of the power MOSFET, turning the MOSFET on. The voltage at the load becomes the same as the source voltage V_s, causing the voltage at the upper capacitor terminal to be $2V_s$. This drive circuit is called a *bootstrap* circuit.

MOSFET gate drivers are available as integrated-circuit (IC) packages. An example is the International Rectifier IR2117 shown in Fig. 10-5*a*. The IC with an external capacitor and diode provides the bootstrap circuit for the MOSFET. Another example is the International Rectifier IR2110 that is designed to drive both high-side and low-side switches (Fig. 10-5*b*). Half-bridge and full-bridge converters are applications where both high-side and low-side drivers are required.

Electrical isolation between the MOSFET and the control circuit is often desirable because of elevated voltage levels of the MOSFET, as in the upper transistors in a full-bridge circuit or a buck converter. Magnetically coupled and optically coupled circuits are commonly used for electrical isolation. Figure 10-6*a* shows a control and power circuit electrically isolated by a transformer. The capacitor on the control side prevents a dc offset in the transformer. A typical switching waveform is shown in Fig. 10-6*b*. Since the volt-second product must be the same on the transformer primary and secondary, the circuit works best when the duty ratio is around 50 percent. A basic optically isolated drive circuit is shown in Fig. 10-6*c*.

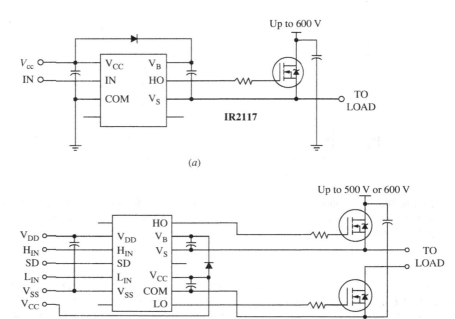

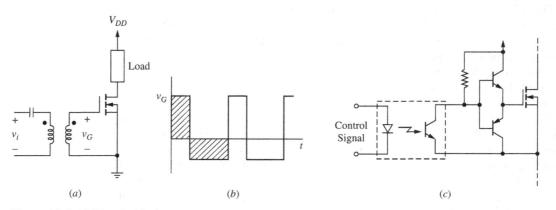

Figure 10.5 (*a*) International Rectifier IR2117 high-side driver; (*b*) International Rectifier IR2110 high- and low-side driver. (*Courtesy of International Rectifier Corporation.*)

Figure 10.6 (*a*) Electrical isolation of control and power circuits; (*b*) Transformer secondary voltage; (*c*) Optically isolated control and power circuits.

10.3 BIPOLAR TRANSISTOR DRIVE CIRCUITS

The bipolar junction transistor (BJT) has largely been replaced by MOSFETs and IGBTs. However, BJTs can be used in many applications. The BJT is a current-controlled device, requiring a base current to maintain the transistor in the conducting state. Base current during the on state for a collector current I_C must be at least I_C/β. The turn-on time depends on how rapidly the required stored charge can be delivered to the base region. Turn-on switching speeds can be decreased by initially applying a large spike of base current and then reducing the current to that required to keep the transistor on. Similarly, a negative current spike at turnoff is desirable to remove the stored charge, decreasing transition time from on to off.

Figure 10-7a shows a circuit arrangement that is suitable for BJT drives. When the input signal goes high, R_2 is initially bypassed by the uncharged capacitor. The initial base current is

$$I_{B_1} = \frac{V_i - v_{BE}}{R_1} \tag{10-1}$$

As the capacitor charges, the base current is reduced and reaches a final value of

$$I_{B_2} = \frac{V_i - v_{BE}}{R_1 + R_2} \tag{10-2}$$

The desired charging time of the capacitor determines its value. Three to five time constants are required to charge or discharge the capacitor. The charging time constant is

$$\tau = R_E C = \left(\frac{R_1 R_2}{R_1 + R_2}\right) C \tag{10-3}$$

The input signal goes low at turnoff, and the charged capacitor provides a negative current spike as the base charge is removed. Figure 10-7b shows the base current waveform.

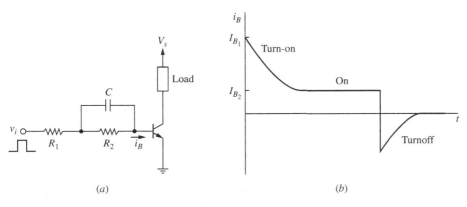

Figure 10.7 (a) Drive circuit for a bipolar transistor; (b) Transistor base current.

EXAMPLE 10.2

Bipolar Transistor Drive Circuit

Design a BJT base drive circuit with the configuration of Fig. 10-7a that has a spike of 1 A at turn-on and maintains a base current of 0.2 A in the on state. The voltage v_i is a pulse of 0 to 15 V with a 50 percent duty ratio, and the switching frequency is 100 kHz. Assume that v_{BE} is 0.9 V when the transistor is on.

■ Solution

The value of R_1 is determined from the initial current spike requirement. Solving for R_1 in Eq. (10-1),

$$R_1 = \frac{V_i - v_{BE}}{I_{B_1}} = \frac{15 - 0.9}{1} = 14.1 \: \Omega$$

The steady-state base current in the on state determines R_2. From Eq. (10-2),

$$R_2 = \frac{V_i - v_{BE}}{I_{B_2}} - R_1 = \frac{15 - 0.9}{0.2} - 14.1 = 56.4 \: \Omega$$

The value of C is determined from the required time constant. For a 50 percent duty ratio at 100 kHz, the transistor is on for 5 μs. Letting the on time for the transistor be five time constants, $\tau = 1$ μs. From Eq. (10-3),

$$\tau = R_E C = \left(\frac{R_1 R_2}{R_1 + R_2}\right) C = 11.3 \: C = 1 \: \mu s$$
$$C = 88.7 \: nF$$

EXAMPLE 10.3

PSpice Simulation for a BJT Drive Circuit

Use PSpice to simulate the circuit of Fig. 10-8a with $V_s = 80$ V, a 10-Ω load resistor, and the base drive components from Example 10-2: (a) with the base capacitor omitted and (b) with the base drive capacitor included. Determine the power absorbed by the transistor for each case. Use the 2n5686 PSpice model from ON Semiconductor.

■ Solution

The circuit of Fig. 10-8a is created using VPULSE for the control voltage source. The transistor model is obtained from the ON Semiconductor website, and the model is copied and pasted into the QbreakN transistor model by choosing *Edit, PSpice Model*.

The resulting switching waveforms are shown in Fig. 10-8. Note the significant difference in switching times with and without the base drive capacitance. Power absorbed by the transistor is determined by entering AVG(W(Q1)) which yields results of 30 W without the base capacitor and 5 W with the capacitor.

Switching times can be reduced by keeping the transistor in the quasi-saturation region, which is just past the linear region but not in hard saturation. This is

10.3 Bipolar Transistor Drive Circuits 439

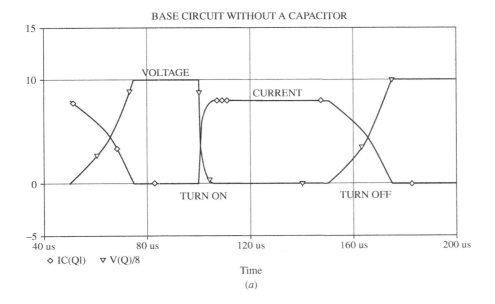

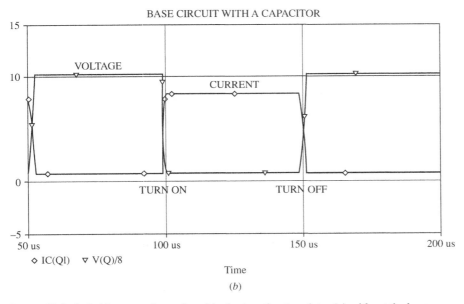

Figure 10.8 Switching waveforms for a bipolar junction transistor (*a*) without the base capacitor and (*b*) with the base capacitor. The voltage is scaled by $\frac{1}{8}$.

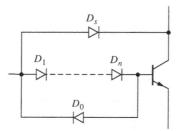

Figure 10.9 Baker's clamp to control the degree of BJT saturation.

controlled by preventing v_{CE} from going too low. However, on-state conduction losses for the BJT are larger than if the transistor were further into saturation where the collector-to-emitter voltage is lower.

A clamping circuit such as the Baker's clamp of Fig. 10-9 can keep the transistor in quasi-saturation by limiting the collector-to-emitter voltage. There are n diodes in series with the base, and a shunting diode D_s is connected from the drive to the collector. The on-state collector-to-emitter voltage is determined from Kirchhoff's voltage law as

$$v_{CE} = v_{BE} + nv_D - v_{D_s} \tag{10-4}$$

The desired value of v_{CE} is determined by the number of diodes in series with the base. Diode D_o allows reverse base current during turnoff.

10.4 THYRISTOR DRIVE CIRCUITS

Thyristor devices such as SCRs require only a momentary gate current to turn the device on, rather than the continuous drive signal required for transistors. The voltage levels in a thyristor circuit may be quite large, requiring isolation between the drive circuit and the device. Electrical isolation is accomplished by magnetic or optical coupling. An elementary SCR drive circuit employing magnetic coupling is shown in Fig. 10-10*a*. The control circuit turns on the transistor and establishes a voltage across the transformer primary and secondary, providing the gate current to turn on the SCR.

The simple gate drive circuit of Fig. 10-10*b* can be used in some applications where electrical isolation is not required. The circuit is a single-phase voltage controller (Chap. 5) of the type that might be used in a common light dimmer. An SCR could be used in place of the triac T_1 to form a controlled half-wave rectifier (Chap. 3). The delay angle is controlled by the RC circuit connected to the gate through the diac T_2. The diac is a member of the thyristor family that operates as a self-triggered triac. When the voltage across the diac reaches a specified value, it begins to conduct and triggers the triac. As the sinusoidal source voltage goes positive, the capacitor begins to charge. When the voltage across the capacitor reaches the diac trigger voltage, gate current is established in the triac for turn-on.

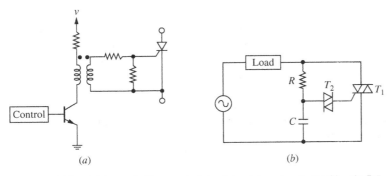

Figure 10.10 (a) Magnetically coupled thyristor drive circuit; (b) Simple *RC* drive circuit.

10.5 TRANSISTOR SNUBBER CIRCUITS

Snubber circuits reduce power losses in a transistor during switching (although not necessarily total switching losses) and protect the device from the switching stresses of high voltages and currents.

As discussed in Chap. 6, a large part of the power loss in a transistor occurs during switching. Figure 10-11a shows a model for a converter that has a large inductive load which can be approximated as a current source I_L. The analysis of switching transitions for this circuit relies on Kirchhoff's laws: the load current must divide between the transistor and the diode; and the source voltage must divide between the transistor and the load.

In the transistor on state, the diode is off and the transistor carries the load current. As the transistor turns off, the diode remains reverse-biased until the

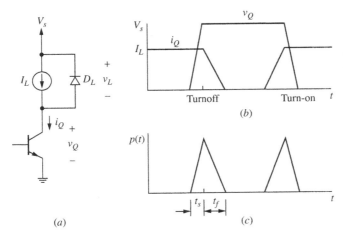

Figure 10.11 (a) Converter model for switching inductive loads; (b) Voltage and current during switching; (c) Instantaneous power for the transistor.

transistor voltage v_Q increases to the source voltage V_s and the load voltage v_L decreases to zero. After the transistor voltage reaches V_s, the diode current increases to I_L while the transistor current decreases to zero. As a result, there is a point during turnoff when the transistor voltage and current are high simultaneously (Fig. 10-11b), resulting in a triangularly shaped instantaneous power waveform $p_Q(t)$, as in Fig. 10-11c.

In the transistor off state, the diode carries the entire load current. During turn-on, the transistor voltage cannot fall below V_s until the diode turns off, which is when the transistor carries the entire load current and the diode current is zero. Again, there is a point when the transistor voltage and current are high simultaneously.

A snubber circuit alters the transistor voltage and current waveforms to an advantage. A typical snubber circuit is shown in Fig. 10-12a. The snubber provides another path for load current during turnoff. As the transistor is turning off

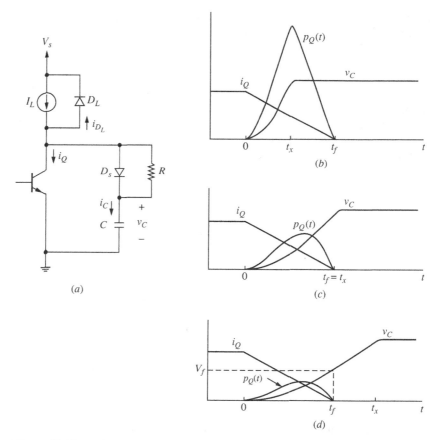

Figure 10.12 (a) Converter with a transistor snubber circuit; (b–d) Turnoff waveforms with a snubber with increasing values of capacitance.

and the voltage across it is increasing, the snubber diode D_s becomes forward-biased and the capacitor begins to charge. The rate of change of transistor voltage is reduced by the capacitor, delaying its voltage transition from low to high. The capacitor charges to the final off-state voltage across the transistor and remains charged while the transistor is off. When the transistor turns on, the capacitor discharges through the snubber resistor and transistor.

The size of the snubber capacitor determines the rate of voltage rise across the switch at turnoff. The transistor carries the load current prior to turnoff, and during turnoff the transistor current decreases approximately linearly until it reaches zero. The load diode remains off until the capacitor voltage reaches V_s. The snubber capacitor carries the remainder of the load current until the load diode turns on. The transistor and snubber-capacitor currents during turnoff are expressed as

$$i_Q(t) = \begin{cases} I_L\left(1 - \dfrac{t}{t_f}\right) & \text{for } 0 \leq t < t_f \\ 0 & t \geq t_f \end{cases} \qquad (10\text{-}5)$$

$$i_C(t) = \begin{cases} I_L - i_Q(t) = \dfrac{I_L t}{t_f} & 0 \leq t < t_f \\ I_L & t_f \leq t < t_x \\ 0 & t \geq t_x \end{cases} \qquad (10\text{-}6)$$

where t_x is the time at which the capacitor voltage reaches its final value, which is determined by the source voltage of the circuit. The capacitor (and transistor) voltage is shown for different values of C in Fig. 10-12b to d. A small snubber capacitor results in the voltage reaching V_s before the transistor current reaches zero, whereas larger capacitance results in longer times for the voltage to reach V_s. Note that the energy absorbed by the transistor (the area under the instantaneous power curve) during switching decreases as the snubber capacitance increases.

The capacitor is chosen on the basis of the desired voltage at the instant the transistor current reaches zero. The capacitor voltage in Fig. 10-12d is expressed as

$$v_c(t) = \begin{cases} \dfrac{1}{C}\displaystyle\int_0^t \dfrac{I_L t}{t_f} dt = \dfrac{I_L t^2}{2Ct_f} & 0 \leq t \leq t_f \\ \dfrac{1}{C}\displaystyle\int_{t_f}^t I_L \, dt + v_c(t_f) = \dfrac{I_L}{C}(t - t_f) + \dfrac{I_L t_f}{2C} & t_f \leq t \leq t_x \\ V_s & t \geq t_x \end{cases} \qquad (10\text{-}7)$$

If the switch current reaches zero before the capacitor fully charges, the capacitor voltage is determined from the first part of Eq. (10-7). Letting $v_c(t_f) = V_f$,

$$V_f = \frac{I_L(t_f)^2}{2Ct_f} = \frac{I_L t_f}{2C}$$

Solving for C,

$$\boxed{C = \frac{I_L t_f}{2V_f}} \quad (10\text{-}8)$$

where V_f is the desired capacitor voltage when the transistor current reaches zero ($V_f \leq V_s$). The capacitor is sometimes selected such that the switch voltage reaches the final value at the same time that the current reaches zero, in which case

$$C = \frac{I_L t_f}{2V_s} \quad (10\text{-}9)$$

where V_s is the final voltage across the switch while it is open. Note that the final voltage across the transistor may be different from the dc supply voltage in some topologies. The forward and flyback converters (Chap. 7), for example, have off-state switch voltages of twice the dc input.

The power absorbed by the transistor is reduced by the snubber circuit. The power absorbed by the transistor before the snubber is added is determined from the waveform of Fig. 10-11c. Turnoff power losses are determined from

$$P_Q = \frac{1}{T}\int_0^T p_Q(t)\,dt \quad (10\text{-}10)$$

The integral is evaluated by determining the area under the triangle for turnoff, resulting in an expression for turnoff power loss without a snubber of

$$P_Q = \frac{1}{2}I_L V_s(t_s + t_f)f \quad (10\text{-}11)$$

where $t_s + t_f$ is the turnoff switching time and $f = 1/T$ is the switching frequency.

Power absorbed by the transistor during turnoff after the snubber is added is determined from Eqs. (10-5), (10-7), and (10-10).

$$P_Q = \frac{1}{T}\int_0^T v_Q i_Q\,dt = f\int_0^{t_f}\left(\frac{I_L t^2}{2Ct_f}\right)I_L\left(1 - \frac{t}{t_f}\right)dt = \frac{I_L^2 t_f^2 f}{24C} \quad (10\text{-}12)$$

The above equation is valid for the case when $t_f \leq t_x$, as in Fig. 10-12c or d.

The resistor is chosen such that the capacitor is discharged before the next time the transistor turns off. A time interval of three to five time constants is necessary for capacitor discharge. Assuming five time constants for complete discharge, the on time for the transistor is

$$t_{on} > 5RC$$

or

$$R < \frac{t_{on}}{5C} \qquad (10\text{-}13)$$

The capacitor discharges through the resistor and the transistor when the transistor turns on. The energy stored in the capacitor is

$$W = \tfrac{1}{2}CV_s^2 \qquad (10\text{-}14)$$

This energy is transferred mostly to the resistor during the on time of the transistor. The power absorbed by the resistor is energy divided by time, with time equal to the switching period:

$$P_R = \frac{\tfrac{1}{2}CV_s^2}{T} = \tfrac{1}{2}CV_s^2 f \qquad (10\text{-}15)$$

where f is the switching frequency. Equation (10-15) indicates that power dissipation in the snubber resistor is proportional to the size of the snubber capacitor. *A large capacitor reduces the power loss in the transistor [Eq. (10-12)], but at the expense of power loss in the snubber resistor.* Note that the power in the snubber resistor is independent of its value. The resistor value determines the discharge rate of the capacitor when the transistor turns on.

The power absorbed by the transistor is lowest for large capacitance, but the power absorbed by the snubber resistor is largest for this case. The total power for transistor turnoff is the sum of the transistor and snubber powers. Figure 10-13 shows the relationship among transistor, snubber, and total losses. The use of the

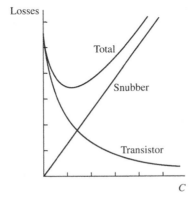

Figure 10.13 Transistor, snubber, and total turnoff losses as a function of snubber capacitance.

snubber can reduce the total switching losses, but perhaps more importantly, the snubber reduces the power loss in the transistor and reduces the cooling requirements for the device. The transistor is more prone to failure and is harder to cool than the resistor, so the snubber makes the design more reliable.

EXAMPLE 10.4

Transistor Snubber Circuit Design

The converter and snubber in Fig. 10-12a has $V_s = 100$ V and $I_L = 5$ A. The switching frequency is 100 kHz with a duty ratio of 50 percent, and the transistor turns off in 0.5 μs. (a) Determine the turnoff losses without a snubber if the transistor voltage reaches V_s in 0.1 μs. (b) Design a snubber using the criterion that the transistor voltage reaches its final value at the same time that the transistor current reaches zero. (c) Determine the transistor turnoff losses and the resistor power with the snubber added.

■ Solution

(a) The turnoff voltage, current, and instantaneous power waveforms without the snubber are like those of Fig. 10-11. Transistor voltage reaches 100 V while the current is still at 5 A, resulting in a peak instantaneous power of (100 V)(5 A) = 500 W. The base of the power triangle is 6 μs, making the area 0.5(500 W)(0.6 μs) = 150 μJ. The switching period is $1/f = 1/100{,}000$ s, so the turnoff power loss in the transistor is $W/T = (150)(10^6)(100{,}000) = 15$ W. Equation (10-11) yields the same result:

$$P_Q = \frac{1}{2} I_L V_s (t_s + t_f) f = \frac{1}{2}(5)(100)(0.1 + 0.5)(10^{-6})(10^5) = 15 \text{ W}$$

(b) The snubber capacitance value is determined from Eq. (10-9):

$$C = \frac{I_L t_f}{2 V_s} = \frac{(5)(0.5)(10^{-6})}{(2)(100)} = 1.25(10^{-8}) = 0.0125 \text{ μF} = 12.5 \text{ nF}$$

The snubber resistor is chosen using Eq. (10-13). The switching frequency is 100 kHz corresponding to a switching period of 10 μs. The on time for the transistor is approximately one-half of the period, or 5 μs. The resistor value is then

$$R < \frac{t_{on}}{5C} = \frac{5 \text{ μs}}{5(0.0125 \text{ μF})} = 80 \text{ }\Omega$$

The resistance value is not critical. Since five time constants is a conservative design criterion, the resistance need not be exactly 80 Ω.

(c) The power absorbed by the transistor is determined from Eq. (10-12):

$$P_Q = \frac{I_L^2 t_f^2 f}{24 C} = \frac{5^2 [(0.5)(10^{-6})]^2 (10^5)}{24(1.25)(10^{-8})} = 2.08 \text{ W}$$

Power absorbed by the snubber resistor is determined from Eq. (10-15):

$$P_R = \frac{1}{2} C V_s^2 f = \frac{0.0125(10^{-6})(100^2)(100{,}000)}{2} = 6.25 \text{ W}$$

Total power due to turnoff losses with the snubber is 2.08 + 6.25 = 8.33 W, reduced from 15 W without the snubber. The losses in the transistor are significantly reduced by the snubber, and total turnoff losses are also reduced in this case.

The other function of the snubber circuit is to reduce voltage and current stresses in the transistor. The voltage and current in a transistor must not exceed the maximum values. Additionally, the transistor temperature must be kept within allowable limits. High current at a high voltage must also be avoided in a bipolar transistor because of a phenomenon called *second breakdown*. Second breakdown is the result of nonuniform distribution of current in the collector-base junction when both voltage and current are large, resulting in localized heating in the transistor and failure.

The forward-bias safe operating area (SOA or FBSOA) of a BJT is the area enclosed by the voltage, current, thermal, and second breakdown limits, as shown in Fig. 10-14a. The FBSOA indicates the capability of the transistor when

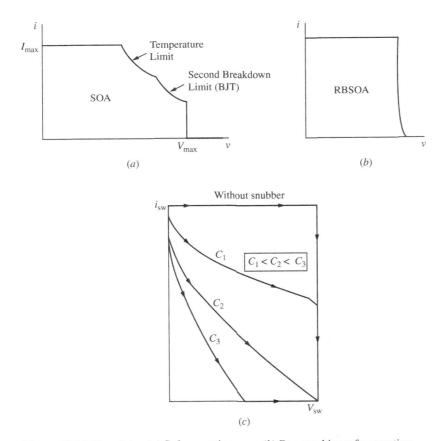

Figure 10.14 Transistor. (*a*) Safe operation area; (*b*) Reverse-bias safe operating area; (*c*) Switching trajectories for different snubber capacitance.

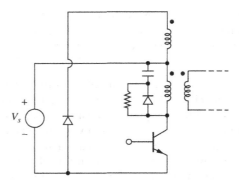

Figure 10.15 Alternate placement of a snubber for the forward converter

the base-emitter junction is forward-biased. The FBSOA indicates maximum limits for steady-state and for turn-on. The SOA can be expanded vertically for pulsed operation. That is, current can be greater if it is intermittent rather than continuous. In addition, there is a reverse-bias safe operating area (RBSOA), shown in Fig. 10-14b. *Forward bias* and *reverse bias* refer to the biasing of the base-emitter junction. The voltage-current trajectory of the switching waveforms of Fig. 10-12 is shown in Fig. 10-14c. A snubber can alter the trajectory and prevent operation outside of the SOA and RBSOA. Second breakdown does not occur in a MOSFET.

Alternative placements of the snubber circuit are possible. The forward converter is shown in Fig. 10-15 with a snubber connected from the transistor back to the positive input supply rather than to ground. The snubber functions like that of Fig. 10-12, except that the final voltage across the capacitor is V_s rather than $2V_s$.

One source of voltage stress in a transistor switch is the energy stored in the leakage inductance of a transformer. The flyback converter model of Fig. 10-16, for example, includes the leakage inductance L_l, which was neglected in the analysis of the converters in Chap. 7 but is important when analyzing the stresses on the switch. The leakage inductance carries the same current as the transistor switch when the transistor is on. When the transistor turns off, the current in the leakage inductance cannot change instantaneously. The large di/dt from the rapidly falling current can cause a large voltage across the transistor.

The snubber circuit of Fig. 10-12 can reduce the voltage stress across the transistor in addition to reducing transistor losses. The diode-capacitor-resistor combination provides a parallel current path with the transistor. When the transistor turns off, the current maintained by the transformer leakage inductance forward-biases the diode and charges the capacitor. The capacitor absorbs energy that was stored in the leakage inductance and reduces the voltage spike that would appear across the transistor. This energy is dissipated in the snubber resistor when the transistor turns on.

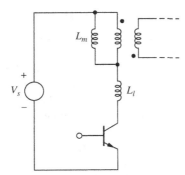

Figure 10.16 Flyback converter with transformer leakage inductance included.

Turn-on snubbers protect the device from simultaneously high voltage and current during turn-on. As with the turnoff snubber, the purpose of the turn-on snubber is to modify the voltage-current waveforms to reduce power loss. An inductor in series with the transistor slows the rate of current rise and can reduce the overlap of high current and high voltage. A turn-on snubber is shown in Fig. 10-17. The snubber diode is off during turn-on. During turnoff, the energy stored in the snubber inductor is dissipated in the resistor.

If a turnoff snubber is also used, the energy stored in the turn-on snubber inductor can be transferred to the turnoff snubber without the need for the additional diode and resistor. Leakage or stray inductance that inherently exists in circuits may perform the function of a turn-on snubber without the need for an additional inductor.

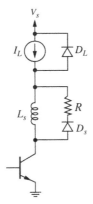

Figure 10.17 Transistor turnon snubber.

10.6 ENERGY RECOVERY SNUBBER CIRCUITS

Snubber circuits reduce the power dissipated in the transistor, but the snubber resistor also dissipates power that is lost as heat. The energy stored in the snubber capacitance is eventually transferred to the snubber resistor. If the energy stored in the snubber capacitor can be transferred to the load or back to the source, the snubber resistor is not necessary, and the losses are reduced.

One method for energy recovery in a snubber is shown in Fig. 10-18. Both D_s and C_s act like the snubber of Fig. 10-12a at turnoff: C_s charges to V_s and delays the voltage rise across the transistor. At turn-on, a current path consisting of Q, C_s, L, D_1, and C_1 is formed, and an oscillatory current results. The charge initially stored in C_s is transferred to C_1. At the next turnoff, C_1 discharges through D_2 into the load while C_s charges again. Summarizing, the energy stored in C_s at turnoff is first transferred to C_1 and is then transferred to the load.

10.7 THYRISTOR SNUBBER CIRCUITS

The purpose of a thyristor snubber circuit is mainly to protect the device from large rates of change of anode-to-cathode voltage and anode current. If *dv/dt* for the thyristor is too large, the device will begin to conduct without a gate signal present. If *di/dt* is too large during turn-on, localized heating will result from the high current density in the region of the gate connection as the current spreads out over the whole junction.

Thyristor snubber circuits can be like those used for the transistor, or they may be of the unpolarized type shown in Fig. 10-19. The series inductor limits *di/dt*, and the parallel *RC* connection limits *dv/dt*.

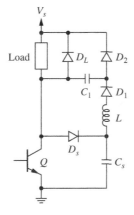

Figure 10.18 Snubber circuit with energy recovery.

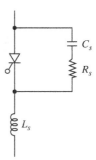

Figure 10.19 Thyristor snubber circuit.

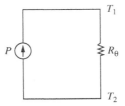

Figure 10.20 An electric circuit equivalent to determine temperature difference.

10.8 HEAT SINKS AND THERMAL MANAGEMENT

Steady-State Temperatures

As discussed throughout this textbook, conduction and switching losses occur in electronic devices. Those losses represent electrical-energy being converted to thermal energy, and removal of thermal energy is essential in keeping the internal temperature of the device below its maximum rated value.

In general, the temperature difference between two points is a function of thermal power and thermal resistance. *Thermal resistance* is defined as

$$R_\theta = \frac{T_1 - T_2}{P} \quad (10\text{-}16)$$

where R_θ = thermal resistance, °C/W (also listed as K/W on some datasheets)
$T_1 - T_2$ = temperature difference, °C
P = thermal power, W

A useful electric circuit analog for steady-state thermal calculations that fits Eq. (10-16) uses P as a current source, R_θ as electrical resistance, and voltage difference as temperature difference, as illustrated in Fig. 10-20.

The internal temperature of an electronic switching device is referred to as the *junction* temperature. Although devices such as MOSFETs do not have a junction per se when conducting, the term is still used. In an electronic device without a heat sink, the junction temperature is determined by thermal power and the junction-to-ambient thermal resistance $R_{\theta,JA}$. The ambient temperature is that of the air in contact with the case. Manufacturers often include the value of $R_{\theta,JA}$ on the datasheet for the device.

EXAMPLE 10.5

MOSFET Maximum Power Absorption

A MOSFET manufacturer's datasheet lists the junction-to-ambient thermal resistance $R_{\theta,JA}$ as 62°C/W. The maximum junction temperature is listed at 175°C, but the designer wishes for it not to exceed 150°C for increased reliability. If the ambient temperature is 40°C, determine the maximum power that the MOSFET can absorb.

■ Solution
From Eq. (10-16),

$$P = \frac{T_1 - T_2}{R_\theta} = \frac{T_J - T_A}{R_{\theta,JA}} = \frac{150 - 40}{62} = 1.77 \text{ W}$$

In many instances, the power absorbed by a device results in an excessive junction temperature, and a heat sink is required. A heat sink reduces the junction temperature for a given power dissipation in a device by reducing the overall thermal resistance from junction to ambient. The case of the device is often attached to the heat sink with a thermal compound to fill the small voids between the imperfect surfaces of the case and sink. Heat sinks are available in all sizes, ranging from small clip-on devices to massive extruded aluminum shapes. Typical heat sinks are shown in Fig. 10-21.

For an electronic device with a heat sink, thermal power flows from the junction to the case, from the case to the heat sink, and then from the heat sink to ambient. The corresponding thermal resistances are $R_{\theta,JC}$, $R_{\theta,CS}$, and $R_{\theta,SA}$, as shown in Fig. 10-22.

The temperature at the heat sink near the mounting point of the electronic device is

$$T_S = PR_{\theta,SA} + T_A \tag{10-17}$$

the temperature at the device case is

$$T_C = PR_{\theta,CS} + T_S = P(R_{\theta,CS} + R_{\theta,SA}) + T_A \tag{10-18}$$

and the temperature at the device junction is

$$T_J = PR_{\theta,JC} + T_C = P(R_{\theta,JC} + R_{\theta,CS} + R_{\theta,SA}) + T_A \tag{10-19}$$

Semiconductor manufacturers' datasheets list the junction-to-case thermal resistance and often list the case-to-heat sink thermal resistance assuming a greased surface. The heat sink-to-ambient thermal resistance is obtained from the heat sink manufacturer.

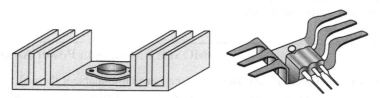

Figure 10.21 Power transistors mounted on heat sinks.

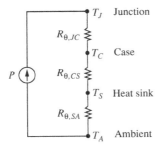

Figure 10.22 The electrical circuit equivalent for a transistor mounted to a heat sink.

EXAMPLE 10.6

MOSFET Junction Temperature with a Heat sink

The datasheet for the MOSFET in Example 10-5 lists the thermal resistance from the junction to case as 1.87°C/W and the thermal resistance from the case to the heat sink as 0.50°C/W. (*a*) If the device is mounted on a heat sink that has a thermal resistance of 7.2°C/W, determine the maximum power that can be absorbed without exceeding a junction temperature of 150°C when the ambient temperature is 40°C. (*b*) Determine the junction temperature when the absorbed power is 15 W. (*c*) Determine $R_{\theta,SA}$ of a heat sink that would limit the junction temperature to 150°C for 15 W absorbed.

■ **Solution**
(*a*) From Eq. (10-19),

$$P = \frac{T_J - T_A}{R_{\theta,JC} + R_{\theta,CS} + R_{\theta,SA}} = \frac{150 - 40}{1.87 + 0.50 + 7.2} = \frac{110}{9.57} = 11.5 \text{ W}$$

Comparing this result with that of Example 10-5, including a heat sink reduces the junction-to-ambient thermal resistance from 62 to 9.57°C/W and enables more power to be absorbed by the device without exceeding a temperature limit. If the MOSFET absorbs 1.77 W as in Example 10-5, the junction temperature with this heat sink will be

$$T_J = P(R_{\theta,JC} + R_{\theta,CS} + R_{\theta,SA}) + T_A$$
$$= 1.77(1.87 + 0.50 + 7.2) + 40 = 56.9°\text{C}$$

compared to 150°C without the heat sink.

(*b*) Also from Eq. (10-19), the junction temperature for 15 W is

$$T_J = P(R_{\theta,JC} + R_{\theta,CS} + R_{\theta,SA}) + T_A = 15(1.87 + 0.50 + 7.2) + 40 = 184°\text{C}$$

(c) Solving Eq. (10-19) for $R_{\theta,SA}$ for a heat sink that would limit the junction temperature to 150°C,

$$R_{\theta,SA} = \frac{T_J - T_A}{P} - R_{\theta,JC} - R_{\theta,CS} = \frac{150 - 40}{15} - 1.87 - 0.50 = 4.96°C/W$$

Time-Varying Temperatures

Temperatures resulting from a time-varying thermal power source are analyzed using an equivalent circuit like that of Fig. 10-23a. The capacitors represent thermal energy storage, resulting in exponential changes in temperatures for a step change in the power source, as shown in Fig. 10-23b and c.

This *RC* model can represent the entire device-case-heat-sink system with T_1, T_2, T_3, and T_4 representing the junction, case, heat sink, and ambient temperature, respectively. The model could also represent just one of those components

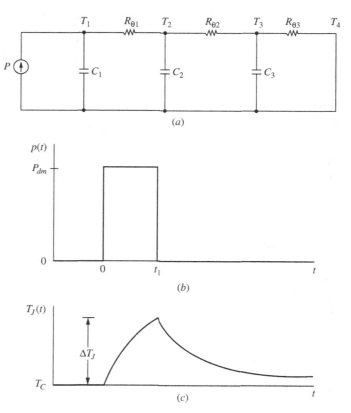

Figure 10.23 (a) An equivalent circuit representation for a time-varying thermal power source; (b) A momentary power pulse; (c) The temperature response due to a power pulse.

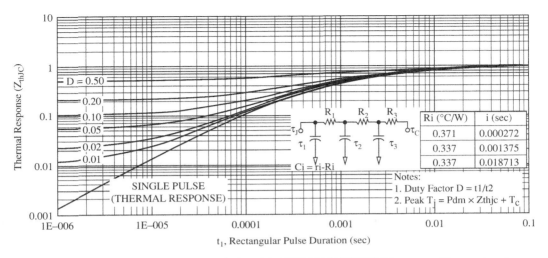

Figure 10.24 Thermal impedance characteristics of the IRF4104 MOSFET. (*Courtesy of International Rectifier Corporation*).

that has been subdivided into multiple sections. For example, it could represent just the junction to case of the device divided into three sections.

Transient thermal impedance from the junction to the case $Z_{\theta,JC}$ is used to determine the change in junction temperature due to momentary changes in absorbed power. Manufacturers typically supply transient thermal impedance information on datasheets. Figure 10-24 shows a graphical representation of $Z_{\theta,JC}$ as well as the *RC* equivalent circuit representation for the junction to case for the IRF4104 MOSFET. Transient thermal impedance is also denoted as Z_{th}.

First, consider the increase in junction temperature due to a single power pulse of amplitude P_{dm} lasting for a duration t_1, as shown in Fig. 10-23b. The thermal model of Fig. 10-23a produces an exponential junction temperature variation like that of Fig. 10-23c. The change in the temperature of the junction in the time interval 0 to t_1 is determined from

$$\Delta T_J = P_{dm} Z_{\theta,JC} \tag{10-20}$$

where $Z_{\theta,JC}$ is the transient thermal impedance from the device junction to case. The maximum junction temperature is ΔT_J plus the case temperature.

$$T_{J,\max} = P_{dm} Z_{\theta,JC} + T_C \tag{10-21}$$

EXAMPLE 10.7

Transient Thermal Impedance

A single power pulse of 100 W with a 100-μs duration occurs in a MOSFET that has the transient thermal resistance characteristics shown in Fig. 10-24. Determine the maximum change in junction temperature.

■ Solution

The bottom curve on the graph gives the thermal impedance for a single pulse. For 100 μs (0.0001 s), $Z_{\theta,JC}$ is approximately 0.11°C/W. Using Eq. (10-20), the increase in junction temperature is

$$\Delta T_J = P_{dm} Z_{\theta,JC} = 100(0.11) = 11°C$$

Next, consider the pulsed power waveform shown in Fig. 10-25a. Junction temperature increases during the power pulse and decreases when the power is zero. After an initial start up interval, the junction temperature reaches equilibrium where thermal energy absorbed in one period matches the thermal energy transferred. Maximum junction temperature $T_{J,max}$ is found using Eq. (10-21) and $Z_{\theta,JC}$ from Fig. 10-24. The horizontal axis is t_1, the time duration of the pulse in each period. The value of $Z_{\theta,JC}$ is read from the curve corresponding to the duty ratio t_1/t_2. The temperature of the case is assumed constant and can be determined from Eq. (10-18) using the average power for P.

If the power pulse is at a high frequency, such as the switching frequency of a typical power converter, the fluctuation in the temperature waveform of Fig. 10-25b becomes small, and temperatures can be analyzed by using $R_{\theta,JC}$ in Eq. (10-19) with P equal to the average power.

EXAMPLE 10.8

Maximum Junction Temperature for Periodic Pulsed Power

The power absorbed by a MOSFET is the pulsed-power waveform like that of Fig. 10-25a with P_{dm} = 100 W, t_1 = 200 μs, and t_2 = 2000 μs. (a) Determine the peak temperature difference between the junction and the case, using the transient thermal impedance from

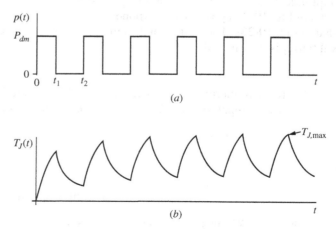

Figure 10.25 (a) A pulsed power waveform; (b) The temperature variation at the junction.

Fig. 10-24. Assume that the case temperature is a constant 80°C. (b) The thermal resistance $R_{\theta,JC}$ for this MOSFET is 1.05°C/W. Compare the result in (a) with a calculation based on the average MOSFET and $R_{\theta,JC}$.

■ **Solution**
(a) The duty ratio of the power waveform is

$$D = \frac{t_1}{t_2} = \frac{200\,\mu s}{2000\,\mu s} = 0.1$$

Using the graph in Fig. 10-24, the transient thermal impedance $Z_{\theta,JC}$ for $t_1 = 200\,\mu s$ and $D = 0.1$ is approximately 0.3°C/W. The maximum temperature difference between the junction and the case is determined from Eq. 10-20 as

$$\Delta T_J = P_{dm} Z_{\theta,JC} = 100(0.3) = 30°$$

making the maximum junction temperature

$$T_{J,\max} = P_{dm} Z_{\theta,JC} + T_C = 30 + 80 = 110°C$$

(b) Using the average power only, the temperature difference from the junction to case would be calculated as $\Delta T_J = P R_{\theta,JC} = (P_{dm} D) R_{\theta,JC} = (10\,W)(1.05°C/W) = 10.5°C$. Therefore, a temperature calculation based on the average power greatly underestimates the maximum temperature difference between the junction and the case. Note that a period of 2000 μs corresponds to a frequency of only 500 Hz. For much higher frequencies (e.g., 50 kHz), the temperature difference based on $R_{\theta,JC}$ and average power is sufficiently accurate.

10.9 Summary

The switching speed of a transistor is determined not only by the device but also by the gate or base drive circuit. The double emitter-follower drive circuit for the MOSFET (or IGBT) significantly reduces the switching time by sourcing and sinking the required gate currents to supply and remove the stored charge in the MOSFET rapidly. A base drive circuit that includes large current spikes at turn-on and turnoff for the bipolar transistor significantly reduces switching times.

Snubber circuits reduce power losses in the device during switching and protect the device from the switching stresses of high voltages and currents. Transistor switching losses are reduced by snubbers, but total switching losses may or may not be reduced because power is dissipated in the snubber circuit. Energy recovery snubber circuits can further reduce the switching losses by eliminating the need for a snubber resistor.

Heat sinks reduce the internal temperature of an electronic device by reducing the total thermal resistance between the device junction and ambient. Equivalently, a heat sink enables a device to absorb more power without exceeding a maximum internal temperature.

10.10 Bibliography

M. S. J. Asghar, *Power Electronics Handbook*, edited by M. H. Rashid, Academic Press, San Diego, Calif., 2001, Chapter 18.

L. Edmunds, "Heatsink Characteristics," International Rectifier Application Note AN-1057, 2004, http://www.irf.com/technical-info/appnotes/an-1057.pdf.

Fundamentals of Power Semiconductors for Automotive Applications, 2d ed., Infineon Technologies, Munich, Germany, 2008.

"HV Floating MOS-Gate Driver ICs," Application Note AN-978, International Rectifier, Inc., El Segunda, Calif., July 2001. http://www.irf.com/technical-info/ appnotes/an-978.pdf.

A. Isurin and A. Cook, "Passive Soft-Switching Snubber Circuit with Energy Recovery," *IEEE Applied Power Electronics Conference*, austin, Tex., 2008.

S. Lee, "How to Select a Heat Sink," Aavid Thermalloy, http://www.aavidthermalloy.com/technical/papers/pdfs/select.pdf

W. McMurray, "Selection of Snubber and Clamps to Optimize the Design of Transistor Switching Converters," *IEEE Transactions on Industry Applications*, vol. IAI6, no. 4, 1980, pp. 513–523.

N. Mohan, T. M. Undeland, and W. P. Robbins, *Power Electronics: Converters, Applications, and Design,* 3d ed., Wiley, New York, 2003.

M. H. Rashid, *Power Electronics: Circuits, Devices, and Systems,* 3d ed., Prentice-Hall, Upper Saddle River, N.J., 2004.

R. E. Tarter, *Solid-State Power Conversion Handbook*, Wiley, New York, 1993.

Problems

MOSFET DRIVE CIRCUITS

10-1 (*a*) Run the PSpice simulation of the circuits of Example 10-1 and use Probe to determine the turnoff and turn-on power loss separately. The restrict data option will be useful. (*b*) From the PSpice simulations, determine the peak, average, and rms values of the MOSFET gate current for each simulation.

10-2 Repeat the PSpice simulation in Example 10-1 for the MOSFET drive circuit of Fig. 10-1*a*, using $R_1 = 75$, 50, and 25 Ω. What is the effect of reducing the drive circuit output resistance?

BIPOLAR TRANSISTOR DRIVE CIRCUITS

10-3 Design a bipolar transistor drive circuit like that shown in Fig. 10-7 with an initial base current of 5 A at turn-on which reduces to 0.5 A to maintain the collector current in the on state. The switching frequency is 100 kHz and the duty ratio is 50 percent.

10-4 Design a bipolar transistor drive circuit like that shown in Fig. 10-7 with an initial base current of 3 A at turn-on which reduces to 0.6 A to maintain the collector current in the on state. The switching frequency is 120 kHz and the duty ratio is 30 percent.

SNUBBER CIRCUITS

10-5 For the snubber circuit of Fig. 10-12*a*, $V_s = 50$, $I_L = 4$ A, $C = 0.05$ μF, $R = 5$ Ω., and $t_f = 0.5$ μs. The switching frequency is 120 kHz, and the duty ratio is 0.4.

(a) Determine expressions for i_Q, i_c, and v_c during transistor turnoff. (b) Graph the i_Q and v_C waveforms at turnoff. (c) Determine the turnoff losses in the switch and the snubber.

10-6 Repeat Prob. 10-5, using $C = 0.01$ μF.

10-7 Design a turnoff snubber circuit like that of Fig. 10-12a for $V_s = 150$ V, $I_L = 10$ A, and $t_f = 0.1$ μs. The switching frequency is 100 kHz, and the duty ratio is 0.4. Use the criteria that the switch voltage should reach V_s when the switch current reaches zero and that five time constants are required for capacitor discharge when the switch is open. Determine the turnoff losses for the switch and snubber.

10-8 Repeat Prob. 10-7, using the criterion that the switch voltage must reach 75 V when the switch current reaches zero.

10-9 Design a turnoff snubber circuit like that of Fig. 10-12a for $V_s = 170$ V, $I_L = 7$ A, and $t_f = 0.5$ μs. The switching frequency is 125 kHz, and the duty ratio is 0.4. Use the criteria that the switch voltage should reach V_s when the switch current reaches zero and that five time constants are required for capacitor discharge when the switch is open. Determine the turnoff losses for the switch and snubber.

10-10 Repeat Prob. 10-9, using the criterion that the switch voltage must reach 125 V when the switch current reaches zero.

10-11 A switch has a current fall time t_f of 0.5 μs and is used in a converter that is modeled as in Fig. 10-11a. The source voltage and the final voltage across the switch are 80 V, the load current is 5 A, the switching frequency is 200 kHz, and the duty ratio is 0.35. Design a snubber circuit to limit the turnoff loss in the switch to 1 W. Determine the power absorbed by the snubber resistor.

10-12 A switch has a current fall time t_f of 1.0 μs and is used in a converter that is modeled as in Fig. 10-11a. The source voltage and the final voltage across the switch are 120 V, the load current is 6 A, the switching frequency is 100 kHz, and the duty ratio is 0.3. Design a snubber circuit to limit the turnoff loss in the switch to 2 W. Determine the power absorbed by the snubber resistor.

HEAT SINKS

10-13 A MOSFET with no heat sink absorbs a thermal power of 2.0 W. The thermal resistance from junction to ambient is 40°C/W, if the ambient temperature is 30°C. (a) Determine the junction temperature. (b) If the maximum junction temperature is 150°C, how much power can be absorbed without requiring a heat sink?

10-14 A MOSFET with no heat sink absorbs a thermal power of 1.5 W. The thermal resistance from junction to ambient is 55°C/W, if the ambient temperature is 25°C. (a) Determine the junction temperature. (b) If the maximum junction temperature is 175°C, how much power can be absorbed without requiring a heat sink?

10-15 A MOSFET mounted on a heat sink absorbs a thermal power of 10 W. The thermal resistances are 1.1°C/W from the junction to the case, 0.9°C/W for the case to the heat sink, and 2.5°C/W for the heat sink to ambient. The ambient temperature is 40°C. Determine the junction temperature.

10-16 A MOSFET mounted on a heat sink absorbs a thermal power of 5 W. The thermal resistances are 1.5°C/W from the junction to the case, 1.2°C/W for the case to the heat sink, and 3.0°C/W for the heat sink to ambient. The ambient temperature is 25°C. Determine the junction temperature.

10-17 A MOSFET mounted on a heat sink absorbs a thermal power of 18 W. The thermal resistances are 0.7°C/W from the junction to the case and 1.0°C/W for the case to the heat sink. The ambient temperature is 40°C. Determine the maximum thermal resistance from the heat sink to ambient such that the junction temperature does not exceed 110°C.

10-18 A single thermal power pulse of 500 W with 10 μs duration occurs in a MOSFET with the transient thermal impedance characteristic of Fig. 10-24. Determine the change in junction temperature due to this pulse.

10-19 In Example 10-8, the switching frequency is 500 Hz. If the switching frequency is increased to 50 kHz with D remaining at 0.1 and P_{dm} remaining at 100 W, determine the change in junction temperature, (a) using the transient thermal impedance $Z_{\theta,JC}$ from Fig. 10-24 and (b) using $R_{\theta,JC} = 1.05°C/W$ and the average transistor power.

APPENDIX A

Fourier Series for Some Common Waveforms

FOURIER SERIES

The Fourier series for a periodic function $f(t)$ can be expressed in trigonometric form as

$$f(t) = a_0 + \sum_{n=1}^{\infty} [a_n \cos(n\omega_0 t) + b_n \sin(n\omega_0 t)]$$

where

$$a_0 = \frac{1}{T} \int_{-T/2}^{T/2} f(t)\, dt$$

$$a_n = \frac{2}{T} \int_{-T/2}^{T/2} f(t) \cos(n\omega_0 t)\, dt$$

$$b_n = \frac{2}{T} \int_{-T/2}^{T/2} f(t) \sin(n\omega_0 t)\, dt$$

Sines and cosines of the same frequency can be combined into one sinusoid, resulting in an alternative expression for a Fourier series

APPENDIX A Fourier Series for Some Common Waveforms

$$f(t) = a_0 + \sum_{n=1}^{\infty} C_n \cos(n\omega_0 t + \theta_n)$$

where $\quad C_n = \sqrt{a_n^2 + b_n^2} \quad$ and $\quad \theta_n = \tan^{-1}\left(\dfrac{-b_n}{a_n}\right)$

or

$$f(t) = a_0 + \sum_{n=1}^{\infty} C_n \sin(n\omega_0 t + \theta_n)$$

where $\quad C_n = \sqrt{a_n^2 + b_n^2} \quad$ and $\quad \theta_n = \tan^{-1}\left(\dfrac{a_n}{b_n}\right)$

The rms value of $f(t)$ can be computed from the Fourier series.

$$F_{\text{rms}} = \sqrt{\sum_{n=0}^{\infty} F_{n,\text{rms}}^2} = \sqrt{a_0^2 + \sum_{n=1}^{\infty}\left(\dfrac{C_n}{\sqrt{2}}\right)^2}$$

HALF-WAVE RECTIFIED SINUSOID (FIG. A-1)

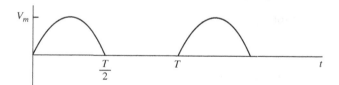

Figure A-1 Half-wave rectified sine wave.

$$v(t) = \dfrac{V_m}{\pi} + \dfrac{V_m}{2}\sin(\omega_0 t) - \sum_{n=2,4,6,\ldots}^{\infty} \dfrac{2V_m}{(n^2-1)\pi}\cos(n\omega_0 t)$$

FULL-WAVE RECTIFIED SINUSOID (FIG. A-2)

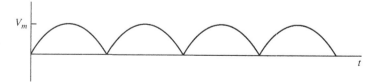

Figure A-2 Full-wave rectified sine wave.

where
$$v_o(t) = V_o + \sum_{n=2,4,\ldots}^{\infty} V_n \cos(n\omega_0 t + \pi)$$

$$V_o = \frac{2V_m}{\pi}$$

and
$$V_n = \frac{2V_m}{\pi}\left(\frac{1}{n-1} - \frac{1}{n+1}\right)$$

THREE-PHASE BRIDGE RECTIFIER (FIG. A-3)

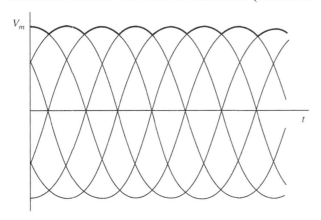

Figure A-3 Three-phase six-pulse bridge rectifier output.

The Fourier series for a six-pulse converter is

$$v_o(t) = V_o + \sum_{n=6,12,18,\ldots}^{\infty} V_n \cos(n\omega_0 t + \pi)$$

$$V_o = \frac{3V_{m,L-L}}{\pi} = 0.955\, V_{m,L-L}$$

$$V_n = \frac{6V_{m,L-L}}{\pi(n^2-1)} \qquad n = 6, 12, 18, \ldots$$

where $V_{m,L-L}$ is the peak line-to-line voltage of the three-phase source, which is $\sqrt{2}V_{L-L,\text{rms}}$.

The Fourier series of the currents in phase a of the ac line (see Fig. 4-17) is

$$i_a(t) = \frac{2\sqrt{3}}{\pi}I_o\left(\cos\omega_0 t - \frac{1}{5}\cos 5\omega_0 t + \frac{1}{7}\cos 7\omega_0 t - \frac{1}{11}\cos 11\omega_0 t + \frac{1}{13}\cos 13\omega_0 t - \ldots\right)$$

which consists of terms at the fundamental frequency of the ac system and harmonics of order $6k \pm 1$, $k = 1, 2, 3, \ldots$.

PULSED WAVEFORM (FIG. A-4)

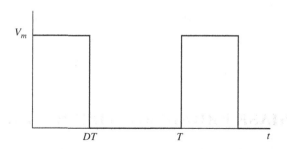

Figure A-4 A pulsed waveform.

$$a_0 = V_m D$$

$$a_n = \left(\frac{V_m}{n\pi}\right) \sin(n2\pi D)$$

$$b_n = \left(\frac{V_m}{n\pi}\right)[1 - \cos(n2\pi D)]$$

$$C_n = \left(\frac{\sqrt{2}V_m}{n\pi}\right)\sqrt{1 - \cos(n2\pi D)}$$

SQUARE WAVE (FIG. A-5)

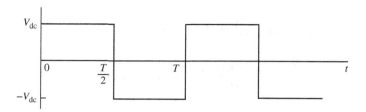

Figure A-5 Square wave.

The Fourier series contains the odd harmonics and can be represented as

$$v_o(t) = \sum_{n \text{ odd}} \left(\frac{4V_{dc}}{n\pi}\right) \sin(n\omega_0 t)$$

MODIFIED SQUARE WAVE (FIG. A-6)

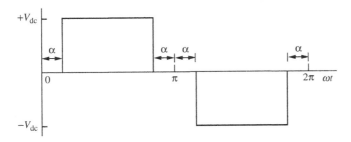

Figure A-6 A modified square wave.

The Fourier series of the waveform is expressed as

$$v_o(t) = \sum_{n \text{ odd}} V_n \sin(n\omega_0 t)$$

Taking advantage of half-wave symmetry, the amplitudes are

$$V_n = \left(\frac{4V_{dc}}{n\pi}\right) \cos(n\alpha)$$

where α is the angle of zero voltage on each end of the pulse.

THREE-PHASE SIX-STEP INVERTER (FIG. A-7)

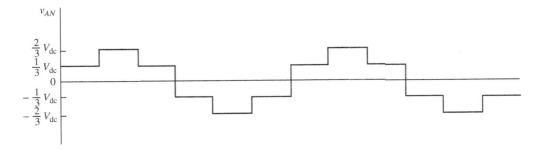

Figure A-7 Three-phase six-step inverter output.

The Fourier series for the output voltage has a fundamental frequency equal to the switching frequency. Harmonic frequencies are of order $6k \pm 1$ for $k = 1, 2, \ldots$ ($n = 5, 7, 11, 13, \ldots$). The third harmonic and multiples of the third do not exist, and even harmonics do not exist. For an input voltage of V_{dc}, the output for an ungrounded wye-connected load (see Fig. 8-17) has the following Fourier coefficients:

$$V_{n,\text{L-L}} = \left| \frac{4V_{dc}}{n\pi} \cos\left(n\frac{\pi}{6}\right) \right|$$

$$V_{n,\text{L-N}} = \left| \frac{2V_{dc}}{3n\pi}\left[2 + \cos\left(n\frac{\pi}{3}\right) - \cos\left(n\frac{2\pi}{3}\right)\right] \right| \qquad n = 1,5,7,11,13,\ldots$$

APPENDIX B

State-Space Averaging

The results of the following development are used in Sec. 7.13 on control of dc power supplies. A general method for describing a circuit that changes over a switching period is called *state-space averaging*. The technique requires two sets of state equations which describe the circuit: one set for the switch closed and one set for the switch open. These state equations are then averaged over the switching period. A state-variable description of a system is of the form

$$\dot{x} = Ac + Bv \qquad \text{(B-1)}$$

$$v_o = C^T x \qquad \text{(B-2)}$$

The state equations for a switched circuit with two resulting topologies are as follows:

$$\begin{array}{cc} \underline{\text{switch closed}} & \underline{\text{switch open}} \\ \dot{x} = A_1 x + B_1 v & \dot{x} = A_2 x + B_2 v \\ v_o = C_1^T x & v_o = C_2^T x \end{array} \qquad \text{(B-3)}$$

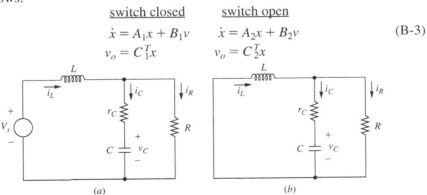

Figure B-1 Circuits for developing the state equations for the buck converter circuit (*a*) for the switch closed and (*b*) for the switch open.

APPENDIX B State-Space Averaging

For the switch closed for the time dT and open for $(1-d)T$, the above equations have a weighted average of

$$\dot{x} = [A_1 d + A_2(1-d)]x + [B_1 d + B_2(1-d)]v \qquad \text{(B-4)}$$

$$v_o = \left[C_1^T d + C_2^T(1-d)\right]x \qquad \text{(B-5)}$$

Therefore, an averaged state-variable description of the system is described as in the general form of Eqs. (B-1) and (B-2) with

$$\begin{aligned} A &= A_1 d + A_2(1-d) \\ B &= B_1 d + B_2(1-d) \\ C^T &= C_1^T d + C_2^T(1-d) \end{aligned} \qquad \text{(B-6)}$$

SMALL SIGNAL AND STEADY STATE

Small-signal and steady-state analyses of the system are separated by assuming the variables are perturbed around the steady-state operating point, namely,

$$\begin{aligned} x &= X + \tilde{x} \\ d &= D + \tilde{d} \\ v &= V + \tilde{v} \end{aligned} \qquad \text{(B-7)}$$

where X, D, and V represent steady-state values and $\tilde{x}$, $\tilde{d}$, and $\tilde{v}$ represent small-signal values. For the steady state, $\dot{x} = 0$ and the small-signal values are zero. Equation (B-1) becomes

or
$$\begin{aligned} 0 &= AX + BV \\ X &= -A^{-1}BV \end{aligned} \qquad \text{(B-8)}$$

$$V_o = -C^T A^{-1} BV \qquad \text{(B-9)}$$

where the matrices are the weighted averages of Eq. (B-6).

The small-signal analysis starts by recognizing that the derivative of the steady-state component is zero.

$$\dot{x} = \dot{X} + \dot{\tilde{x}} = 0 + \dot{\tilde{x}} = \dot{\tilde{x}} \qquad \text{(B-10)}$$

Substituting steady-state and small-signal quantities into Eq. (B-4),

$$\dot{\tilde{x}} = \{A_1(D+\tilde{d}) + A_2[1-(D+\tilde{d})]\} + \{B_1(D+\tilde{d}) + B_2[1-(D+\tilde{d})]\}(V+\tilde{v}) \qquad \text{(B-11)}$$

If the products of small-signal terms $\tilde{x}\tilde{d}$ can be neglected, and if the input is assumed to be constant, $v = V$ and

$$\dot{\tilde{x}} = [A_1 D + A_2(1-D)]\tilde{x} + [(A_1 - A_2)X + (B_1 - B_2)V]\tilde{d} \qquad \text{(B-12)}$$

Similarly, the output is obtained from Eq. (B-5).

$$\tilde{v}_o = \left[C_1^T + C_2^T(1 - D)\right]\tilde{x} + \left[(C_1^T - C_2^T)X\right]\tilde{d} \qquad \text{(B-13)}$$

STATE EQUATIONS FOR THE BUCK CONVERTER

State-space averaging is quite useful for developing transfer functions for switched circuits such as dc-dc converters. The buck converter is used as an example. State equations for the switch closed are developed from Fig. B-1a, and state equations for the switch open are from Fig. B-1b.

Switch Closed

First, the state equations for the buck converter (also for the forward converter) are determined for the switch closed. The outermost loop of the circuit in Fig. B-1a has Kirchhoff's voltage law equation

$$L\frac{di_L}{dt} + i_R R = V_s \qquad \text{(B-14)}$$

Kirchhoff's current law gives

$$i_R = i_L - i_C = i_L - C\frac{dv_C}{dt} \qquad \text{(B-15)}$$

Kirchhoff's voltage law around the left inner loop gives

$$L\frac{di_L}{dt} + i_C r_C + v_C = V_s \qquad \text{(B-16)}$$

which gives the relation

$$i_C = C\frac{dv_C}{dt} = \frac{1}{r_C}\left(V_s - L\frac{di_L}{dt} - v_C\right) \qquad \text{(B-17)}$$

Combining Eqs. (B-14) through (B-17) gives the state equation

$$\frac{di_L}{dt} = -\frac{Rr_C}{L(R + r_C)}i_L - \frac{R}{L(R + r_C)}v_C + \frac{1}{L}V_s \qquad \text{(B-18)}$$

Kirchhoff's voltage law around the right inner loop gives

$$-v_C - i_C r_C + i_R R = 0 \qquad \text{(B-19)}$$

Combining the above equation with Eq. (B-15) gives the state equation

$$\frac{dv_C}{dt} = \frac{R}{C(R + r_C)}i_L - \frac{1}{C(R + r_C)}v_C \qquad \text{(B-20)}$$

Restating Eqs. (B-18) and (B-20) in state-variable form gives

$$\dot{x} = A_1 x + B_1 V_s \tag{B-21}$$

where

$$\dot{x} = \begin{bmatrix} \dot{i}_L \\ \dot{v}_C \end{bmatrix}$$

$$A_1 = \begin{bmatrix} -\dfrac{Rr_C}{L(R+r_C)} & -\dfrac{R}{L(R+r_C)} \\ \dfrac{R}{C(R+r_C)} & -\dfrac{1}{C(R+r_C)} \end{bmatrix} \tag{B-22}$$

$$B_1 = \begin{bmatrix} \dfrac{1}{L} \\ 0 \end{bmatrix}$$

If $r_C \ll R$,

$$A_1 \approx \begin{bmatrix} -\dfrac{r_C}{L} & -\dfrac{1}{L} \\ \dfrac{1}{C} & -\dfrac{1}{RC} \end{bmatrix} \tag{B-23}$$

Switch Open

The filter is the same for the switch closed as for the switch open. Therefore, the A matrix remains unchanged during the switching period.

$$A_2 = A_1$$

The input to the filter is zero when the switch is open and the diode is conducting. State equation (B-16) is modified accordingly, resulting in

$$B_2 = 0$$

Weighting the state variables over one switching period gives

$$\begin{aligned} \dot{x}d &= A_1 x d + B_1 V_s d \\ \dot{x}(1-d) &= A_2 x(1-d) + B_2 V_s(1-d) \end{aligned} \tag{B-24}$$

Adding the above equations and using $A_2 = A_1$,

$$\dot{x} = A_1 x + [B_1 d + B_2(1-d)] V_s \tag{B-25}$$

In expanded form,

$$\begin{bmatrix} \dot{i}_L \\ \dot{v}_C \end{bmatrix} = \begin{bmatrix} -\dfrac{r_C}{L} & -\dfrac{1}{L} \\ \dfrac{1}{C} & -\dfrac{1}{RC} \end{bmatrix} \begin{bmatrix} i_L \\ v_C \end{bmatrix} + \begin{bmatrix} \dfrac{d}{L} \\ 0 \end{bmatrix} V_s \tag{B-26}$$

Equation (B-26) gives the averaged state-space description of the output filter and load of the forward converter or buck converter.

The output voltage v_o is determined from

$$v_o = Ri_R = R(i_L - i_R) = R\left(i_L - \frac{v_o - v_C}{r_C}\right) \qquad \text{(B-27)}$$

Rearranging to solve for v_o,

$$v_o = \left(\frac{Rr_C}{R + r_C}\right)i_L + \left(\frac{R}{R + r_C}\right)v_C \approx r_C i_L + v_C \qquad \text{(B-28)}$$

The above output equation is valid for both switch positions, resulting in $C_1^T = C_2^T = C^T$. In state-variable form

$$v_o = C^T x$$

where

$$C^T = \left[\frac{Rr_C}{R + r_C} \quad \frac{R}{R + r_C}\right] \approx [r_C \quad 1] \qquad \text{(B-29)}$$

and

$$x = \begin{bmatrix} i_L \\ v_C \end{bmatrix} \qquad \text{(B-30)}$$

The steady-state output is found from Eq. (B-9),

$$V_o = -C^T A^{-1} B V_s \qquad \text{(B-31)}$$

where $A = A_1 = A_2$, $B = B_1 D$, and $C^T = C_1^T = C_2^T$. The final result of this computation results in a steady-state output of

$$V_o = V_s D \qquad \text{(B-32)}$$

The small-signal transfer characteristic is developed from Eq. (B-12), which in the case of the buck converter results in

$$\dot{\tilde{x}} = A\tilde{x} + BV_s\tilde{d} \qquad \text{(B-33)}$$

Taking the Laplace transform,

$$s\tilde{x}(s) = A\tilde{x}(s) + BV_s\tilde{d}(s) \qquad \text{(B-34)}$$

Grouping $\tilde{x}(s)$

$$[sI - A]\tilde{x}(s) = BV_s\tilde{d}(s) \qquad \text{(B-35)}$$

where I is the identity matrix. Solving for $\tilde{x}(s)$,

$$\tilde{x}(s) = [sI - A]^{-1} BV_s\tilde{d}(s) \qquad \text{(B-36)}$$

Expressing $\tilde{v}_o(s)$ in terms of $\tilde{x}(s)$,

$$\tilde{v}_o(s) = C^T \tilde{x}(s) = C^T[sI - A]^{-1} B V_s \tilde{d}(s) \qquad \text{(B-37)}$$

Finally, the transfer function of output to variations in the duty ratio is expressed as

$$\frac{\tilde{v}_o(s)}{\tilde{d}(s)} = C^T[sI - A]^{-1} B V_s \qquad \text{(B-38)}$$

Upon substituting for the matrices in the above equation, a lengthy evaluation process results in the transfer function

$$\frac{\tilde{v}_o(s)}{\tilde{d}(s)} = \frac{V_s}{LC} \left[\frac{1 + s r_C C}{s^2 + s(1/RC + r_C/L) + 1/LC} \right] \qquad \text{(B-39)}$$

The above transfer function was used in the section on control of dc power supplies in Chap. 7.

Bibliography

S. Ang and A. Oliva, *Power-Switching Converters*, 2d ed., Taylor & Francis, Boca Raton, Fla., 2005.

R. D. Middlebrook and S. Ćuk, "A General Unified Approach to Modelling Switching—Converter Power Stages," *IEEE Power Electronics Specialists Conference Record*, 1976.

N. Mohan, T. M. Undeland, and W. P. Robbins, *Power Electronics: Converters, Applications, and Design*, 3d ed., Wiley, New Yorks, 2003.

INDEX

A

Ac voltage controller, 171, 172
Ac-ac converter, 171
Active power, 22
Adjustable-speed motor drives, 349
Amplitude control, 346
    inverters, 342
    resonant converter, 404
Amplitude modulation ratio, 360
Average power, 22
Averaged circuit model, 254

B

Battery charger, 24, 120
Bipolar junction transistor, 9, 437
    Darlington, 10
Bipolar PWM inverter, 361
Body diode, 10, 207
Boost converter, 211, 244, 301
Bridge rectifier, 111, 114, 131, 160
    three phase, 463
Buck converter, 198, 310
    control, 303
    design, 207, 208, 210
Buck-boost converter, 221

C

Capacitors, 25
    average current, 26
    average power, 26
    ESR, 206
    stored energy, 25
Capture, 13

Carrier signal, 357
Charge pump, 247
Chopper, dc, 197
Class D amplifiers, 366
Commutation, 103, 160
Compensation, 308, 317
Conduction angle, 97, 189
Continuous current, 120, 126, 198
Control, 302
Control loop, 297, 303
Controlled full-wave rectifier, 131
Controlled half-wave rectifier, 94
Converter
    ac-ac, 2
    ac-dc, 2
    classification, 1
    dc-ac, 2
    dc-dc, 2
    selection, 298
Crest factor, 50
Cross-over frequency, 304
Ćuk converter, 226
Current-fed converter, 294

D

Darlington, 10
Dc link, 382
Dc link resonant converter, 422
Dc power supplies, 265
    complete, 325
    off line, 326
Dc power transmission, 1, 156
Dc-dc converter
    boost, 211

473

Dc-dc converter —(*Cont.*)
   buck, 198
   buck-boost, 221
   Cuk, 226
   current-fed, 294
   double-ended forward, 285
   flyback, 267
   forward, 277
   full-bridge, 291
   half-bridge, 291
   multiple outputs, 297
   push-pull, 287
   SEPIC, 231
   switched capacitor, 247
Delay angle, 94, 131
Design
   boost converter, 216
   buck converter, 207, 208, 210
   Ćuk converter, 230
   flyback converter, 274
   forward converter, 284
   half-wave rectifier, 74
   inverter, 344, 364
   type 2 error amplifier, 311
   type 3 error amplifier, 318
Diode, 6
   fast-recovery, 7
   freewheeling, 81
   ideal, 17, 72
   MOSFET body, 9
   reverse recovery, 7
   Schottky, 7
Discontinuous current, 198
Displacement power factor, 49
Distortion factor, 49
Distortion volt-amps, 50
Double-ended forward converter, 285
Drive circuits
   BJT, 437
   high side, 433
   low side, 431
   MOSFET, 431
   PSpice, 17
   thyristor, 440
   transistor, 8
Duty ratio, 35, 198

E

Electric arc furnaces, 192
Electronic switch, 5, 65
Energy, 22
Energy recovery, 27, 32
Equivalent series resistance (ESR), 206, 273, 307, 309, 323
Error amplifier, 303, 307, 308, 311
Extinction angle, 70, 72, 77, 96

F

Fast Fourier transform (FFT), 55
Feedback, 302
Filter
   capacitor, 88, 122
   L-C, 126, 323, 404
   transfer function, 306
Flyback converter, 267
Forced response, 67, 76
Form factor, 50
Forward converter, 277
Fourier series, 4, 43, 45
   amplitude control, 343
   common waveforms, 461
   controlled rectifier, 136
   full-wave rectified sine wave, 115
   half-wave rectified sine wave, 82
   multilevel inverter, 349
   PSpice, 54
   PWM inverter, 361
   square-wave inverter, 337
   three-phase rectifier, 146
Freewheeling diode, 81, 86, 103
Frequency modulation ratio, 360
Fuel injector, 27
Full-bridge converter, 291, 331

G

Gate turnoff thyristor (GTO), 7

H

Half-bridge converter, 291
Half-wave rectifier, 65
   controlled, 94, 95, 99

Heat sinks, 450
  steady-state temperatures, 450
  time-varying temperatures, 454
High-side drivers, 433

I

Induction motor speed control, 379
Inductors, 25
  average power, 25
  average voltage, 25
  stored energy, 25, 30
Insulated gate bipolar transistor (IGBT), 9, 336, 432
Interleaved converters, 237
International Rectifier
  IR2110, 435
  IR2117, 435
  IRF150, 16
  IRF4104, 455
  IRF9140, 16
Inverter, 2, 142, 331
  amplitude control, 342
  full bridge, 331
  half bridge, 346
  harmonic control, 342
  multilevel, 348
  PWM, 357
  six-step, 373
  square wave, 333

K

K factor, 312, 318

L

Light dimmer, 192
Linear voltage regulator, 196
Low-side drivers, 431

M

MOS-controlled thyristor (MCT), 7
MOSFET, 9
  drive circuits, 431
  on-state resistance, 10

Multilevel inverters, 348
  diode clamped, 354
  independent dc sources, 349
  pattern swapping, 353
  three phase, 378

N

National Semiconductor
  LM2743 control circuit, 323
Natural response, 67, 76

O

Orthogonal functions, 40

P

Parallel dc-dc resonant converter, 415
Passive sign convention, 21
Phase control, 171
Phase margin, 304, 311
Power
  apparent, 42
  average, 22, 46, 70, 77, 79
  complex, 44
  computations, 21
  dc source, 24
  factor, 43, 96
  instantaneous, 21
  reactive, 44
  real, 22
Power factor correction, 299
Probe, 13, 52, 72
PSpice, 13
  average power, 52
  control loop, 311, 315
  controlled rectifier, 100
  convergence, 18
  dc power supplies, 301
  default diode, 17
  energy, 52
  Fourier analysis, 54
  half-wave rectifier, 72
  ideal diode, 17
  instantaneous power, 52
  power computations, 51

PSpice —(*Cont.*)
   rms, 54
   Sbreak switch, 14
   SCR, 18, 100
   THD, 56
   voltage-controlled switch, 14
Pulse-width modulation, 307, 357
Push-pull converter, 287
PWM control circuits, 323

R

Rectifier
   filter capacitor, 88, 122
   half-wave, 65
   three-phase, 144
Reference voltage, 361
Resonant converter, 387
   comparison, 421
   dc link, 422
   parallel resonant dc-dc, 415
   series resonant dc-dc, 407
   series resonant inverter, 401
   series-parallel dc-dc, 418
   zero-current switching, 387
   zero-voltage switching, 394
Reverse recovery, 7
Ripple voltage
   boost converter, 215
   buck converter, 204
   buck-boost converter, 225
   Cuk converter, 228
   effect of ESR, 206
   flyback converter, 273
   Forward converter, 282
   full-wave rectifier, 124
   half-wave rectifier, 90, 91
   push-pull converter, 289
   SEPIC, 234
Rms, 34
   PSpice, 54
   pulse waveform, 35
   sinusoids, 36
   sum of waveforms, 40
   triangular waveform, 41

S

Safe operating area (BJT), 447
Schottky diode, 7, 207
Series resonant dc-dc converter, 407
Series resonant inverter, 401
Series-parallel dc-dc converter, 418
Silicon controlled rectifier (SCR), 7
Single-ended primary inductance converter
   (SEPIC), 231
Six-pulse rectifier, 145
Six-step three-phase inverter, 373
Small-signal analysis, 304
Snubber circuits
   energy recovery, 449
   thyristor, 450
   transistor, 441
Solenoid, 27
Solid-state relay, 179
SPICE, 13
Stability, 157, 303, 307, 311, 317
State-space averaging, 307, 467
Static VAR control, 191
Stepped parameter, 73
Switch selection, 11
Switched-capacitor converter
   inverting, 249
   step-down, 250
   step-up, 247
Switching losses, 240, 241
Synchronous rectification, 207

T

Thermal impedance, 455
Thermal resistance, 451
Three phase
   controlled rectifier, 149
   inverter, 154, 373
   neutral conductor, 38
   rectifiers, 144
   voltage controller, 183
Thyristor, 7
   drive circuits, 440
   snubber circuit, 450
Time constant, 69, 93

Total harmonic distortion (THD), 49, 339
Transfer function
  filter, 306
  PWM, 307
  switch, 305
Transformer
  center tapped, 114
  dot convention, 266
  leakage inductance, 267
  magnetizing inductance, 266
  models, 265
Transient thermal impedance, 455
Transistor switch, 27
Transistors, 8
Triac, 7, 8
Twelve-pulse rectifiers, 151
Type 2 compensated error amplifier, 308
Type 3 compensated error amplifier, 317
  placement of poles and zeros, 323

U

Uninterruptible power supplies (UPS), 331
Unipolar PWM inverter, 365

V

Voltage doubler, 125
Vorperian's model, 259

Z

Zero-current switching, 387
Zero-voltage switching, 394